Autodesk Revit 2024 Architecture Certified Professional Exam Study Guide

Elise Moss

Trademarks
The following are registered trademarks of Autodesk, Inc.: AutoCAD, AutoCAD Architecture, Revit, AutoDesk, AutoCAD Design Center, Autodesk Device Interface, VizRender, and HEIDI. Microsoft, Windows, Word, and Excel are either registered trademarks of trademarks of Microsoft Corporation.

All other trademarks are trademarks of their respective holders.

SDC Publications
P.O. Box 1334
Mission, KS 66222
913-262-2664
www.SDCpublications.com
Publisher: Stephen Schroff

ISBN-13: 978-1-63057-597-7
ISBN-10: 1-63057-597-6

Printed and bound in the United States of America.

Preface

This book is geared towards users who have been using Revit for at least six months and are ready to pursue their official Autodesk certification. You can locate the closest testing center on Autodesk's website. Check with your local reseller about special Exam Days when certification exams can be taken at a discount. Autodesk has restructured their certification exams so they can be taken in your home if you have the proper equipment (web cam and microphone). This edition of the textbook has been restructured as well to align with the new certification exams.

I wrote this book because I taught a certification preparation class at SFSU. I also teach an introductory Revit class using my Revit Basics text. I heard many complaints from students who had taken my Revit Basics class when they had to switch to the Autodesk AOTC courseware for the exam preparation class. Students preferred the step by step easily accessible instruction I use in my texts.

This textbook includes exercises which simulate the knowledge users should have in order to pass the certification exam. I advise my students to do each exercise two or three times to ensure that they understand the user interface and can perform the task with ease.

I have endeavored to make this text as easy to understand and as error-free as possible…however, errors may be present. Please feel free to email me if you have any problems with any of the exercises or questions about Revit in general.

I have posted videos for some of the lessons from this text on YouTube – search for Moss Designs and you will find any videos for my book on that channel.

Exercise files can be accessed and downloaded from the publisher's website at:
www.SDCpublications.com/downloads/978-1-63057-597-7

I have included a sample practice exam for the Professional certification exam. It may only be downloaded using a special access code printed on the inside cover of the textbook. The practice exam is free and does not require any special software, but it is meant to be used on a computer to simulate the same environment as taking the test at an Autodesk testing center.

Please feel free to email me if you have any questions or problems with accessing files or regarding any of the exercises in this text.

Acknowledgements

A special thanks to Rick Rundell, Gary Hercules, Christie Landry, and Steve Burri, as well as numerous Autodesk employees who are tasked with supporting and promoting Revit.

Additional thanks to Gerry Ramsey, Will Harris, Scott Davis, James Cowan, James Balding, Rob Starz, and all the other Revit users out there who provided me with valuable insights into the way they use Revit.

Thanks to Zach Werner for the cover artwork for this textbook along with his technical assistance. Thanks to Karla Werner for her help with editing and formatting to ensure this textbook looks proper for our readers.

My eternal gratitude to my life partner, Ari, my biggest cheerleader throughout our years together.

Elise Moss
Elise_moss@mossdesigns.com

Table of Contents

Lesson Two
Documentation

Lesson Three
Collaboration and Coordination

Lesson Four
Project Standards and Setup

Lesson Five
Information Analysis

Introduction
FAQs on Getting Certified in Revit

The first day of class students are understandably nervous and they have a lot of questions about getting certification. Throughout the class, I am peppered with the same or similar questions.

Certification is done through a website. You can learn more at https://home.pearsonvue.com/autodesk. The website currently has exams for Revit Architecture and Revit Structure. They are only offering the Professional exam at this time.

In the past, the exams required you to use Revit during the exam to walk through problems. The exam is now entirely – 100% – browser-based. All the questions are multiple choice, fill in the blank, true/false, or point and click. You do not have any other software open beside the browser. This means you have to be really familiar with Revit's dialog boxes and options as you are relying on your memory for most of the exam.

You now have the option to take the exam at a Pearson Vue testing center or in the comfort of your own home. You can go on-line to determine the location of the testing center closest to your location and schedule the exam.

If you opt to take the exam in your home, you still need to "schedule" the exam as a proctor will be monitoring you remotely during the exam. In order to take the exam at home, you need a laptop, workstation or tablet device connected to the internet equipped with a microphone and a webcam. Prior to the exam, you will be required to test your system to ensure the microphone and webcam are functioning properly. You also need to upload your identification and take a "selfie" so the proctor can verify that you are the one taking the exam.

When I took the exam at home, I had to clear off my desk, turn off and unplug all monitors except for the laptop I was using for the exam. The proctor made me pick up my laptop and rotate the laptop 360 degrees so she could see the area I was taking the exam and verify that I had no papers on my desk and nothing that could help me during the exam. You are not allowed to take notes or refer to any reference material during the exam. The only device you are allowed to use is the device you are using to take the exam. Your cell phone is to be stored away from your work area.

The Professional certification requires about 1,200 hours or two years of experience with the software.

The Autodesk Certified Professional Exam is 45 questions and 120 minutes. This means you have an average of two and a half minutes per question. If you have a slow internet connection, there may be a lag and this will slow you down. So, make sure you have a good internet connection prior to scheduling an exam outside a testing center.

In the past, you could be certified as a User or as a Professional. In order to be certified as a Professional, you needed to pass both the user and the professional exams. Then, you could opt to take one or the other exam. Now, you can only be certified as a Professional. The user exam is no longer being offered.

What are the exams like?

The first time you take the exam, you will create an account with a login. Be sure to write down your login name and password. You will need this regardless of whether you pass or fail.

If you fail and decide to retake the exam, you want to be able to log in to your account.

If you pass, you want to be able to log in to download your certificate and other data.

Autodesk certification tests use a "secure" browser, which means that you cannot cut and paste or copy from the browser. You cannot take screenshots of the browser.

If you go on Autodesk's site, there is a list of topics covered for each exam. I have added a note next to each exercise in the Table of Contents to indicate which domain is covered by that exercise. You may find this helpful if you failed to pass the exam. At the end of the exam, you are provided a list of domains where you missed the question, so you can use the domain references to help you figure out which exercises to practice.

Expect questions on modeling, documentation, collaboration, project standards, and information analysis. The topics include creating and managing Revit families, linking and monitoring files, as well as more complex wall families.

Exams are timed. This means you only have one to three minutes for each question. You have the ability to "mark" a question to go back if you are unsure. This is a good idea because a question that comes later on in the exam may give you a clue or an idea on how to answer a question you weren't sure about.

Exams pull from a question "bank" and no two exams are exactly alike. Two students sitting next to each other taking the same exam will have entirely different experiences and an entirely different set of questions. However, each exam covers specific topics. For example, you will get at least one question about family parameters. You will probably not get the same question as your neighbor.

At the end of the exam, the browser will display a screen listing the question numbers and indicate any marked or incomplete questions. Any questions where you forgot to select an answer will be marked incomplete. Questions that you marked will display as answered. You can click on those answers and the browser will link you back directly to those questions so you can review them and modify your answers.

I advise my students to mark any questions where they are struggling and move forward, then use any remaining time to review those questions. A student could easily spend ten to fifteen minutes pondering a single question and lose valuable time on the exam.

Once you have completed your review, you will receive a prompt to END the exam. Some students find this confusing as they think they are quitting the exam and not receiving a score. Once you end the exam, you may not change any of your answers. There will be a brief pause and then you will see a screen where you will be notified whether you passed or failed. You will also see your score in each section. Again, you will not see any of the actual questions. However, you will see the topic, so you might see that you scored poorly on Documentation, but you won't know which questions you missed.

Results	100	200	300	400	500	600	700	800	900	1000
Required Score: 700										
Your Score: 752										

Section analysis		Outcome	
Modeling	Meets expectation	**Pass**	✓
Documentation	Meets expectation		
Collaboration and Coordination	Meets expectation		
Project Standards and Setup	Meets expectation		
Information Analysis	Meets expectation		

If you failed the test, you do want to note where you scored poorly. Review those topics both in this guide and in the software to help prepare you to retake the exam. You can sign into the Pearson Vue website and download the report of how you did on the exam and use that as a study guide.

How many times can I take the test?

You can take any test up to three times in a 12-month period. There is a 24 hour waiting period between re-takes, so if you fail the exam on Tuesday, you can come back on Wednesday and try again. Of course, this depends on the testing center where you take the exam. If you do not pass the exam on the second re-take, you must wait five days before retaking the exam again. There is no limit on the number of re-takes. Some testing centers may provide a free voucher for a re-take of the exam. You should check the policy of your testing center and ask if they will provide a free voucher for a re-take in case you fail.

Do a lot of students have to re-take the test or do they pass on the first try?

About half of my students pass the exam on the first try, but it definitely depends on the student. Some people are better at tests than others. My youngest son excels at "multiple guess" style exams. He can pretty much ace any multiple guess exam you give him regardless of the topic. Most students are not so fortunate. Some students find a timed test extremely stressful. For this reason, I have created a simulated version of the exam for my students simply so they can practice taking an online timed exam. This has the effect of "conditioning" their responses, so they are less stressed taking the actual exam.

Some students find the multiple-choice style extremely confusing, especially if they are non-native English speakers. There are exams available in many languages, so if you are not a native English speaker, check with the testing center about the availability of an exam in your native language.

Why take a certification exam?

The competition for jobs is steep and employers can afford to be picky. Being certified provides employers with a sense of security knowing that you passed a difficult exam that requires a basic skill set. It is important to note that the certification exam does not test your ability as a designer or drafter. The certification exam tests your knowledge of the Revit software. This is a fine distinction, but it is an important one.

If you pass the exam, you have the option of having a badge displayed on your LinkedIn profile verifying that you are certified in the Revit software. This may help you convince a recruiter or prospective employer to grant you an interview.

How long is the certification good for?

Your certification is good for three years. It used to be that certification was specific to a release. In other words, you would be certified in Revit 2024 and be required to take the exam using Revit 2024. Now, since the exam is entirely browser-based and doesn't require you to even use the software, it is no longer tied to a specific release.

Most employers want you to be certified within a couple of years of the most current release, so if you wish to maintain your "competitive edge" in the employment pool, expect that you will have to take the certification exam every three years. I recommend students take an "update" class from an Autodesk Authorized Training Center before they take the exam to improve their chances of passing. Autodesk is constantly tweaking and changing the exam format. Check with your testing center about the current requirements for certification.

How much does it cost?

It costs $200 USD to take the Professional exam.

Do I need to be able to use the software to pass the exam?

This sounds like a worse question than intended. Some of my students have taken the Revit classes but have not actually gotten a job using Revit yet. They are in that catch-22 situation where an employer requires experience or certification to hire them, but they can't pass the exam because they aren't using the software every day. For those students, I advise some self-discipline where they schedule at least six hours a week for a month where they use the software – even if it is on a "dummy" project – before they take the exam. That will boost the odds in their favor. However, you do not have to have Revit installed or operating on your computer in order to take the exam. You do need to be very familiar with the software.

What happens if I pass?

You want to be sure you wrote down your login information. That way, you can log in to the certification center website and download your certificate. You also can download a logo which shows you are a Certified Professional that you can post on your website or print on your business card. Autodesk now offers a badge which you can add to your LinkedIn profile.

How many times can I log into Autodesk's testing center?

You can log in as often as you like. The tests you have taken will be listed as well as whether you passed or failed.

Can everybody see that I failed the test?

Autodesk is kind enough to keep it a secret if you failed the exam. Nobody knows unless you tell them. If you passed the test, people only see that information if you selected the option to post that result. Regardless, only you and Pearson VUE will know whether you passed or failed unless *you* choose to share that information.

What if I need to go to the bathroom during the exam or take a break?

You are allowed to "pause" the test. This will stop the clock. You then alert the proctor that you need to leave the room for a break. When you return, you will need the proctor to approve you to re-enter the test and start the clock again. Even if you take the exam at home, you are monitored during the exam. Just open a chat window and tell the proctor you need to pause the test for a bathroom break.

How many questions can I miss?

A passing score is 700. Because the test is constantly changing, the number of questions can change and the amount of points each question is worth can also vary.

How much time do I have for the exam?

The Professional exam is 120 minutes.

Can I ask for more time?

You can make arrangements for more time if you are a non-English speaker or have problems with tests. Be sure to speak with the proctor about your concerns. Most proctors will provide more time if you truly need it. However, my experience has been that most students are able to complete the exam with time to spare. I have only had one or two students that felt they "ran out of time."

What happens if the computer crashes during the test?

Don't worry. Your answers will be saved, and the clock will be stopped. Simply reboot your system. Let the proctor know when you are ready to start the exam again, so you can re-enter the testing area in your browser.

Can I have access to the practice exams you set up for your students?

I am including a Professional practice exam in the exercise files with this text.

What sort of questions do you get in the exams?

Many students complain that the questions are all about Revit software and not about building design or the uniform building code. Keep in mind that this test is to determine your knowledge about Revit software. This is not an exam to see if you are a good architect or designer.

The exams have several question types:

One best answer – this is a multiple-choice style question. You can usually arrive at the best answer by figuring out which answers do NOT apply.

Select all that apply – this can be a confusing question for some users because unless they know how *many* possible correct answers there are, they aren't sure. The test tells you to select 2/3/4 correct answers out of 5 and will prompt you if you select too many or not enough.

Point and click – this has a java-style interface. You will be presented with a picture, and then asked to pick a location on the picture to simulate a user selection. When you pick, a mark will be left on the image to indicate your selection. Each time you pick in a different area, the mark will shift to the new location. You do not have to pick an exact point…a general target area is all that is required.

Matching format – you are probably familiar with this style of question from elementary school. You will be presented with two columns. One column may have assorted terms and the second column the definitions. You then are expected to drag the terms to the correct definition to match the items.

True/False – you will be provided a set of sentences and asked to determine which ones are true or false.

Order of Operation – There is usually one question where you have to organize the order you would perform tasks in order to achieve a goal. For example, you may be asked the steps required to set up a collaborative environment with work sets or the steps required to set up and select a preferred design option. This can be a challenging question when you don't have access to the software, so practice getting comfortable with these types of tasks.

Any tips?

I suggest you read every question at least twice. Some of the wording on the questions is tricky.

Be well rested and be sure to eat before the exam. Most testing centers do not allow food, but they may allow water. Keep in mind that you can take a break if you need one. If you are taking the test at home, you can have water on your desk or work area, but that is it.

Relax. Maintain perspective. This is a test. It is not fatal. If you fail, you will not be the first person to have failed this exam. Failing does not mean you are a bad designer or architect or even a bad person. It just means you need to study the software more.

Remember to write down your login name and password for your account. The proctor will not be able to help you if you forget.

Practice Exams

I have created practice exams at the end of each lesson. Do not memorize the answers. The questions on the practice exams will not be the same as the questions on the actual exam, but they may be similar. Some users have complained that they were marked wrong when they gave the correct answer, but they still were able to pass the exam. Some of this has to do with the way questions are worded. Some of the wording of the exam questions is vague or misleading, so be sure to read each question carefully.

Autodesk also provides preparatory courseware, including videos, and practice exercises to help you tune up your skills for the exam. This is an excellent free resource.

Lesson

01

Modeling and Materials

This lesson addresses the following certification exam questions:

- Create and Modify Architectural Components
  This includes walls, curtain walls, roofs, floor, ceilings, stairs, railings, and columns
- Create, Configure and Apply Materials
  Understand how to create a custom material, how to apply a material to a family and how to use the Split Face and Paint tools
- Configure Rooms
  Modify room parameters and understand how to assign room boundaries
- 3D Parametric Families
  Understand how to select the correct template when creating a family, how reference lines and planes are used to control geometry, how to manage visibility/graphic overrides, how nested families operate, how to define and leverage parameters
- Topography
  Revit 2024 has updated how topography is created and modified. However, the exam may still use older terminology and methods. You need to be familiar with toposurfaces and toposolids. You also need to know how to create topography from a linked file.
- Model Groups
  Demonstrate how to create and modify model groups

Users should be able to understand the difference between a hosted and non-hosted component. A hosted component is a component that must be placed or constrained to another element. For example, a door or window is hosted by a wall. You should be able to identify what components can be hosted by which elements. Walls are non-hosted. Whether or not a component is hosted is defined by the template used for creating the component. A wall, floor, ceiling or face can be a host.

Some components are level-based, such as furniture, site components, plumbing fixtures, casework, roofs and walls. When you insert a level-based component, it is constrained to that level and can only be moved within that infinite plane.

Components must be loaded into a project before they can be placed. Users can pre-load components into a template, so that they are available in every project.

Users should be familiar with how to use Element and Type Properties of components in order to locate and modify information.

There are three kinds of families in Revit Architecture:
- system families
- loadable families
- in-place families

System families are walls, ceilings, stairs, floors, etc. These are families that can only be created by using an existing family, duplicating, and redefining. These families are loaded into a project using a project template. You can copy system families from one project to another using the Transfer Project Standards tool.

Loadable families are external files. These include doors, windows, furniture, and plants.

In-place families are components that are created inside of a project and are unique to that project.

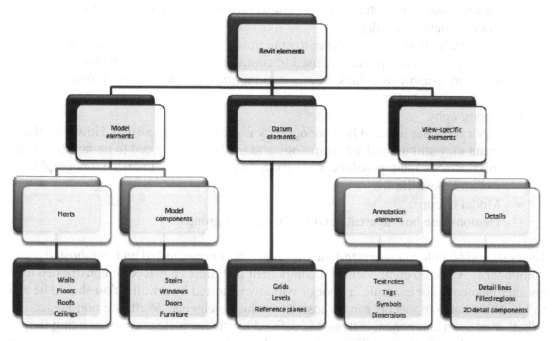

Revit elements are separated into three different types of elements: Model, Datum and View-specific. Users are expected to know if an element is model, datum or view specific.

Model elements are broken down into categories. A category might be a wall, window, door, or floor. If you look in the Project Browser, you will see a category called Families. If you expand the category, you will see the families for each category in the current project. Each family may contain multiple types.

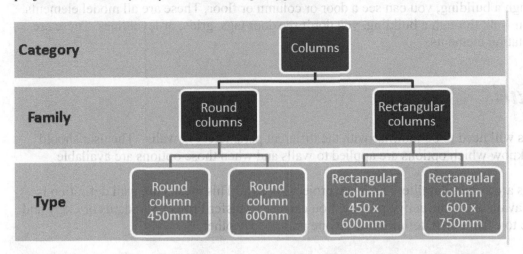

Every Revit file is considered a Project. A Revit project consists of the Project Environment, components, and views. The Project Environment is managed in the Project Browser.

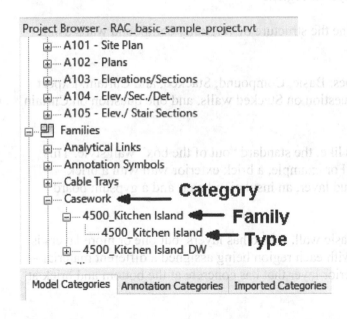

There may be one question on the exam asking you to describe the hierarchy of elements in a Revit project.

Elements are organized by Category, then Family, then Type, then Instance. This is easier to visualize if you look in the Project Browser.

If you look in the Visibility/Graphics Overrides dialog, elements are divided into Model Categories and Annotation Categories.

You should be able to identify whether an object is a model element or an annotation element.

One way to think of it is that model elements are actual physical items. If you walk through a building, you can see a door or column or floor. These are all model elements. If you walk through a building, you don't see door tags, grids, or level lines. These are annotation elements.

Walls

Users will need to be familiar with the different parameters in walls. The user should also know which options are applied to walls and when those options are available.

Walls are system families. They are project-specific. This means the wall definition is only available in the active project. You can use Transfer Project Standards or Copy and Paste to copy a wall definition from one project to another.

On the Professional exam, you may be shown an image of a wall and asked to identify different wall properties.

Just as roofs, floors, and ceilings can consist of multiple horizontal layers, walls can consist of more than one vertical layer or region.

You can modify a wall type to define the structure of vertically compound walls using layers or regions.

Revit has several different wall types: Basic, Compound, Stacked, and Curtain. Expect one question on Basic walls, one question on Stacked walls, and one question on Curtain walls.

A Basic Wall is just what it sounds like, the standard "out of the box" wall style. This wall type may have several layers. For example, a brick exterior wall with a brick exterior layer, an air gap layer, a stud layer, an insulation layer, and a gypsum board layer.

A Compound wall is similar to a Basic wall. It also has layers, but one or more layers is divided into one or more regions, with each region being assigned a different material—for example, a wall that has an exterior layer that has concrete at the bottom and brick at the top.

A Stacked wall is two or more basic and/or compound walls that are stacked on top of each other. While Basic and Compound walls have a uniform thickness or width defined by the layers, a Stacked wall can have a variable thickness or width.

A Curtain wall is defined by a curtain grid. Mullions can be placed at the grid lines. Panels are placed in the spaces between the grid lines.

Exercise 1-1

Wall Options

Drawing Name: **i_firestation_basic_plan.rvt**
Estimated Time to Completion: 10 Minutes

Scope
Exploring the different wall options

Solution

1. Views (all)
    Floor Plans
    Ground Floor
    Lower Roof
    Main Floor
    Main Roof
    Site
    T.O. Footing
    T.O. Parapet

    Activate the **Ground Floor** floor plan.

2. 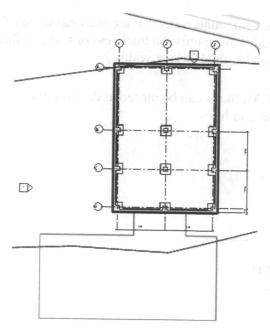 Zoom into the area where the green polygon is.

3. 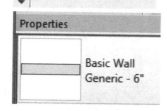 Select **Wall** from the Architecture ribbon.

4. Set the Wall Type to **Generic – 6″** in the Properties pane.

5.  Set the Location Line to **Core Face: Exterior**.

6. 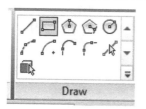 Select the **Rectangle** tool on the Draw panel.

7.

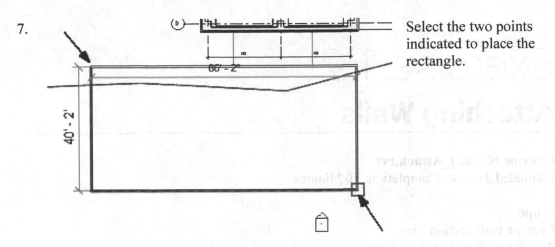

Select the two points indicated to place the rectangle.

8.

New Walls	⌄ 🔲 Edit Type
Constraints	⌃ ⌃
Location Line	Core Face: Exterior
Base Constraint	Ground Floor
Base Offset	0' 0"
Base is Attached	☐
Base Extension Distance	0' 0"
Top Constraint	Unconnected
Unconnected Height	20' 0"
Top Offset	0' 0"
Top is Attached	☐

Notice that if the Top Constraint is set to Unconnected, you can define the Unconnected Height.

9. Close the file without saving.

Attaching Walls

After placing a wall, you can override its initial top and base constraints by attaching its top or base to another element in the same vertical plane. By attaching a wall to another element, you avoid the need to manually edit the wall profile when the design changes.

The other element can be a floor, a roof, a ceiling, a reference plane, or another wall that is directly above or below. The height of the wall then increases or decreases as necessary to conform to the boundary represented by the attached element.

You can detach walls from elements as well. If you want to detach selected walls from all other elements at once, click Detach All on the Options Bar.

Exercise 1-2
Attaching Walls

Drawing Name: **i_Attach.rvt**
Estimated Time to Completion: 10 Minutes

Scope
Create a wall section view.
Attach a wall to a roof or floor.

Solution

1. 📂   Open *i_Attach.rvt*.

2. 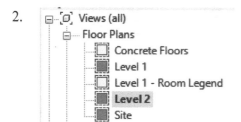   Activate Level 2 Floor Plan.

3. 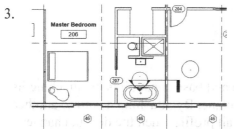   *Place a wall section as shown.*

    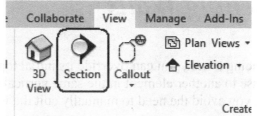

    Go to the **View** ribbon.
    Select the **Section** tool.

4. Set the view type to **Wall Section**.

5.

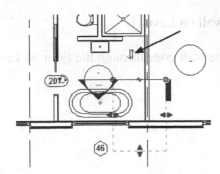

The first point selected will place the section head.
The second point selected will place the section tail.

Use the Flip controls if needed to orient the section head to face down/south.

Double left click on the section head to open the section view.

6.

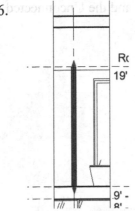

Select the wall on Level 2.

7.

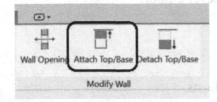

Select **Attach Top/Base** from the ribbon.

8.

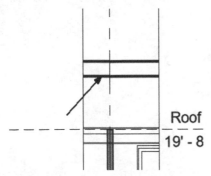

Select the roof.

9.

What is the volume of the wall after it is attached to the roof?

Select the wall and then go to the Properties panel to determine the correct volume.
It should be **88.83 CF**.

10.

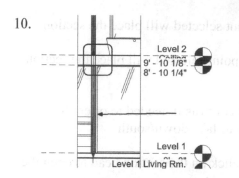

Select the wall on Level 1.

Note that the wall goes through the floor on Level 2.

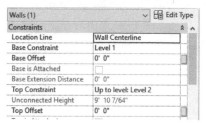

Check the Properties palette and note that the Top Constraint is set to Level 2 and the Unconnected Height is grayed out.

Walls (1)	∨	Edit Type
Constraints		
Location Line	Wall Centerline	
Base Constraint	Level 1	
Base Offset	0' 0"	
Base is Attached	☐	
Base Extension Distance	0' 0"	
Top Constraint	Up to level: Level 2	
Unconnected Height	9' 10 7/64"	
Top Offset	0' 0"	

11.

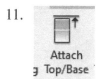

Select **Attach Top/Base** on the ribbon.

12.

Select the Level 2 floor.

13.

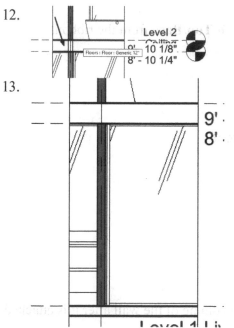

The wall adjusts.

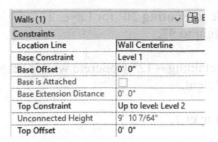

Note that the Top Constraint and Unconnected Height values did not change.

Exercise 1-3

Stacked Walls

Drawing Name: **i_Footing.rvt**
Estimated Time to Completion: 10 Minutes

Scope
Defining a stacked wall.

A stacked wall uses more than one wall type. Stacked walls may have varying widths.

Solution

1. 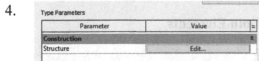 Open *i_Footing.rvt*.

2.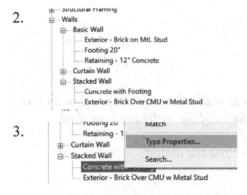

    In the Project Browser, locate the Walls family category.
    Expand the Stacked wall section.

    Select the **Concrete with Footing** wall type.

3. Right click and select **Type Properties**.

4. Select **Edit** next to Structure.

5. Select the **Insert** button.

6.

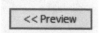

	Name	Height	Offset	Top	Base	Flip
1	Footing 20'	0' 9"	0' 0"			☐
2	Retaining - 12"	Variable	0' 0"	0' 0"	0' 0"	☐

Set the type to **Footing 20'** for Layer 1.
Set the Height to **9"**.

Note the Retaining – 12" Concrete wall is set to a Variable Height.

7.

Types

	Name	Height	Offset	Top	Base	Flip
1	Retaining - 12"	Variable	0' 0"	0' 0"	0' 0"	☐
2	Footing 20'	0' 9"	0' 0"			☐

Highlight Layer 1.

Use the Down button to move the Footing 20' below the Retaining – 12" Concrete.

Down

8.

<< Preview

Select the Preview button to expand the dialog and see what the wall looks like.

9.

Family: Stacked Wall
Type: Concrete with Footing
Offset: Core Centerline ▾
  Wall Centerline
  Core Centerline
  Finish Face: Exterior
  Finish Face: Interior
  Core Face: Exterior
  Core Face: Interior

Types

Set the Offset to **Finish Face: Interior**.

10.

Notice how the wall adjusts in the preview so that the interior faces of the walls are now flush.

Click **OK** to close the dialog.

11.

Autodesk Revit 2024
Warning
Highlighted elements are joined but do not intersect.

Show   More Info   Expand >>

Unjoin Elements   OK   Cancel

Click **Unjoin Elements**.

Click **OK**.

12.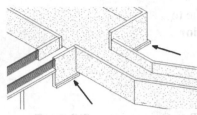

Notice how the stacked walls updated in the model.

Exercise 1-4

Placing a Cut in a Wall

Drawing Name: **walls.rvt**
Estimated Time to Completion: 20 Minutes

Scope
Placing a horizontal or vertical cut in a wall using Reveal.
Creating a cut in a wall using Edit Profile.

Solution

1. 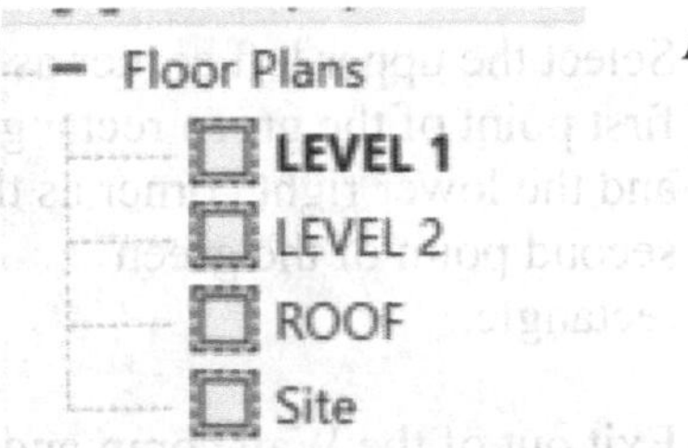 Activate **Level 1** Floor Plan.

   — Floor Plans
   ☐ **LEVEL 1**
   ☐ LEVEL 2
   ☐ ROOF
   ☐ Site

   A floor, some model lines and some grid lines are already placed in the drawing.

2.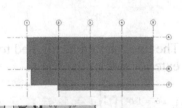

   Select the **Wall** tool from the Architecture ribbon.

3.

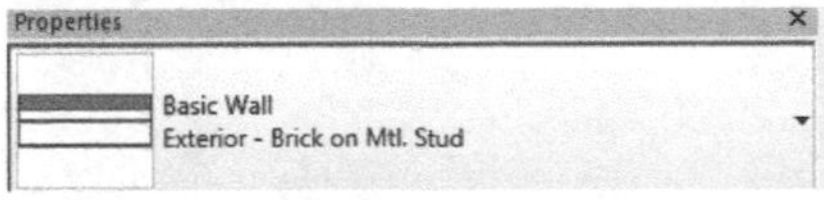

   Set the wall type to **Exterior - Brick on Mtl. Stud** using the Type Selector on the Properties pane.

4. 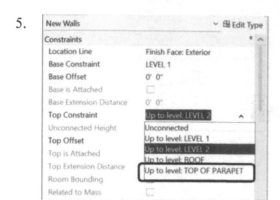 Set the Location Line to **Finish Face: Exterior**.

5. 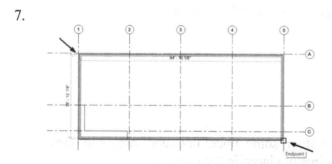 Set the Top Constraint to **TOP OF PARAPET**.

Set the Top Offset to **0"**.

6. 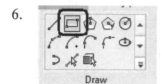 Select the **Rectangle** tool from the Draw panel.

7. 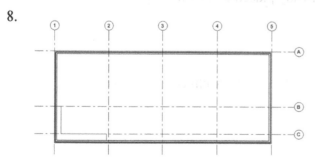 Select the upper left corner as the first point of the green rectangle and the lower right corner as the second point of the green rectangle.

Exit out of the Wall command.

8. The lines should be aligned to the exterior side of the walls.

Set the Detail Level to **Medium**.

9. Select a wall.

Note if the exterior side of the wall is oriented correctly. If it needs to be re-oriented, use the Flip Arrows to position it correctly.

10.

Elevations (Building Elevation)
- East
- North
- South
- West

Switch to a South elevation.

11.

Edit Profile

Select the South wall.

Click **Edit Profile** on the ribbon.

12. ns   Modify | Walls > Edit P

Select the **Line** tool from the Draw Panel.

Draw

13.

Start the line using the endpoint located to the left of Grid 2.

Draw the line vertically and select an endpoint above Level 2.

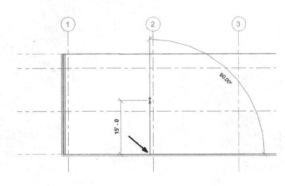

14.

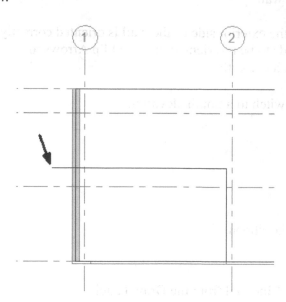

Drag the mouse to the left to place a horizontal line and click outside the building to select the endpoint for the horizontal line.

Exit the line command.

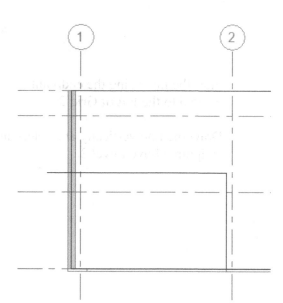

Your profile should look like this.

15.

Select the horizontal line.

Modify the temporary dimension to **20' 8"**.

Note that you don't need to use the feet or inches symbols, just place a space between the units.

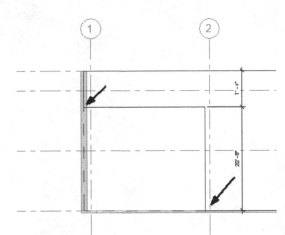

16. Select the **TRIM** tool from the ribbon.

17. Trim the two corners to create a single continuous wall boundary.

18. Select the **Green Check** to exit the edit profile mode.

19. Switch to a **West** elevation.

Elevations (Building Elevation)
- East
- North
- South
- **West**

20. 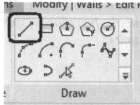 Select the West wall.

    Click **Edit Profile** on the ribbon.

21. 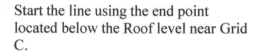 Select the **Line** tool from the Draw Panel.

22. 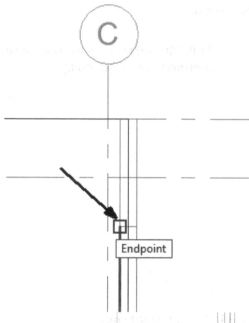 Start the line using the end point located below the Roof level near Grid C.

23. Draw a horizontal line across Grid B.

24.

Select the endpoint at Level 1 and Grid B and draw a vertical line up to intersect with the horizontal line.

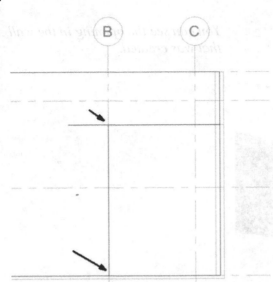

25. Select the **TRIM** tool from the ribbon.

26.

Trim the three corners to create a single continuous wall boundary.

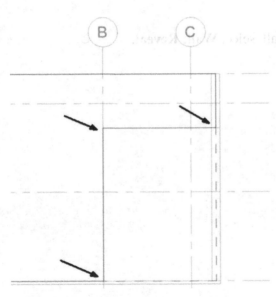

27. Select the **Green Check** to exit the edit profile mode.

28. 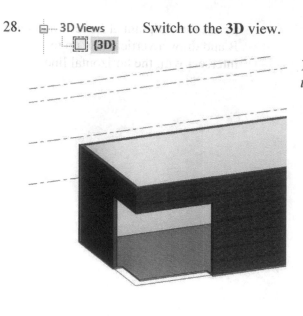 Switch to the **3D** view.

You can see the opening in the wall that was created.

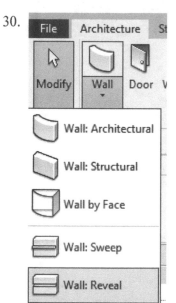

29. 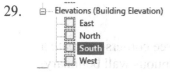 Activate the **South** elevation.

30. Under Wall, select **Wall: Reveal**.

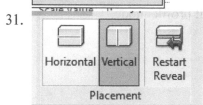

31. Enable **Vertical** on the ribbon.

32.

Click on Grid 3 to place the reveal.

Exit the command.

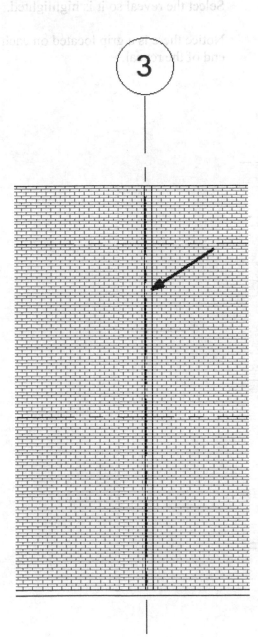

33.

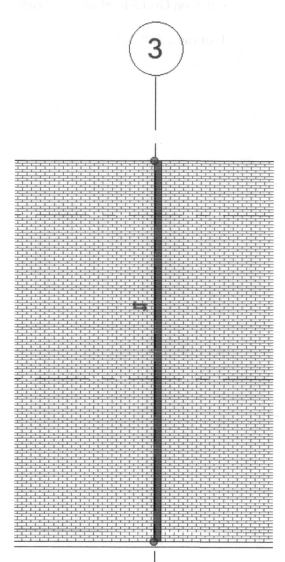

Select the reveal so it is highlighted.

Notice there is a grip located on each end of the reveal.

34.

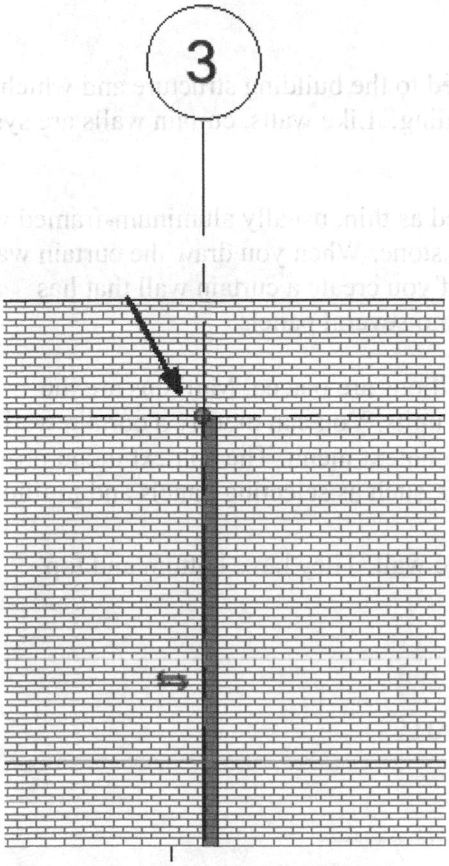

Drag the top end of the reveal to the intersection of Grid 3 and Roof Level.

35.

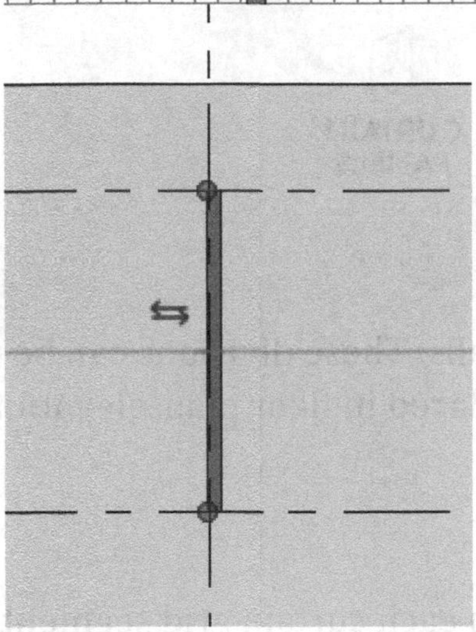

Drag the bottom end of the reveal to the intersection of Grid 3 and Level 2.

Close the drawing file.

Curtain Walls

A curtain wall is any exterior wall that is attached to the building structure and which does not carry the floor or roof loads of the building. Like walls, curtain walls are system families.

In common usage, curtain walls are often defined as thin, usually aluminum-framed walls containing in-fills of glass, metal panels, or thin stone. When you draw the curtain wall, a single panel is extended the length of the wall. If you create a curtain wall that has automatic curtain grids, the wall is subdivided into several panels.

In a curtain wall, grid lines define where the mullions are placed. Mullions are the structural elements that divide adjacent window units. You can modify a curtain wall by selecting the wall and right-clicking to access a context menu. The context menu provides several choices for manipulating the curtain wall, such as selecting panels and mullions.

Curtain Walls contains most properties of a basic wall. They have bottom and top constraints and their profile can be modified.

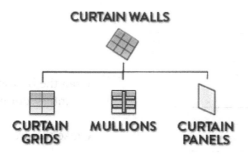

CURTAIN GRIDS
Curtain grids are divisions created on the walls. These divisions can be horizontal or vertical. Curtain grids can be placed in floor plan, elevations, and 3D views.

MULLIONS
Mullions are elements that can be created on each curtain grid segment, as well as on each curtain wall extremity.

CURTAIN PANELS
Curtain panels are rectangular elements located between each curtain grid.

Exercise 1-5

Curtain Walls

Drawing Name: **linear_curtain_wall.rvt**
Estimated Time to Completion: 30 Minutes

Scope
Placing a Curtain Walls
Adding and removing Curtain Wall Grids
Adding a Curtain Wall Door
Adding and removing mullions

Solution

1.
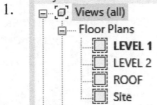
Activate **Level 1** Floor Plan.

2.
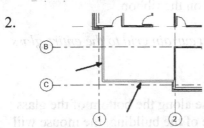
Select the two walls located between Grids 1 and 2 and Grids B and C.

Hint: Hold down the CTL key to select more than one element.

3.
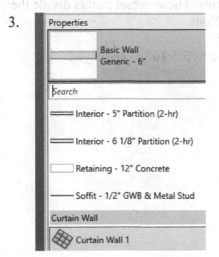
Use the Type Selector to assign **Curtain Wall 1** to the selected walls.

4. 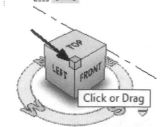   Switch to a **3D** View.

5.   Use the Viewcube to orient the view to a SW perspective view.

6.   Zoom into the Southwest corner of the building model.

7.   Select **Curtain Grid** from the ribbon.

Curtain
Grid

8.   Enable **All Segments** on the ribbon.

All segments applies a curtain grid to the entire glass panel.

9.   If you hover the mouse along the bottom of the glass panel on the west side of the building, the mouse will snap to different points. These preset points divide the panel into equal sections.

Click when the value **5' 5 3/16"** is displayed.

10.

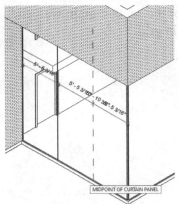

MIDPOINT OF CURTAIN PANEL

Click a second time to place a curtain grid at the midpoint of the curtain panel.

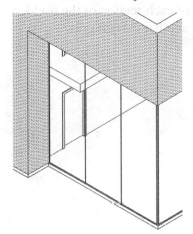

If you cancel out of the command, you see the glass panel has been divided into three equal panes.

11.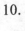

Curtain
Grid

Select **Curtain Grid** from the ribbon.

12.

Add-Ins   Modify | Place Curtain Grid

All   One   All Except   Restart
Segments Segment   Picked   Curtain Grid

Placement

Enable **All Segments** on the ribbon.

All segments applies a curtain grid to the entire glass panel.

13.

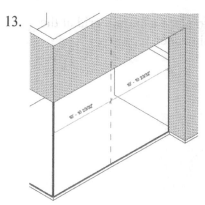

Locate the midpoint of the south curtain wall and select to place a curtain grid.

14.

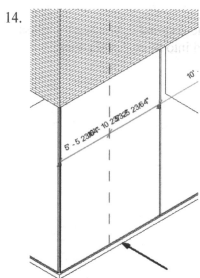

Locate the midpoint of each section of the divided curtain wall to place another curtain grid.

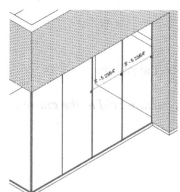

Cancel out of the command so you can inspect your curtain wall.

The south curtain wall should have four vertical glass panels of equal width of 5' 5-23/64".

15.

Select **Curtain Grid** from the ribbon.

Curtain N
Grid

16.

Enable **All Segments** on the ribbon.

All segments applies a curtain grid to the entire glass panel.

17.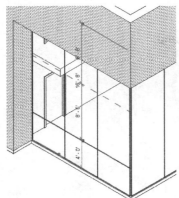

Place a horizontal curtain grid by hovering over the vertical edge of the curtain wall.

Locate the bottom grid 4' above Level 1.

18.

Place a second horizontal grid 8' above first horizontal grid.

19.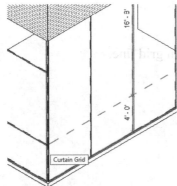

Move over to the west curtain wall.

The curtain grid tool should automatically snap to the horizontal grids placed on the west wall.

Add two horizontal grids at the same distances on the south curtain wall.

The curtain wall should look as shown.

We will add a door in the area indicated by the arrow, but first we need to define a panel the correct size.

20. Select the **Curtain Grid** tool from the ribbon.

21. 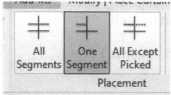 Set the Placement to **One Segment**.

22. 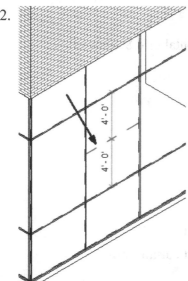 Place the segment at the midpoint of the middle glass panel.

ESC out of the command.

23. 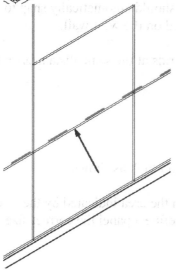 Select the lower curtain grid line.

24. 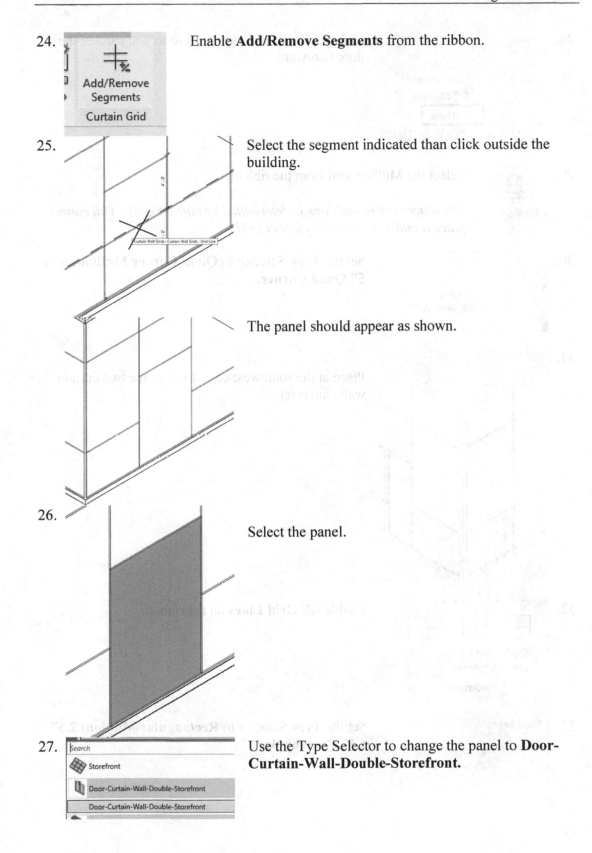 Enable **Add/Remove Segments** from the ribbon.

Add/Remove
Segments

Curtain Grid

25. Select the segment indicated than click outside the building.

Curtain Wall Grids : Curtain Wall Grids : Grid Line

The panel should appear as shown.

26. Select the panel.

27. Use the Type Selector to change the panel to **Door-Curtain-Wall-Double-Storefront.**

Search

Storefront

Door-Curtain-Wall-Double-Storefront

Door-Curtain-Wall-Double-Storefront

28. Change the Detail Level to **Fine** so you can see the door hardware.

29. Select the Mullion tool from the ribbon.

The placement of mullions is determined by curtain grids. You cannot place a mullion without a curtain grid.

30.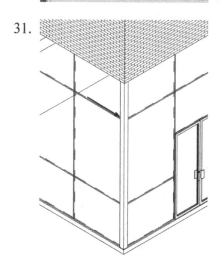

Set the Type Selector to **Quad Corner Mullion: 5" x 5" Quad Corner**.

31. Place at the southwest corner where the two curtain walls intersect.

32.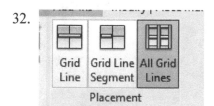

Enable **All Grid Lines** on the ribbon.

33.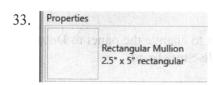

Set the Type Selector to **Rectangular Mullion: 2.5" x 5" Rectangular**.

34.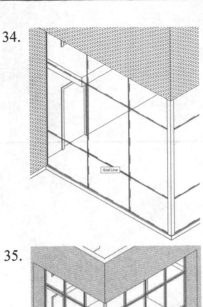
Select the Grid Line on the west curtain wall.

35.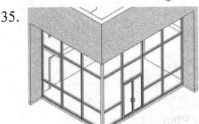
Select the grid lines on the south curtain wall.

ESC out of the command.

36.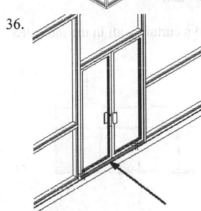
Use the TAB key to cycle through selections until the mullion below the door is selected.

Delete the mullion by clicking the DELETE key on the keyboard or using the Delete tool on the ribbon.

37.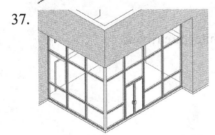
The curtain wall should appear as shown.

Save as *ex1-5.rvt*.

Exercise 1-6
Embedded Curtain Walls

Drawing Name: **embedded curtain wall.rvt**
Estimated Time to Completion: 20 Minutes

Scope
Placing a Curtain Wall inside an existing wall
Adding and removing Curtain Wall Grids
Adding a Curtain Wall Door
Adding and removing mullions

Solution

1.

   Activate **Level 1** Floor Plan.

   We are going to embed a curtain wall in the indicated south wall.

   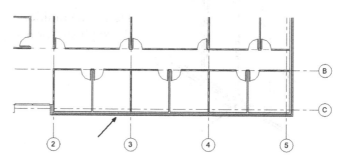

2.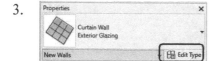

   Select the **Wall** tool from the Architecture ribbon.

3. Select the **Curtain Wall: Exterior Glazing** from the Type Selector.

   Click **Edit Type**.

4.

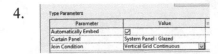

Enable **Automatically Embed**.

Set the Curtain Panel to **System Panel: Glazed**.
Set the Join Condition to **Vertical Grid Continuous**.
Under Vertical Grid:
Set the Layout to **None**.

5.

Vertical Grid	
Layout	None
Spacing	2' 0"
Adjust for Mullion Size	☐
Horizontal Grid	
Layout	None
Spacing	4' 0"
Adjust for Mullion Size	☐
Vertical Mullions	
Interior Type	None
Border 1 Type	None
Border 2 Type	None
Horizontal Mullions	
Interior Type	None
Border 1 Type	None
Border 2 Type	None

Under Horizontal Grid:
Set the Layout to **None**.

Set the Vertical and Horizontal Mullions to **None**.

Click **OK**.

6.

In the Properties palette:

Set the Base Constraint to **LEVEL 1**.
Set the Base Offset to **2' 8"**.
Set the Top Constraint to **LEVEL 1**.
Set the Top Offset to **8' 8"**.

7.

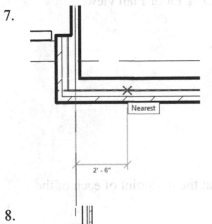

Set the start point at **2' 6"** to the right of Grid 2.

8.

Set the end point at Grid 5.

ESC to exit the command.

9.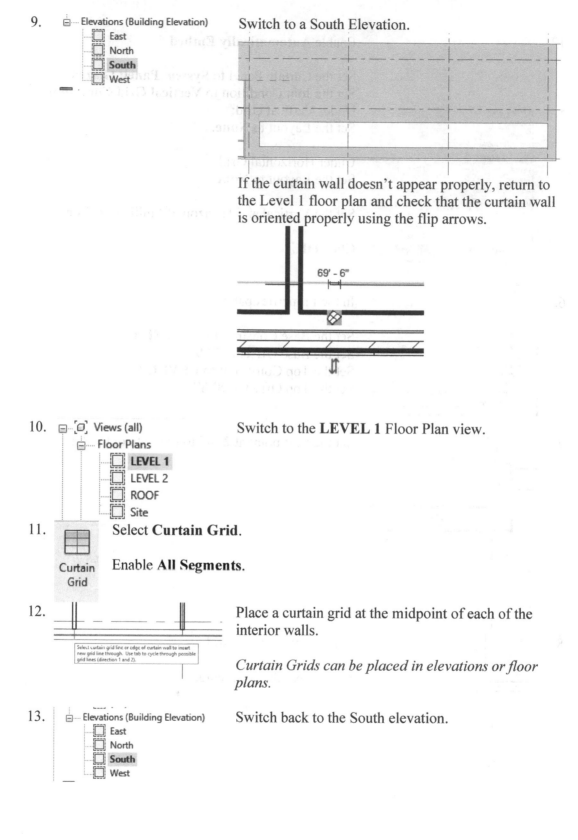

Elevations (Building Elevation)
- East
- North
- **South**
- West

Switch to a South Elevation.

If the curtain wall doesn't appear properly, return to the Level 1 floor plan and check that the curtain wall is oriented properly using the flip arrows.

69' - 6"

10.

Views (all)
- Floor Plans
  - **LEVEL 1**
  - LEVEL 2
  - ROOF
  - Site

Switch to the **LEVEL 1** Floor Plan view.

11.

Curtain Grid

Select **Curtain Grid**.

Enable **All Segments**.

12.

Select curtain grid line or edge of curtain wall to insert new grid line through. Use tab to cycle through possible grid lines (direction 1 and 2).

Place a curtain grid at the midpoint of each of the interior walls.

Curtain Grids can be placed in elevations or floor plans.

13.

Elevations (Building Elevation)
- East
- North
- **South**
- West

Switch back to the South elevation.

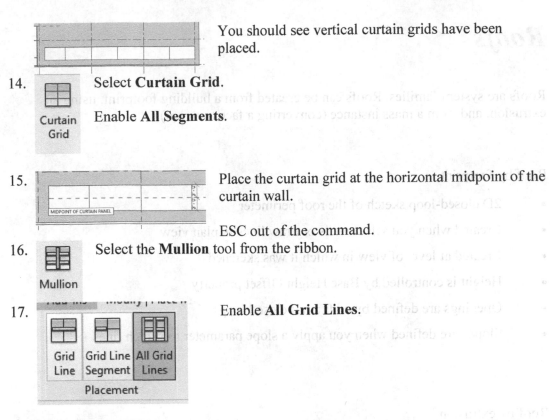

You should see vertical curtain grids have been placed.

14. Select **Curtain Grid**.

Enable **All Segments**.

15. Place the curtain grid at the horizontal midpoint of the curtain wall.

ESC out of the command.

16. Select the **Mullion** tool from the ribbon.

17. Enable **All Grid Lines**.

18. Select the **Rectangular Mullion: 1" Square** from the Type Selector.

19.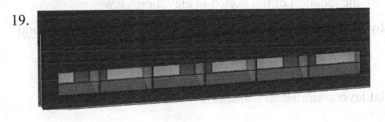

Click on one of the curtain grids to place the mullions.

Switch to a 3D view to inspect your embedded curtain wall.

20. If the embedded curtain wall does not create an opening in the south wall, select the south wall, click Edit Profile and then place a rectangle at the location of the curtain wall.

Save as *ex1-6.rvt*.

Roofs

Roofs are system families. Roofs can be created from a building footprint, using an extrusion, and from a mass instance (converting a face to a roof).

Roof by footprint

- 2D closed-loop sketch of the roof perimeter

- Created when you select walls or draw lines in plan view

- Created at level of view in which it was sketched

- Height is controlled by Base Height Offset property

- Openings are defined by additional closed loops

- Slopes are defined when you apply a slope parameter to sketch lines

Roof by extrusion

- Open-loop sketch of the roof profile

- Created when you use lines and arcs to sketch the profile in an elevation view

- Height is controlled by the location of the sketch in elevation view

- Depth is calculated by Revit based on size of sketch, unless you specify start and end points.

Roofs are defined by material layers, similar to walls.

Exercise 1-7

Creating a Roof by Footprint

Drawing Name: **i_roofs.rvt**
Estimated Time to Completion: 10 Minutes

Scope
Create a roof.

Solution

1. 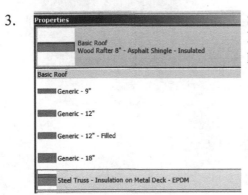 Activate the **T.O. Parapet** floor plan.

2. Select **Roof by Footprint** from the tab.

3. Select **Steel Truss - Insulation on Metal Deck - EPDM** using the Type Selector on the Properties panel.

4.  Select **Pick Walls** mode.

5. 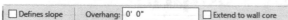 On the Options bar,
uncheck **Defines slope**.
Set the Overhang to **0'-0"**.
Uncheck **Extend to Wall Core**.

6. 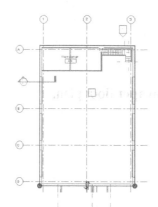 Select all the exterior walls.

7.  Use the **Trim** tool from the Modify panel to
create a closed boundary, if necessary.

8. 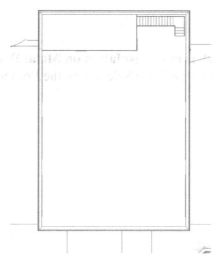 Align the boundary lines with the exterior face
of each wall.

9. 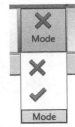 Select the **Green Check** on the Model panel to **Finish Roof**.

10. Close without saving.

Exercise 1-8
Creating a Roof by Extrusion

Drawing Name: **i_roofs_extrusion.rvt**
Estimated Time to Completion: 30 Minutes

Scope
Create a roof by extrusion.
Modify a roof.

Solution

1.  ⊟ 3D Views
        □ {3D}
    Activate the **3D view.**

2.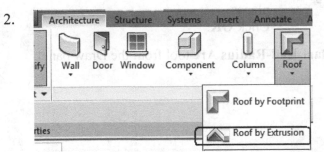
    Activate the Architecture tab.
    Select **Roof by Extrusion** under the Build panel.

3. 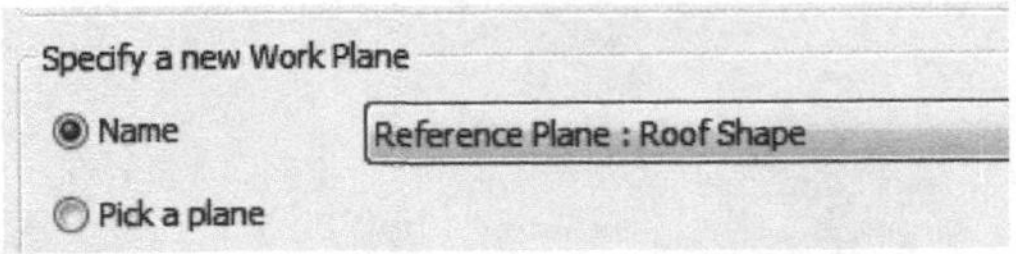 Specify a new Work Plane
    ◉ Name   Reference Plane : Roof Shape
    ○ Pick a plane
    Enable **Name**.
    Select **Roof Shape** from the list of reference planes.
    Click **OK**.

4.

Roof Reference Level and Offset

Level: Upper Roof

Offset: 0' 0"

OK   Cancel

Select **Upper Roof** from the list.
Click **OK**.

5.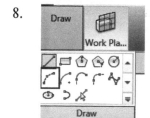

Select the **Show** tool from the Work Plane panel.
This will display the active work plane.

6.

Right click on the ViewCube's ring.
Select **Orient to a Plane**.

Go Home   Home

Save View

Lock to Selection

Set Current View as Home

Set Front to View   ▶

Reset Front

✓ Show Compass

Orient to View   ▶

Orient to a Direction   ▶

Orient to a Plane...

7.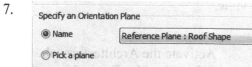

Specify an Orientation Plane

◉ Name   Reference Plane : Roof Shape

◯ Pick a plane

Enable **Name**.
Select **Roof Shape** from the list of reference planes.
Click **OK**.

8.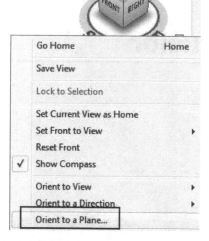

Select the **Start-End-Radius Arc** tool from the Draw panel.

9.

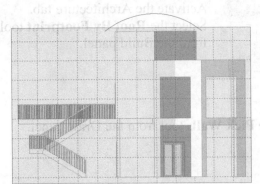

Draw an arc over the building as shown.

10.

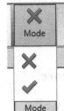

Select the **Green Check** under the Mode panel to finish the roof.

11.

Switch to an isometric 3D view.

12.

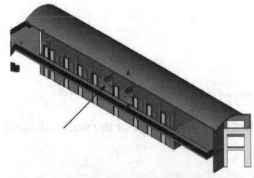

Hover over the wall below the roof.
Click TAB.
All connected walls will be selected.

13.

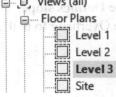

Select **Attach Top/Base** from the Modify Wall panel.
Then select the roof.
The walls will adjust to meet the roof.
Orbit the model to inspect the walls.

14.

Activate the **Level 3** floor plan.

15.  Activate the Architecture tab.
Select the **Roof By Footprint** tool from the Build panel.

16. 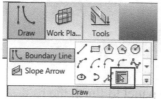 Select the **Pick Walls** tool from the Draw panel.

17. On the Options bar:

Enable **Defines slope**.
Set the Overhang to **2' 0"**.
Enable **Extend to wall core**.

18. 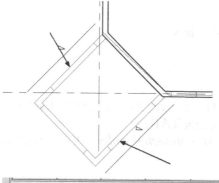 Select the outside edge of the two walls indicated.

19. Disable **Defines slope**.

20. 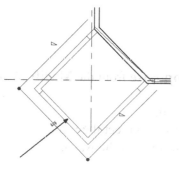 Select the outside edge of the wall indicated.

21. Disable **Defines slope**.
Set the Overhang to **0' 0"**.

22. 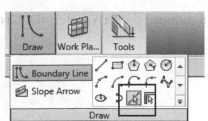 Select the **Pick Line** tool from the Draw panel.

23. 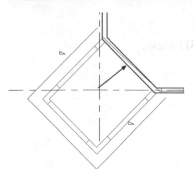 Pick the exterior side of the wall indicated.

24. 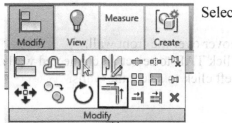 Select the **Trim** tool from the Modify panel.

25. 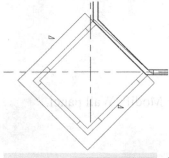 Trim the roof boundaries so they form a closed polygon.

26. Set the Type to **Generic- 9″**. Click **OK**.

27.  Set the Slope to **6″/12″**.

28. Verify that both the back and front boundary lines have no slope assigned.

    If they have a slope symbol, select the back and front boundary and lines and uncheck **Defines Slope** on the Options bar.

29. 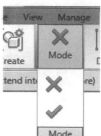 Once the sketch is closed and trimmed properly, select the **Green Check** on the Mode panel.

30. 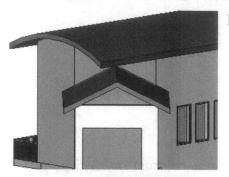 Return to a 3D view.

31. 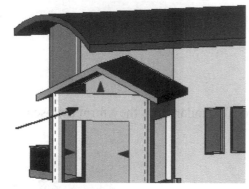 Hover over the front wall.
Click TAB to select all connected walls.
Left click to select the walls.

32.  Select Attach Top/Base from the Modify Wall panel.

33.  Select the roof to attach the wall.

34.

Error - cannot be ignored   --   1 Error,

Can't keep wall and target joined

<<   1 of 2   >>   Show

Unjoin Elements

Click **Unjoin Elements**.

Left click in the window to release the selection.

Orbit around the model to inspect the roofs.

35. Close without saving.

Exercise 1-9
Add Split Lines to a Roof

Drawing Name: **Split Lines Roof.rvt**
Estimated Time to Completion: 20 Minutes

Scope
Create a roof by footprint.
Modify a roof using split lines.

Solution

1.

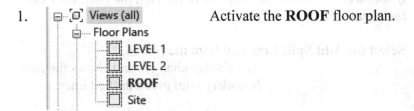

Views (all)
   Floor Plans
     LEVEL 1
     LEVEL 2
     ROOF
     Site

Activate the **ROOF** floor plan.

2.

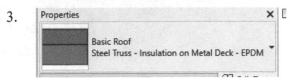

Component   Column   Roof

Roof by Footprint

Select the **Roof by Footprint** tool from the ribbon.

3.

Properties   ×

Basic Roof
Steel Truss - Insulation on Metal Deck - EPDM

Verify that the Type is set to Steel Truss – Insulation on Metal Deck – EPDM.

4. 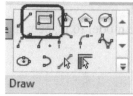 Disable **Defines slope**.
Set the Overhang to **0' 0"**.
Disable **Extend to wall core.**

5. 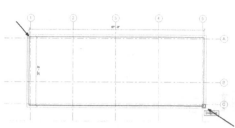 Select the **Rectangle** tool from the Draw panel.

6. Select the two inside corners of the parapet walls to define the rectangle.

7. Select **Green Check** to complete the roof.

8. *Add Split Line*   *The roof should still be highlighted.*
If you accidentally clicked ESC or deselected the roof, click on the roof to select it.

Select the **Add Split Line** tool from the ribbon.

The display changes and shows the roof boundary with green dashed lines.

9. Start the split line at Grid 3 and Grid A.

10.

End the split line at Grid 3 and Grid C.

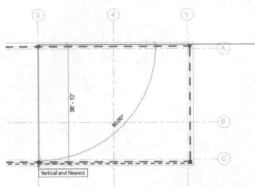

11.

Draw a split line from C1 to B2.

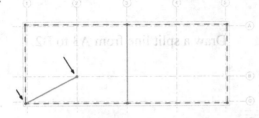

12.

Draw a split line from A1 to B2.

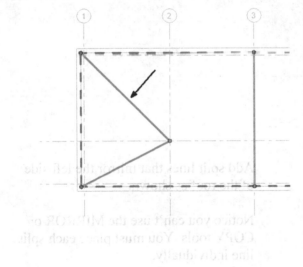

13.   Draw a split line from C3 to B2.

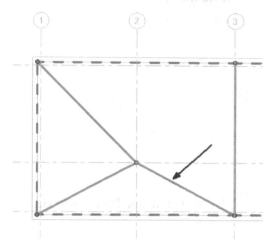

14.   Draw a split line from A3 to B2.

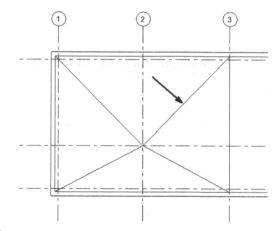

15.   Add split lines that mirror the left side of the roof as shown.

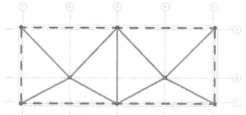

Notice you can't use the MIRROR or COPY tools. You must place each split line individually.

16.   Enable **Modify Sub Elements**.

17.

Select the point at B2.

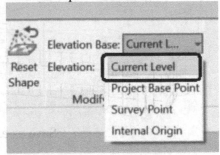

On the ribbon: set the Elevation Base to **Current Level**.

Set the Elevation to **2'-0"**.

Repeat for the point at B4.

18. Select **Edit Type** on the Properties palette.

Properties

Basic Roof
Steel Truss - Insulation on Metal Deck - EPDM

Roofs (1)   Edit Type

19. Click **Edit** next to Structure.

Type Parameters	
Parameter	Value
Construction	
Structure	Edit...
Default Thickness	2' 4 1/2"

20. Note that Layer 2 has Variable enabled.

Layers

	Function	Material	Thickness	Wraps	Variable
1	Finish 1 [4]	Roofing, EPD	0' 0 1/4"		☐
2	Thermal/Air Layer [3]	Rigid insulatio	1' 0"		☑
3	Core Boundary	Layers Above	0' 0"		
4	Structure [1]	Metal Deck	0' 0 1/4"		☐
5	Structure [1]	Structure, Ste	1' 4"		☐
6	Core Boundary	Layers Below W	0' 0"		

This is because we added a slope with split lines.

Click **OK** twice to close the dialog boxes.

21. Click **Modify** on the ribbon to exit the roof editing mode.

Modify
Select ▾

22.

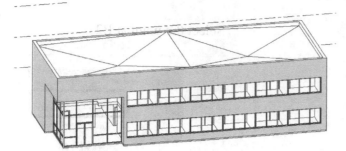

Switch to a 3D view to see how the roof appears.

Change the display to Realistic to see how the material layers display for the roof.

23.   Save as *ex1-9.rvt*.

Exercise 1-10
Add Spot Elevation and Slope Annotations

Drawing Name: **spot elevation.rvt**
Estimated Time to Completion: 10 Minutes

Scope
Add a spot slope annotation to a roof.
Add a spot elevation annotation to a roof.

Solution

1.

> Views (all)
> └─ Floor Plans
>   ├─ LEVEL 1
>   ├─ LEVEL 2
>   ├─ **ROOF**
>   └─ Site

Activate the **ROOF** floor plan.

2.

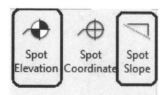

Activate the Annotate ribbon.

Locate the Spot Slope and Spot Elevation tools on the Dimension panel.

Select the **Spot Slope** tool.

3.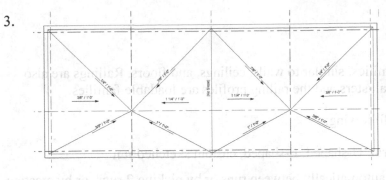

Select each split line.

You will need to select the split line, then click to select which side of the split line you want the label to be placed.

You can also place spot slope labels between the split lines.

4. 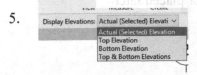 Select the **Spot Elevation** tool.

5.  On the Options bar:

Enable **Actual (Selected) Elevation**.

6. 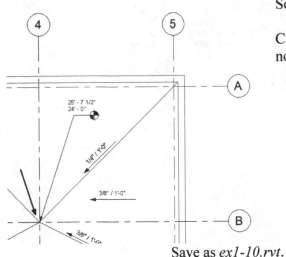 Select the intersection point at B2.

Click to place the leader and elevation note.

7. On the Options bar:

Enable**Top & Bottom Elevation**.

8. Select the intersection point at B4.

Click to place the leader and elevation note.

9. Save as *ex1-10.rvt*.

Stairs

Revit stairs are system families, similar to walls, ceilings, and floors. Railings are also system families, but the balusters and the railing profiles are loadable families.

A stair can consist of the following:

- Runs: straight, spiral, U-shaped, L-shaped, custom sketched run

- Landings: created automatically between runs or by picking 2 runs, or by creating a custom sketched landing

- Supports (side and center): created automatically with the runs or by picking a run or landing edge

- Railings: automatically generated during creation or placed later

Be familiar with all the properties that control stairs.

	BASE AND TOP LEVELS: Stairs are based on selected levels that already exist in the project. You can add an offset on these levels if required.
	DESIRED STAIR HEIGHT: Total distance between the base and the top of the stairs, including offsets.
	DESIRED NUMBER OF RISERS: Automatically calculated by Revit, dividing Stair Height by Maximum Riser Height. You can change this number, which will modify the stair slope.
	ACTUAL NUMBER OF RISERS: The number of risers you modeled so far.
	MAXIMUM RISER HEIGHT: Riser height for your stair will never go above this value. This parameter is set on the stair type. Usually on par with code requirements.
	ACTUAL RISER HEIGHT: This distance is automatically calculated by Revit, dividing the Stair Height by the Desired Number of Risers.
	MINIMUM TREAD DEPTH: On the stair type, specify the minimum tread depth. When you start modeling your stair, you can go above this number, but not below.

	ACTUAL TREAD DEPTH: By default, this value is equal to minimum tread depth set in the stair type. However, you can set a bigger value if you want more depth.
	MINIMUM RUN WIDTH: Set on the stair type, you can specify the minimum run width. This does not include support (stringers).
	ACTUAL RUN WIDTH: By default, this will be the same as the minimum run width. You can set a higher value than the minimum, but a lower value will result in a Warning.

Exercise 1-11

Creating Stairs by Sketch

Drawing Name: **i_stairs.rvt**
Estimated Time to Completion: 20 Minutes

Scope
Place stairs using reference work planes.

Solution

1. 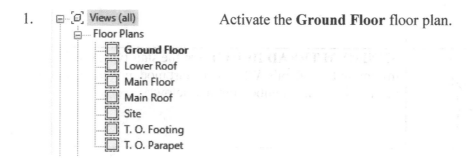 Activate the **Ground Floor** floor plan.

2.   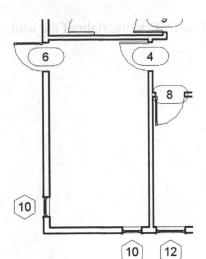 Zoom into the lower left corner of the building.

3.   Select the **Reference Plane** tool from the Work Plane panel on the Architecture ribbon.

4.   Offset: 2' 4"   Set the Offset to **2' 4"** on the Options bar.

5.   Select the **Pick Lines** tool from the Draw panel.

6.   Place a vertical reference plane 2' 4" to the left of the wall where Door 4 is placed.

7. 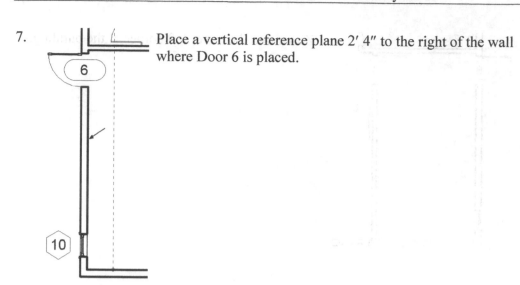 Place a vertical reference plane 2′ 4″ to the right of the wall where Door 6 is placed.

8. Offset: 4′ 4″ ☐ Lock   Set the Offset to **4′ 4″** on the Options bar.

9. 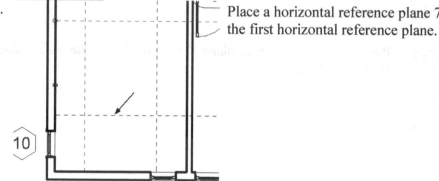 Place a horizontal reference plane 4′ 4″ above the wall where Window 10 is located.

10. Offset: 7′ 4″ ☐   Set the Offset to **7′ 4″** on the Options bar.

11. Place a horizontal reference plane 7′ 4″ above the first horizontal reference plane.

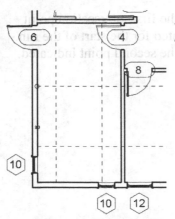

Four reference planes – two horizontal and two vertical should be placed in the room.

12. Activate the Architecture ribbon.
Select the **Stair** tool from the Circulation panel.

Stair

13. Edit Type   Select **Edit Type** from the Properties pane.

14.

Type Parameters	
Parameter	
Calculation Rules	
Maximum Riser Height	0' 8"
Minimum Tread Depth	0' 11"
Minimum Run Width	4' 0"
Calculation Rules	

Change the Maximum Riser Height to **8"**.
Set the stair width to **4' 0"**.
Click **OK**.

15.

Dimensions	
Desired Number of Risers	16
Actual Number of Risers	1
Actual Riser Height	0' 7 1/2"
Actual Tread Depth	0' 11"
Tread/Riser Start Number	1

Set the desired number of risers to **16**.

16.

Run		
Landing		
Support		
Components		

Select the **Run** tool from the Draw panel on the ribbon.

17. Location Line: Run: Center   Set the Location line to **Run: Center** on the Options bar.

18.

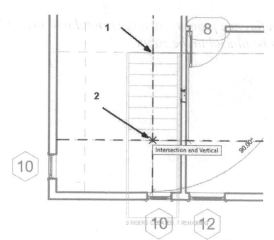

Pick the first intersection point indicated for the start of the run. Pick the second point indicated.

19.

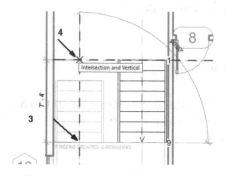

Moving clockwise, select the lower intersection point #3 and then the upper intersection point #4.

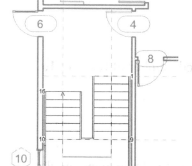

U-shaped stairs are placed.

20. Select the **Green Check** on the Mode panel to **Finish Stairs**.

21. Close without saving.

Exercise 1-12
Creating Stairs by Component

Drawing Name: **i_stairs_component.rvt**
Estimated Time to Completion: 10 Minutes

Scope
Place stairs using the component method.

Solution

1. 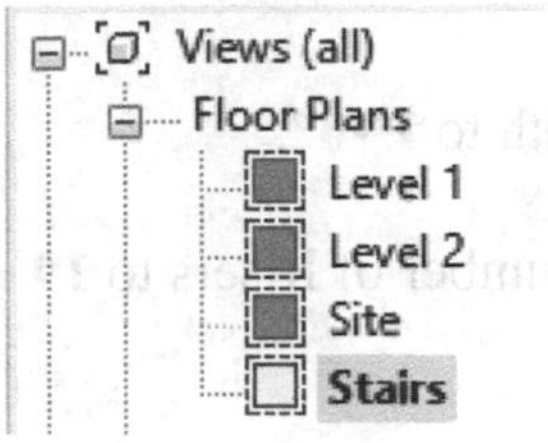 [Views (all) tree showing: Views (all), Floor Plans, Level 1, Level 2, Site, Stairs]   Activate the **Stairs** floor plan.

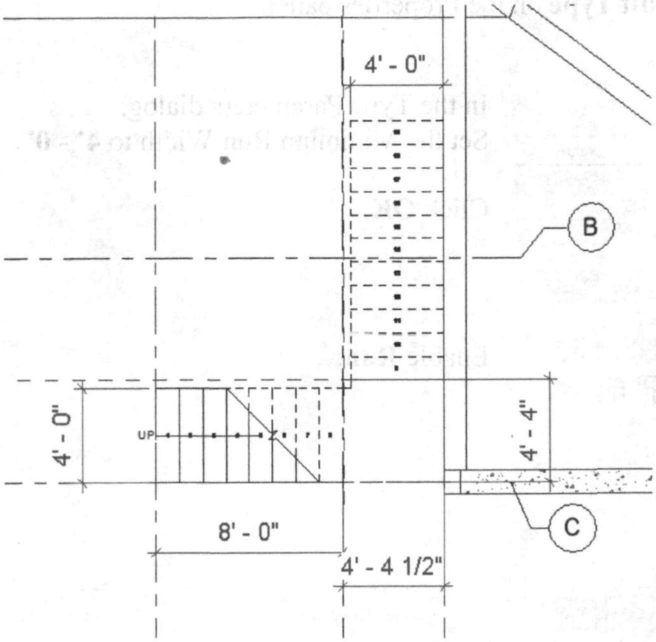

Stairs will be placed using the dimensions shown.

There are eight horizontal risers and 11 vertical risers.

Use the reference planes provided as guides.

Use the Cast in Place Monolithic Stair.

Set the Base Level to Level 1.

Set the Top Level to Level 2.

Set the Tread Depth to 1'-0".

Set the Desired Number of Risers to 19.

2. 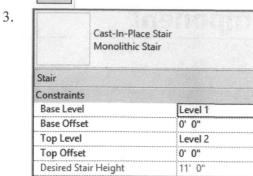 Activate the Architecture ribbon.

Select the **Stair** tool.

3. Set the Stair type to **Cast-In-Place Stair: Monolithic Stair**.

Stair	
Cast-In-Place Stair Monolithic Stair	

Constraints	
Base Level	Level 1
Base Offset	0' 0"
Top Level	Level 2
Top Offset	0' 0"
Desired Stair Height	11' 0"

Set the Base Level to **Level 1**.

Set the Top Level to **Level 2**.

4.

Dimensions	
Desired Number of Risers	19
Actual Number of Risers	1
Actual Riser Height	0' 6 243/256"
Actual Tread Depth	1' 0"
Tread/Riser Start Number	1
Identity Data	

Scroll down.

Set the Tread Depth to **1'-0"**.

Set the Desired Number of Risers to **19**.

5. Edit Type — Select **Edit Type** on the Properties panel.

6. In the Type Parameters dialog:
Set the Minimum Run Width to **4' - 0"**.

Type Parameters	
Parameter	Value
Calculation Rules	
Maximum Riser Height	0' 7 11/128"
Minimum Tread Depth	0' 11 3/128"
Minimum Run Width	4' 0"
Calculation Rules	Edit...

Click **OK.**

7. Modify | Create Stair — Run, Landing, Support, Components — Enable **Run**.

8.

Location Line: Exterior Support: Right ∨	Offset: 0' 0"	Actual Run Width: 4' 0"	☑ Automatic Landing

On the Options bar: Set the Location Line: **Exterior Support Right**
Set the Offset to **0' 0"**. Set the Actual Run Width to **4' 0"**.

9. Start the stairs at C6 and end the run at C8.

This will place 8 risers.

7' - 11 243/256"

10.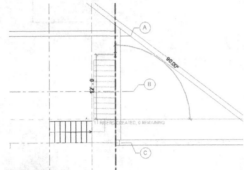

Start the second run at the intersection of the reference plane and Grid line 7.

Drag the cursor straight up vertically and left click to end when you see 11 risers.

11. Green check to complete the stairs.

12. Close without saving.

Landings

Landings are placed automatically when you draw more than one run.

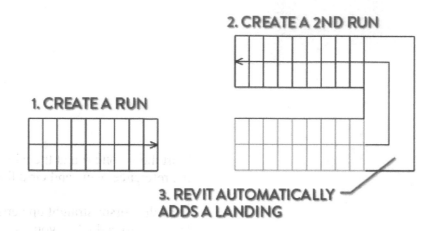

Exercise 1-13
Stair Landings

Drawing Name: **landings.rvt**
Estimated Time to Completion: 10 Minutes

Scope
Create a stair landing

Solution

1. Activate the **Ground Floor – Main Stairs** floor plan.

2. Select **Stair** from the Architecture ribbon.

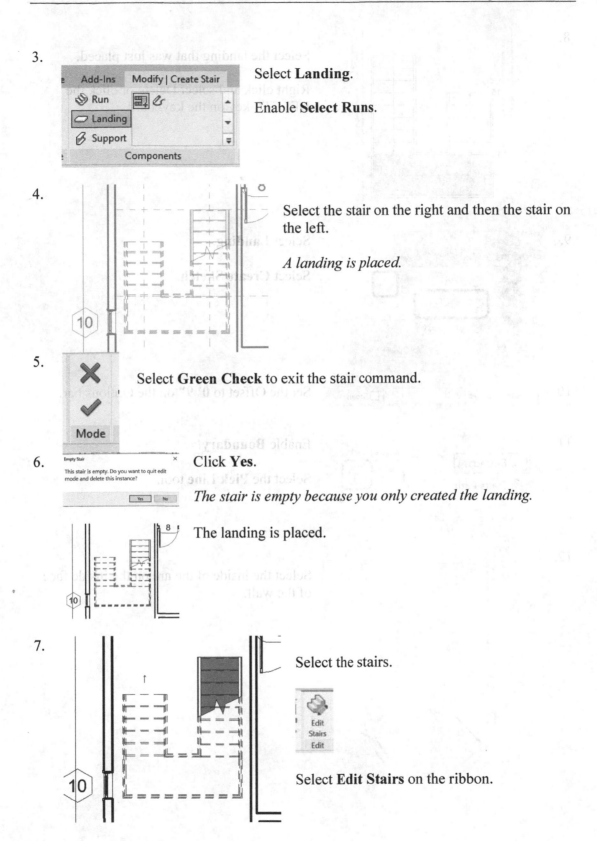

3. Select **Landing**.

Enable **Select Runs**.

4. Select the stair on the right and then the stair on the left.

A landing is placed.

5. Select **Green Check** to exit the stair command.

6. Click **Yes**.

The stair is empty because you only created the landing.

The landing is placed.

7. Select the stairs.

Select **Edit Stairs** on the ribbon.

8.

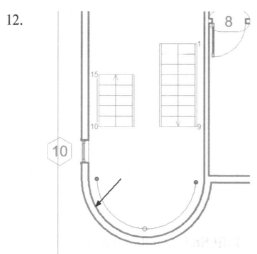

Select the landing that was just placed.

Right click and select Delete or click the DELETE key on the keyboard.

9.

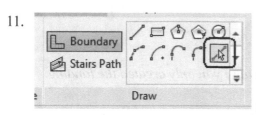

Select **Landing**.

Select **Create Sketch**.

10.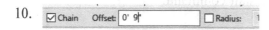

Set the Offset to **0' 9"** on the Options bar.

11.

Enable **Boundary**.

Select the **Pick Line** tool.

12.

Select the inside of the arc on the inside face of the wall.

13.

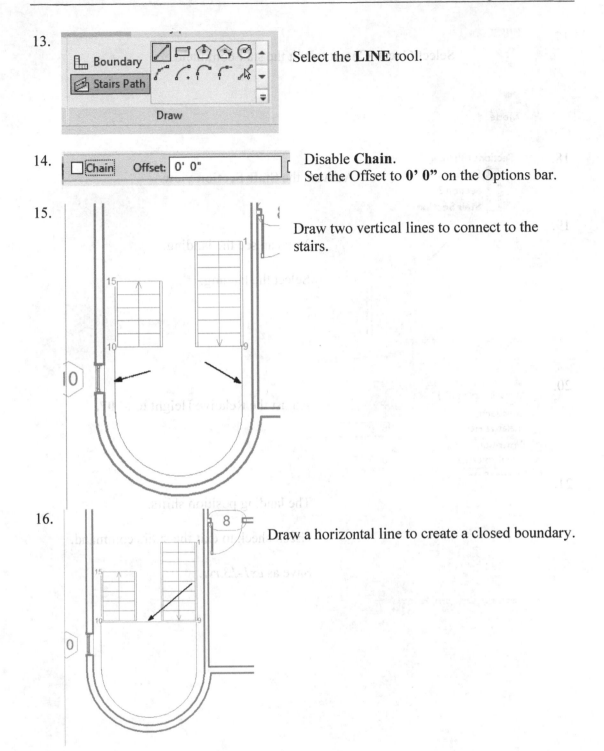

Select the **LINE** tool.

14. Disable **Chain**.
Set the Offset to **0' 0"** on the Options bar.

15. Draw two vertical lines to connect to the stairs.

16. Draw a horizontal line to create a closed boundary.

17.

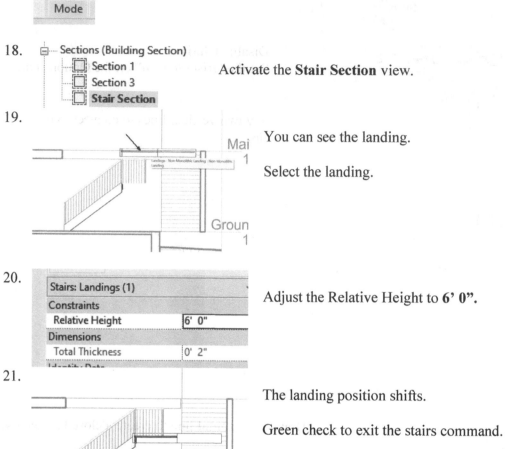

Select **Green Check** to exit the stair command.

18.

─── Sections (Building Section)
    ☐ Section 1
    ☐ Section 3
    ☐ **Stair Section**

Activate the **Stair Section** view.

19.

You can see the landing.

Select the landing.

20.

Stairs: Landings (1)	
Constraints	
Relative Height	6' 0"
Dimensions	
Total Thickness	0' 2"

Adjust the Relative Height to **6' 0"**.

21.

The landing position shifts.

Green check to exit the stairs command.

Save as *ex1-13.rvt*.

Railings

Railings are system families. Railings can be free-standing or they can be hosted. Railing hosts include stairs, ramps, and floors. (This is a possible exam question.)

You need to understand and identify the different components which make up a railing.

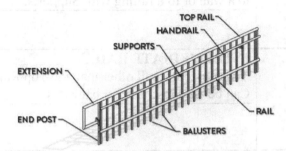

When defining a railing, you should be familiar with the different Type properties.

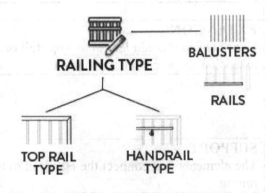

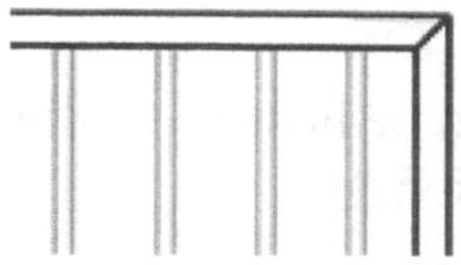	**TOP RAIL** Top rail is the highest horizontal element of a railing. It is created by selecting a 2D profile and a height.
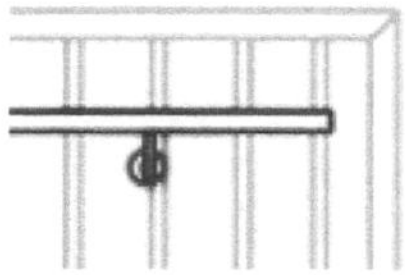	**HANDRAIL** Handrail is an intermediate rail used for hands. They are linked to a wall or to a railing with Supports.
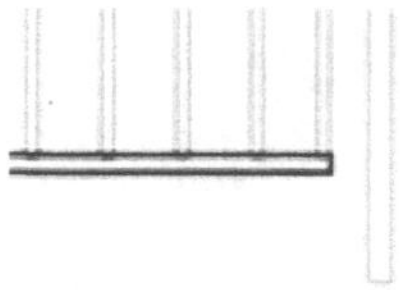	**INTERMEDIATE RAIL** Any horizontal rail other than the Top Rail and the Handrail. Can be used to constrain balusters.
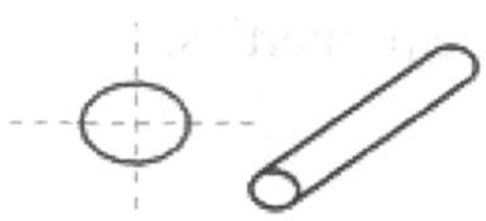	**RAIL 2D PROFILE** Every Rail in Revit is an extrusion from a 2D Profile Family. Use default profiles for simple shapes, or create a custom one for fancy shapes.
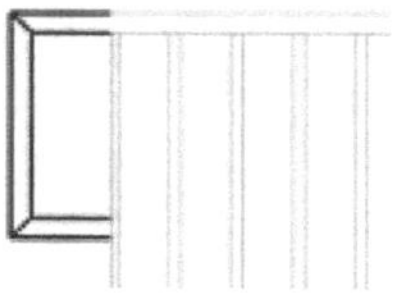	**EXTENSION** Use extension to add length to Top Rail or Handrail. The extension shape can be customized.
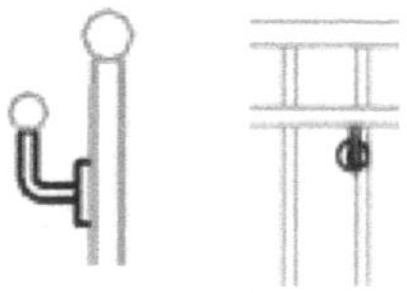	**SUPPORT** The elements that connect the Handrail to the wall or to the railing.

Baluster Elements

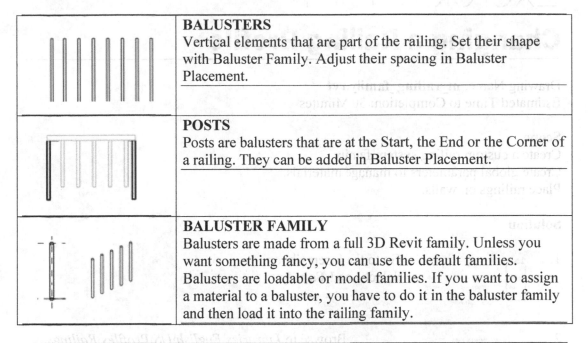

	BALUSTERS Vertical elements that are part of the railing. Set their shape with Baluster Family. Adjust their spacing in Baluster Placement.
	POSTS Posts are balusters that are at the Start, the End or the Corner of a railing. They can be added in Baluster Placement.
	BALUSTER FAMILY Balusters are made from a full 3D Revit family. Unless you want something fancy, you can use the default families. Balusters are loadable or model families. If you want to assign a material to a baluster, you have to do it in the baluster family and then load it into the railing family.

You should be aware of where you click in the Railing's Type Properties to define different components.

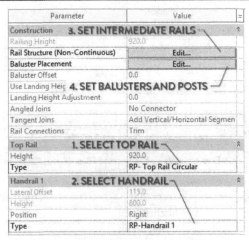

Parameter	Value	=
Construction	**3. SET INTERMEDIATE RAILS**	
Railing Height	920.0	
Rail Structure (Non-Continuous)	Edit...	
Baluster Placement	Edit...	
Baluster Offset	0.0	
Use Landing Heig	**4. SET BALUSTERS AND POSTS**	
Landing Height Adjustment	0.0	
Angled Joins	No Connector	
Tangent Joins	Add Vertical/Horizontal Segmen	
Rail Connections	Trim	
Top Rail	**1. SELECT TOP RAIL**	
Height	920.0	
Type	RP- Top Rail Circular	
Handrail 1	**2. SELECT HANDRAIL**	
Lateral Offset	115.0	
Height	800.0	
Position	Right	
Type	RP-Handrail 1	

Exercise 1-14
Changing a Railing Profile

Drawing Name: **m_railing_family.rvt**
Estimated Time to Completion: 30 Minutes

Scope
Create a custom railing system family.
Create global parameters to manage materials.
Place railings on walls.

Solution

1. Go to the Insert ribbon.
   Select **Load Family**.

2. Browse to *Libraries\English\US\Profiles\Railings*.

3. Select the *M_Decorate Rail* family.
   Click **Open**.

4. Open the **Parking Area** 3D view.

5.

Select one of the railings placed on the concrete walkway.

6.
Select **Edit Type** on the Properties panel.

Properties	×
Railing SH_1100mm	▾
Railings (1)	⊞ Edit Type

7. Duplicate...

Select **Duplicate**.

8.
Name: Railing_Crescent_City

Type **Railing_Crescent_City**.
Click **OK.**

9.

Type Parameters		
Parameter	Value	=
Construction		⌃
Railing Height	800.0	
Rail Structure (Non-Continuous)	Edit...	
Baluster Placement	Edit...	
Baluster Offset	0.0	
Use Landing Height Adjustment	No	
Landing Height Adjustment	0.0	
Angled Joins	Add Vertical/Horizontal Segmen	
Tangent Joins	Extend Rails to Meet	
Rail Connections	Trim	

Select **Edit** next to Rail Structure.

10. Delete all the rails except for **New Rail (1)**. Set the Height to **250.0**. Set the Profile to **M_Rectangular HandRail: 50 x 50 mm.** Set the Material to **Iron, Wrought.**
You will have to import the material into the document.
Click **OK.**

Rails

	Name	Height	Offset	Profile	
1	New Rail(1)	250.0	0.0	M_Rectangular Handrail : 50 x 50mm ▾	Iron, Wrought

11.

Construction	
Railing Height	800.0
Rail Structure (Non-Continuous)	Edit...
Baluster Placement	Edit...
Baluster Offset	0.0
Use Landing Height Adjustment	No
Landing Height Adjustment	0.0
Angled Joins	Add Vertical/Horizontal Segments
Tangent Joins	Extend Rails to Meet
Rail Connections	Trim
Top Rail	

Select **Edit** next to Baluster Placement.

12. Set the Baluster Family to **Baluster –Square: 25 mm**. Set the Host to **Top Rail Element**. Set the Top Offset to **50.0**. Set Distance from Previous to **200.00.**

	Name	Baluster Family	Base	Base offset	Top	Top offset	Dist. from previous	Offset
1	Pattern start	N/A	N/A	N/A	N/A	N/A	N/A	N/A
2	Regular balust	M_Baluster - Square : 25mm	Host	0.0	Top Rail Eleme	50.0	200	0.0
3	Pattern end	N/A	N/A	N/A	N/A	N/A	0.0	N/A

13. Set the Start Post to **None**. Set the Corner Post to **None**. Set the End Post to **None**. Click **OK**.

Posts

	Name	Baluster Family	Base	Base offset	Top	Top offset	Space	Offset
1	Start Post	None	Host	-247.0	Top Rail Ele	0.0	0.0	0.0
2	Corner Post	None	Host	-247.0	Top Rail Ele	0.0	0.0	0.0
3	End Post	None	Host	-247.0	Top Rail Ele	0.0	0.0	0.0

14.  Set the Top Rail Height to **800.00**. Click **OK**.

Rail Connections	Trim
Top Rail	
Use Top Rail	☑
Height	800.0
Type	Rectangular - 50x50mm

15.  Inspect the railing.

16. 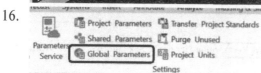 Select the **Manage** ribbon. Select **Global Parameters**.

Parameters Service
Project Parameters | Transfer Project Standards
Shared Parameters | Purge Unused
Global Parameters | Project Units
Settings

17.  Select **New Parameter**.

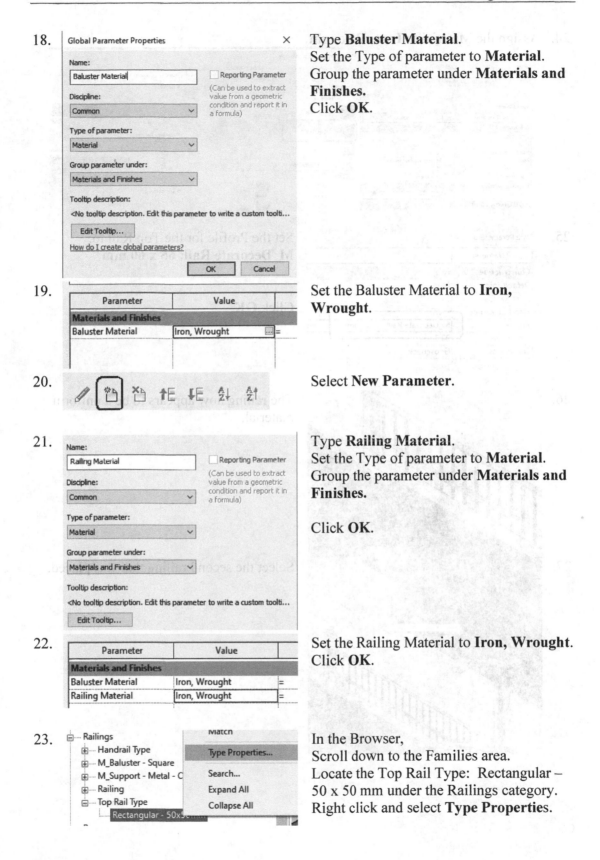

18.

Type **Baluster Material**.
Set the Type of parameter to **Material**.
Group the parameter under **Materials and Finishes.**
Click **OK**.

19.

Set the Baluster Material to **Iron, Wrought**.

20.

Select **New Parameter**.

21.

Type **Railing Material**.
Set the Type of parameter to **Material**.
Group the parameter under **Materials and Finishes.**

Click **OK**.

22.

Set the Railing Material to **Iron, Wrought**.
Click **OK**.

23.

In the Browser,
Scroll down to the Families area.
Locate the Top Rail Type: Rectangular –
50 x 50 mm under the Railings category.
Right click and select **Type Propertie**s.

24. Assign the Material to **Iron, Wrought**.

25.

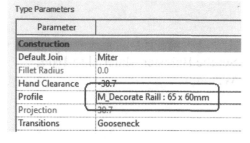

Set the Profile for the Top Rail to:
M_Decorate Rail: 65 x 60 mm.

Click **OK**.

26.

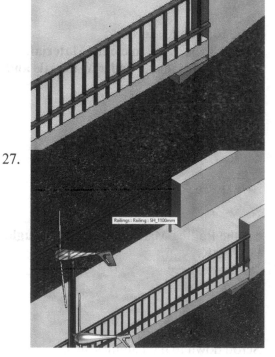

The railing now appears to be a uniform material.

27. Select the second railing that was placed.

28.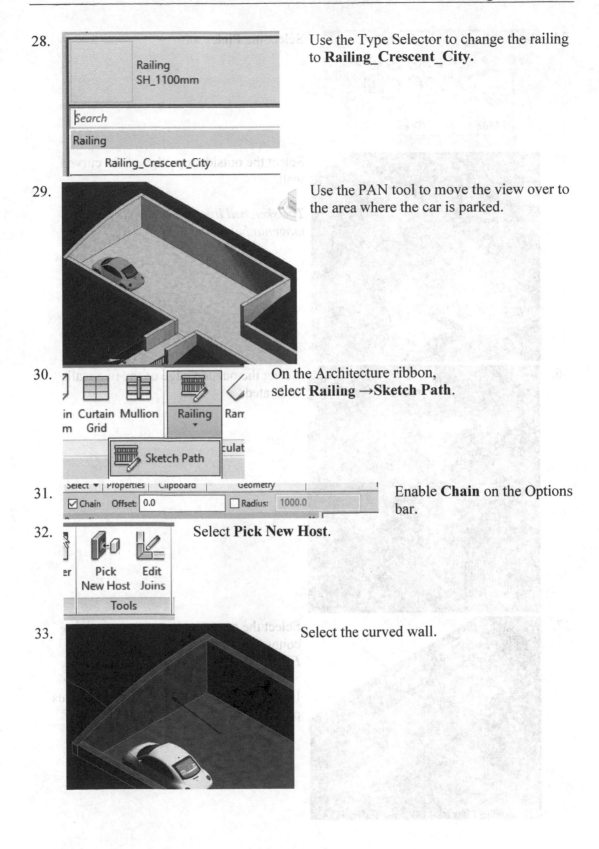
Use the Type Selector to change the railing to **Railing_Crescent_City.**

29. Use the PAN tool to move the view over to the area where the car is parked.

30. On the Architecture ribbon, select **Railing →Sketch Path**.

31. Enable **Chain** on the Options bar.

32. Select **Pick New Host**.

33. Select the curved wall.

34.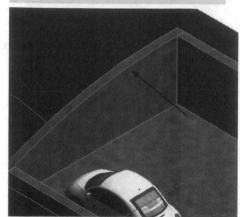

Select the **Pick** tool.

35.

Select the outside top edge of the curved wall.

The sketched line will be displayed in magenta below the wall.

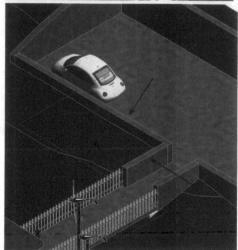

36.

Select the outside edge of the two walls indicated.

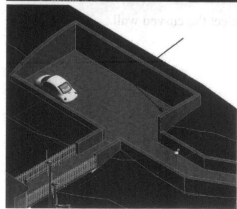

37.

Select the outside edge of the remaining connecting walls.
Do not select the lower walls.

Use the TRIM tool to create a continuous polyline.

38.

> Railing
> SH_1100mm

> Search
> Railing
>   Railing_Crescent_City

Set the Railing Type to
Railing_Crescent_City.

39.

Mode

Click the green check to finish.

40.

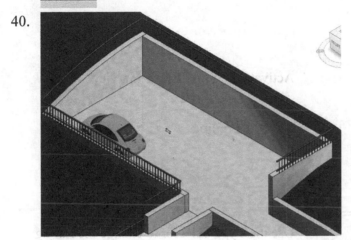

The railing is placed.

Save as *ex1-14.rvt*.

Exercise 1-15
Modify a Railing

Drawing Name: **railing.rvt**
Estimated Time to Completion: 15 Minutes

Scope
Modify an existing railing.

Solution

1.   Sections (Building Section)
      Stair Elevation
          Activate the **Stair Elevation** view.

2.   12'
          Select the railing.

          Railings : Railing : Handrail - Rectangular
          LEVI
          0'

3.   Properties
      Railing
      Handrail - Rectangular
      Search
      Railing
        Glass Panel - Bottom Fill
        Guardrail - Pipe
          Use the Type Selector to change the railing type to **Guardrail – Pipe**.

4.
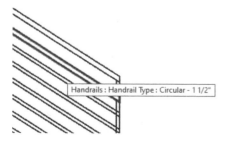
      Handrails : Handrail Type : Circular - 1 1/2"

          Hover your mouse over the second from the top railing.

          Click the TAB key to cycle through selections until the handrail highlights.

          Click to select the handrail.

5.

Properties	×
Handrail Type Circular - 1 1/2"	
Railings: Handrails (1)	∨ [Edit Type]
Dimensions	
Length	36' 2 45/256"

Select **Edit Type** from the Properties palette.

6.

Extension (Beginning/Bottom)	
Extension Style	Wall
Length	1' 0"
Plus Tread Depth	☑

Under Extension (Beginning/Bottom):
Set the Extension Style to **Wall**.
Set the Length to **1'-0"**.
Enable **Plus Tread Depth**.

7.

Extension (End/Top)	
Extension Style	Wall
Length	1' 0"

Under Extension (End/Top):
Set the Extension Style to **Wall**.
Set the Length to **1'-0"**.

To preview, click **Apply**.

Try changing the extension style to Post and then Floor to see how it looks.

The railing will adjust to the new settings.

Click **OK**.

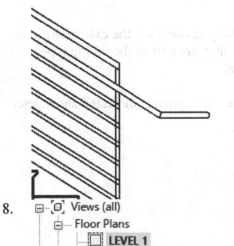

8.

```
⊟ [◌] Views (all)
  ⊟ Floor Plans
      ▢ LEVEL 1
      ▢ LEVEL 2
      ▢ ROOF
      ▢ Site
```

Activate the **LEVEL 1** floor plan view.

9.

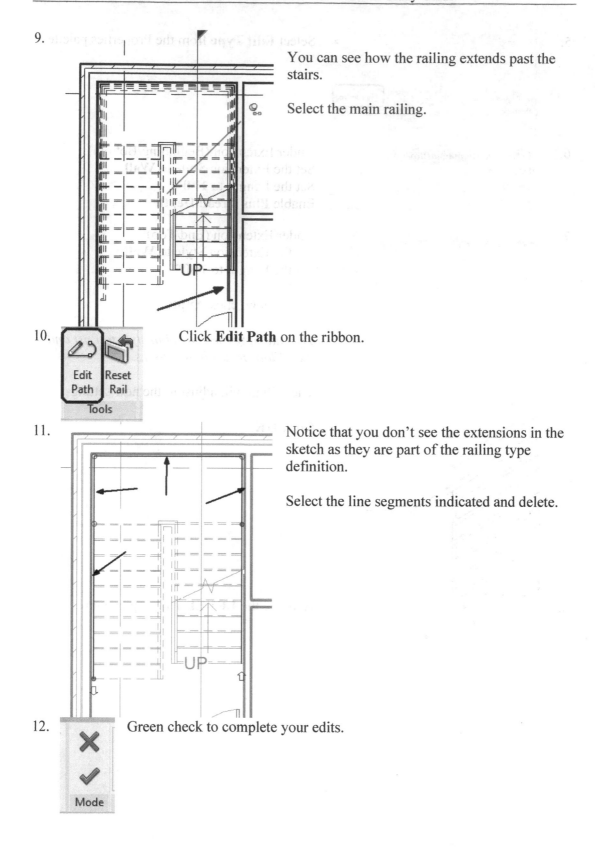

You can see how the railing extends past the stairs.

Select the main railing.

10.

Click **Edit Path** on the ribbon.

11.

Notice that you don't see the extensions in the sketch as they are part of the railing type definition.

Select the line segments indicated and delete.

12.

Green check to complete your edits.

13.

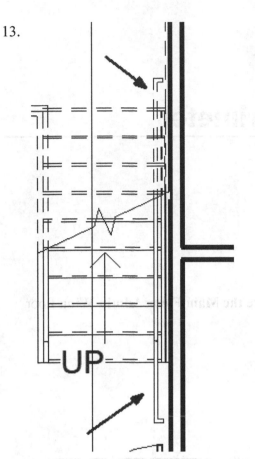

Notice that the extensions are now visible again.

Save as *ex1-15.rvt*.

Floors

Floors are system families. They have layers, just like walls. You create floors by defining their boundaries, either by picking walls or using drawing tools. Floors are placed relative to levels. You do not need walls or a building pad to place a floor element.

Typically, you sketch a floor in a plan view, although you can use a 3D view if the work plane of the 3D view is set to the work plane of a plan view.

Floors are offset downward from the level on which they are sketched.

Exercise 1-16
Modifying a Floor Perimeter

Drawing Name: **i_floors.rvt**
Estimated Time to Completion: 15 Minutes

Scope
Modify a floor.

Solution

1. 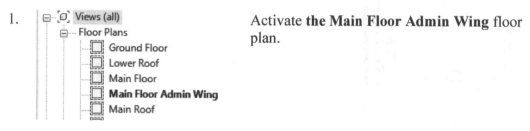   Activate **the Main Floor Admin Wing** floor plan.

2. Window around the area indicated.

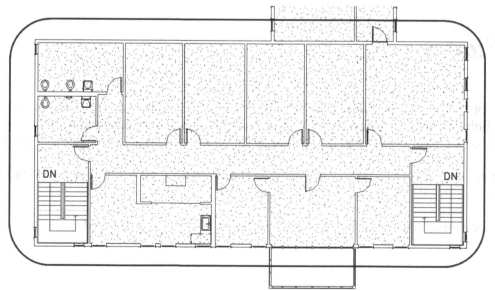

3. Select the **Filter** tool located in the lower right corner of the window.

4.

Category:	Count:
<Room Separation>	1
Casework	5
Curtain Panels	12
Curtain Wall Grids	5
Curtain Wall Mullions	16
Doors	15
☑ Floors	1
Lines (Lines)	16
Plumbing Fixtures	7
Railings	4
Rooms	13
Stairs	2
Walls	27
Windows	18

Total Selected Items: 1

Select **Check None** to disable all the checks.

Then check only the **Floors**.

Click **OK**.

5.

Dimensions	
Slope	
Perimeter	254' 8"
Area	3254.30 SF
Volume	3457.69 CF
Thickness	1' 0 3/4"

In the Properties pane,
Note that the floor has a perimeter of 254' 8".

6. Edit Boundary / Mode — Select **Edit Boundary** under the Mode panel.

7. Boundary Line, Slope Arrow, Span Direction / Draw — Select **Pick Walls** mode under the Draw panel.

8. Offset: 0' 0" ☐ Extend into wall (to core) — Uncheck the Extend into wall (to core) option.

9. Select the three walls indicated.

10. Modify — Select the **Trim** tool on the Modify panel from the tab.

11.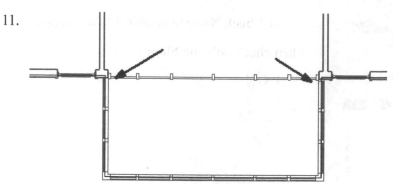

Trim the two corners indicated so that there is no wall in the section between the two vertical walls on the upper ends.

12.

Select the **Green Check** to **Finish Floor** from the tab.

13.

Attaching to floor   ×

Would you like walls that go up to this floor's level to attach to its bottom?

☐ Do not show me this message again   Attach   Don't attach

Select **Don't attach**.

14.

Dimensions	
Slope	
Perimeter	270' 8"
Area	3390.71 SF
Volume	3602.63 CF
Thickness	1' 0 3/4"

Note that the floor has a perimeter of 270′ 8″. Click **OK** to close the dialog.

15. Close without saving.

Exercise 1-17
Modifying Floor Properties

Drawing Name: **floors_i.rvt**
Estimated Time to Completion: 5 Minutes

Scope
Modify a floor using Type Properties
Determine the floor's Elevation at Bottom

Solution

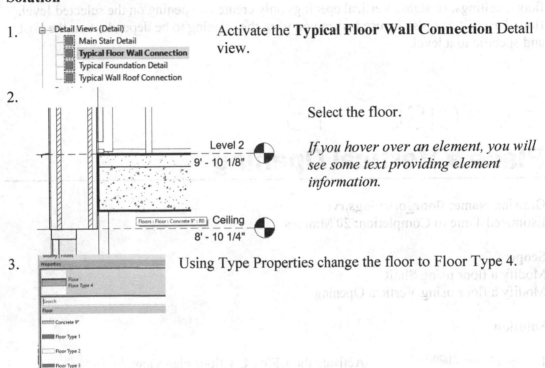

1.   Activate the **Typical Floor Wall Connection** Detail view.

2.   Select the floor.

If you hover over an element, you will see some text providing element information.

3.   Using Type Properties change the floor to Floor Type 4.

4.   What is the Elevation at Bottom of the floor?

It should read 8' 11 23/64"

Vertical Openings

The three methods for creating a vertical opening are:

- Shaft
- Vertical
- Boundary Edit

You should be familiar with the different methods of creating vertical openings, such as shafts and vertical as well as by modifying floor or ceiling boundaries. There may be a question to determine if you understand best practices and when which tool should be applied. In general, shafts are used when the openings go through more than one level/floor/ceiling. Shafts are independent elements. Vertical openings are hosted by floors, ceilings, or slabs. Vertical openings only create an opening on the selected level. Boundary edit can also be used when you want the opening to be dependent on the host and specific to a level.

Exercise 1-18
Place a Vertical Opening

Drawing Name: **floor_openings.rvt**
Estimated Time to Completion: 20 Minutes

Scope
Modify a floor using Shaft
Modify a floor using Vertical Opening

Solution

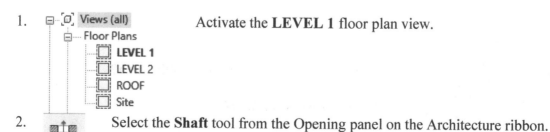

1.   Activate the **LEVEL 1** floor plan view.

2.   Select the **Shaft** tool from the Opening panel on the Architecture ribbon.

3.

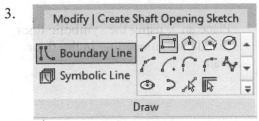

Enable **Boundary Line**.

Select the **Rectangle** tool from the Draw panel.

4.

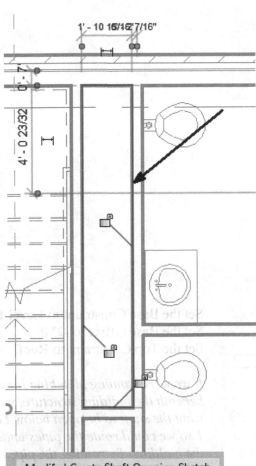

Place the rectangle flush to the finish face of the walls behind the top bathroom area.

This will be a shaft used for routing plumbing lines. We are using a shaft instead of an opening to ensure that the plumbing can travel the entire vertical distance from Level 1 to the roof.

5.

Enable **Symbolic Line**.

Select the **Line** tool from the Draw panel.

6.

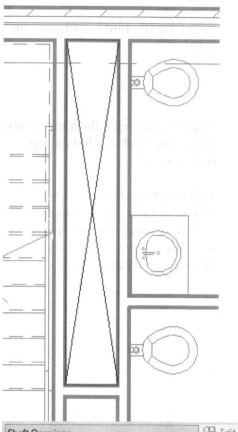

Place an X using the symbolic lines to indicate an opening in the floor plan.

7.

Shaft Openings		∨	Edit Type
Constraints			≪
Base Constraint	LEVEL 1		
Base Offset	-2' 0"		
Top Constraint	Up to level: ROOF		
Unconnected Height	26' 0"		
Top Offset	0' 0"		
Phasing			≪
Phase Created	New Construction		
Phase Demolished	None		

Set the Base Constraint to **Level 1**.
Set the Base Offset to **-2' 0"**.
Set the Top Constraint to **Roof.**

If we were running plumbing lines beneath the building structure, we want the shaft to be open below Level 1 so we could route the pipes under the building. By routing the pipes below the building, we can connect to the city water supply as well as to any sewer or septic lines.

8.

Green check to complete the shaft.

✗ Boundary Line
✓ Symbolic Line
Mode

9.

Sections (Building Section)
    Floor Opening

Switch to the **Floor Opening** Section view.

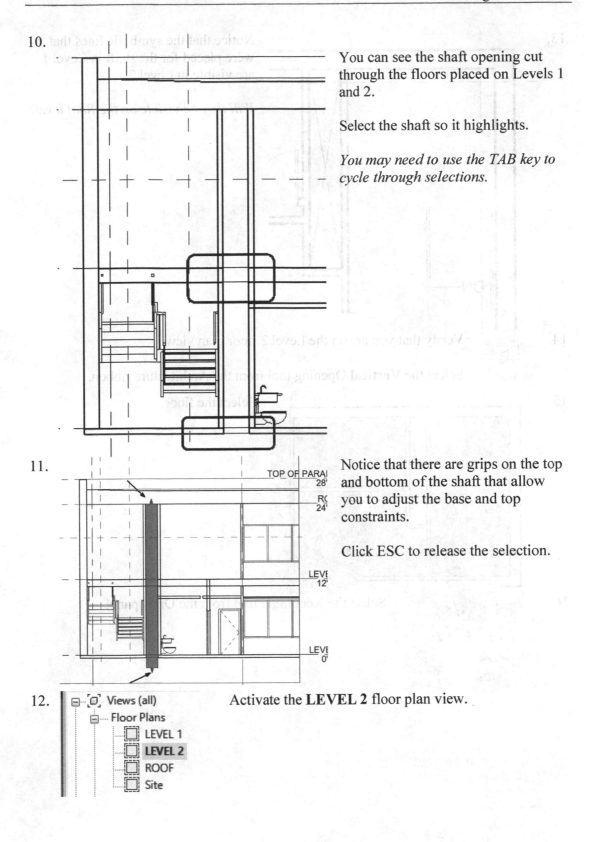

10. You can see the shaft opening cut through the floors placed on Levels 1 and 2.

Select the shaft so it highlights.

You may need to use the TAB key to cycle through selections.

11. Notice that there are grips on the top and bottom of the shaft that allow you to adjust the base and top constraints.

Click ESC to release the selection.

TOP OF PARA
28'

R(
24'

LEVI
12'

LEVI
0'

12. Activate the **LEVEL 2** floor plan view.

Views (all)
  Floor Plans
    LEVEL 1
    LEVEL 2
    ROOF
    Site

13.

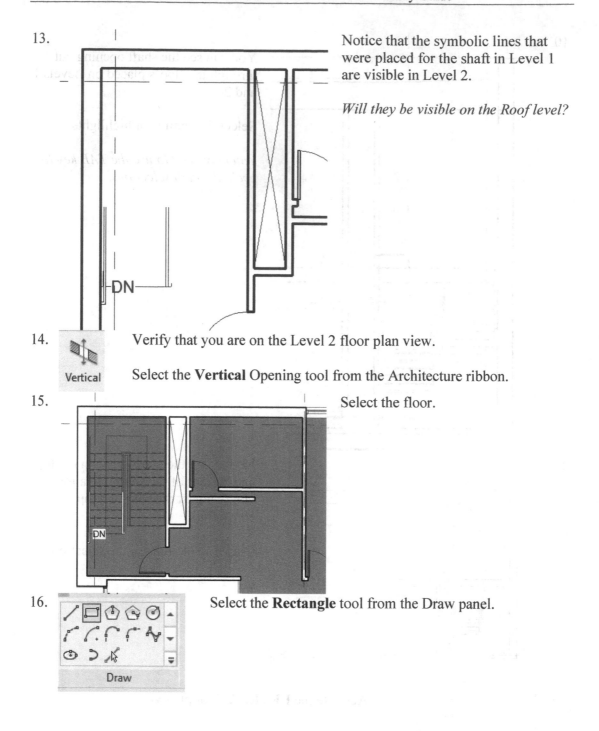

Notice that the symbolic lines that were placed for the shaft in Level 1 are visible in Level 2.

Will they be visible on the Roof level?

DN

14. Verify that you are on the Level 2 floor plan view.

Vertical

Select the **Vertical** Opening tool from the Architecture ribbon.

15. Select the floor.

DN

16. Select the **Rectangle** tool from the Draw panel.

Draw

17.

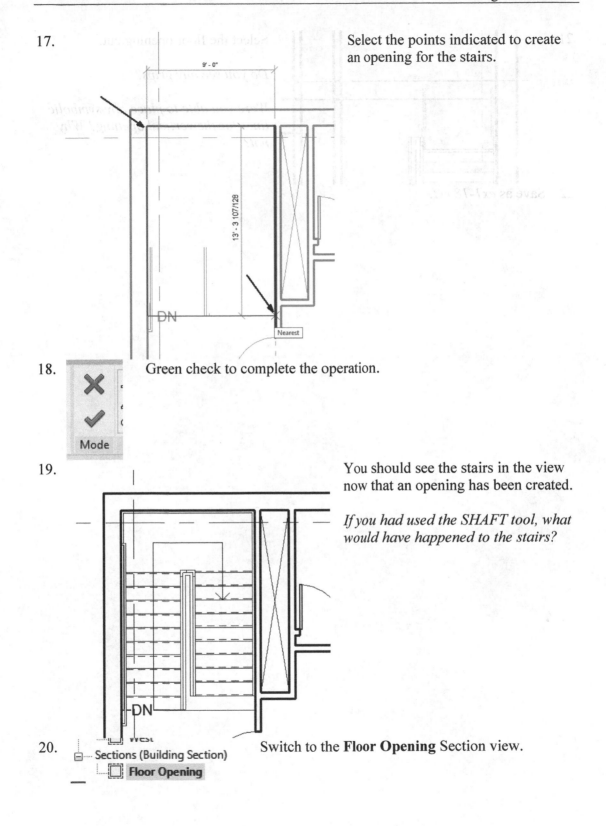

Select the points indicated to create an opening for the stairs.

18. Green check to complete the operation.

19. You should see the stairs in the view now that an opening has been created.

If you had used the SHAFT tool, what would have happened to the stairs?

20. Switch to the **Floor Opening** Section view.

Sections (Building Section)
    Floor Opening

21.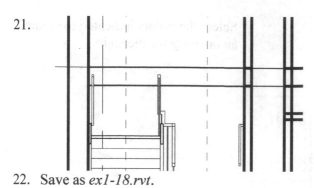

Select the floor opening cut.

Do you see any grips?

Were you able to place any symbolic lines for the vertical opening? Why not?

22. Save as *ex1-18.rvt*.

Placing Columns

Although structural columns share many of the same properties as architectural columns, structural columns have additional properties defined by their configuration and industry standards which provide different behaviors.

Structural elements such as beams, braces, and isolated foundations join to structural columns; they do not join to architectural columns.

In addition, structural columns have an analytical model that is used for data exchange.

Typically, drawings or models received from an architect may contain a grid and architectural columns. You create structural columns by manually placing each column or by using the At Grids tool to add a column to selected grid intersections. In most cases, it is helpful to set up a grid before adding structural columns, as they snap to grid lines.

Structural columns can be created in plan or 3D views.

You can place a slanted structural column, but you cannot place a slanted architectural column.

You can use architectural columns to model column box-outs around structural columns and for decorative applications. Architectural columns inherit the material of other elements to which they are joined.

By default, structural columns extend downward when you place them. For example, if you place a structural column on Level 2 of your project, it will extend from Level 2 down to Level 1.

Architectural columns behave in the opposite manner. If you place one on Level 1 of your project, it will extend up to Level 2.

To extend a structural column up to the next level

1. On the Structural tab of the Design bar, click Structural Column.

2. On the Options bar, select Height from the drop-down menu and then select a Level from the drop-down menu. If you are placing the structural column on Level 1, select Level 2 for the height. The structural column will extend from Level 1 to Level 2.

Exercise 1-19
Placing Columns

Drawing Name: **columns.rvt**
Estimated Time to Completion: 30 Minutes

Scope
Place structural columns at grid lines.
Place an architectural column.
Insert a structural column into an architectural column.

Solution

1. Activate the **Level 1** floor plan.

2. Select **Structural Column** from the drop-down on the Architecture ribbon.

3. Select **W10X33** using the Type Selector.

4. On the Options bar:
   Specify Height.
   Specify Level 2.

   Remember structural columns extend DOWN not up.

5. 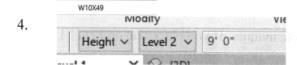 Left click on **At Grids**.

6.

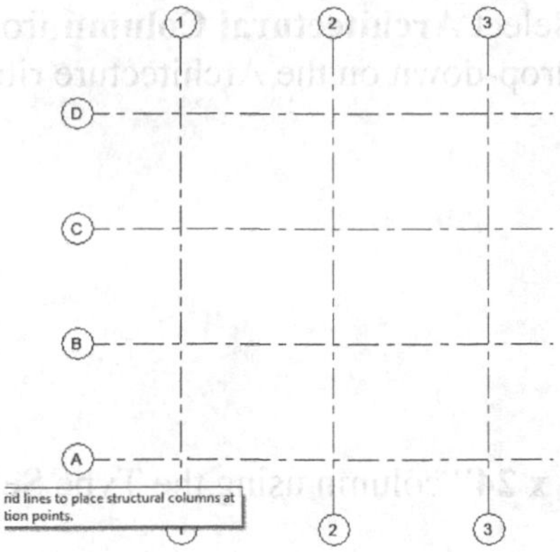

Hold down the CTL key.

Select Grids 1,2, 3, A,B,C, and D.

Select **Finish** on the ribbon.

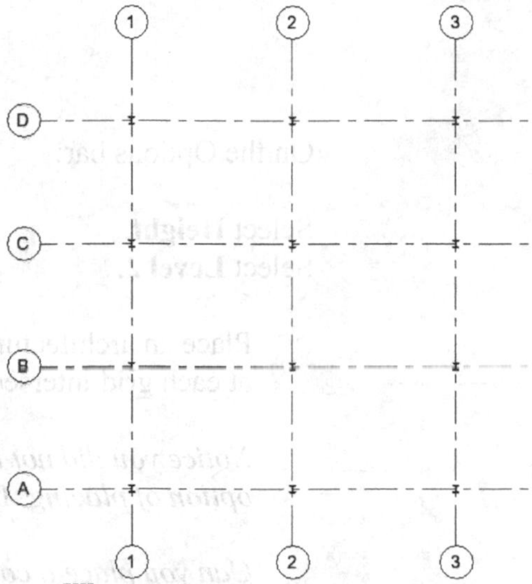

A structural column was placed at each grid intersection.

Click Cancel to exit the command.

7. Elevations (Building Elevation)
   East
   North
   South
   West

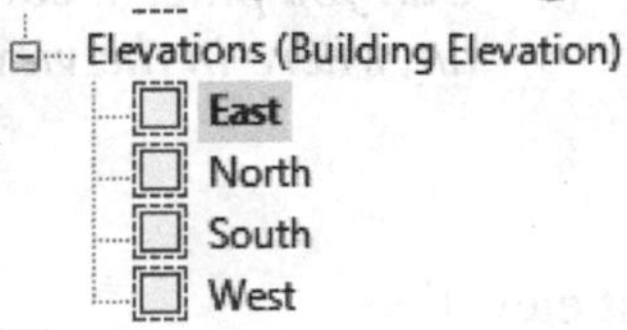

Activate the East elevation.

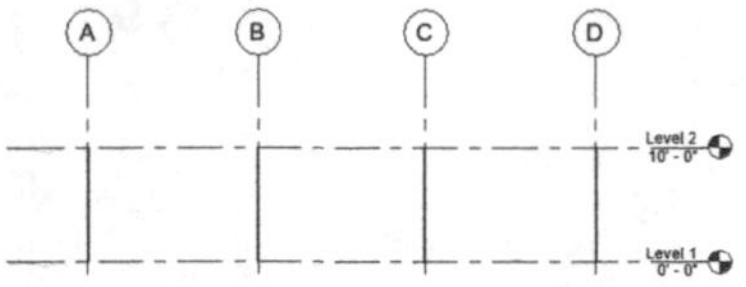

Notice that the structural columns were placed between Level 1 and Level 2.

8. Views (all)
   Floor Plans
   Level 1
   Level 2
   Site

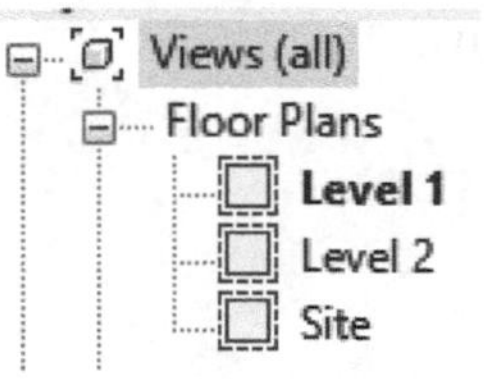

Activate the **Level 1** floor plan.

9.

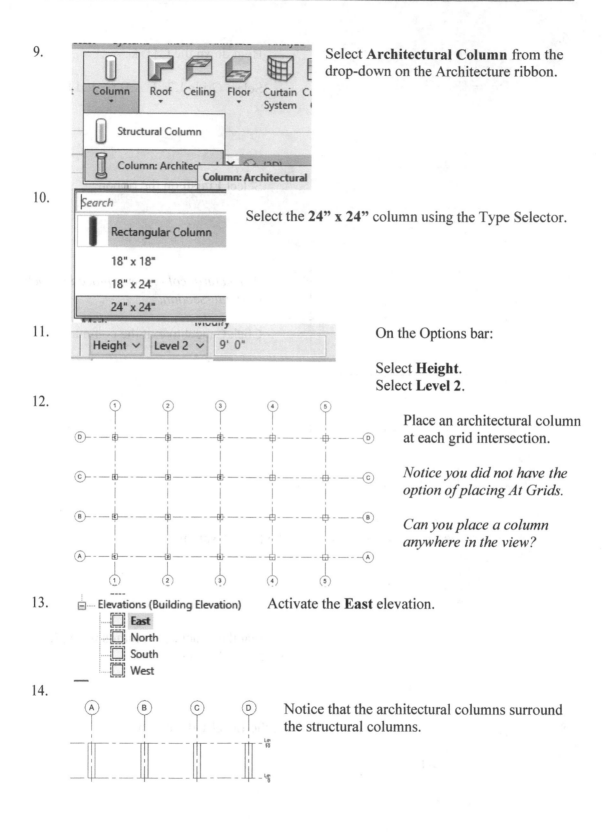

Select **Architectural Column** from the drop-down on the Architecture ribbon.

10.

Select the **24" x 24"** column using the Type Selector.

11.

On the Options bar:

Select **Height**.
Select **Level 2**.

12.

Place an architectural column at each grid intersection.

Notice you did not have the option of placing At Grids.

Can you place a column anywhere in the view?

13.

Activate the **East** elevation.

14.

Notice that the architectural columns surround the structural columns.

15. 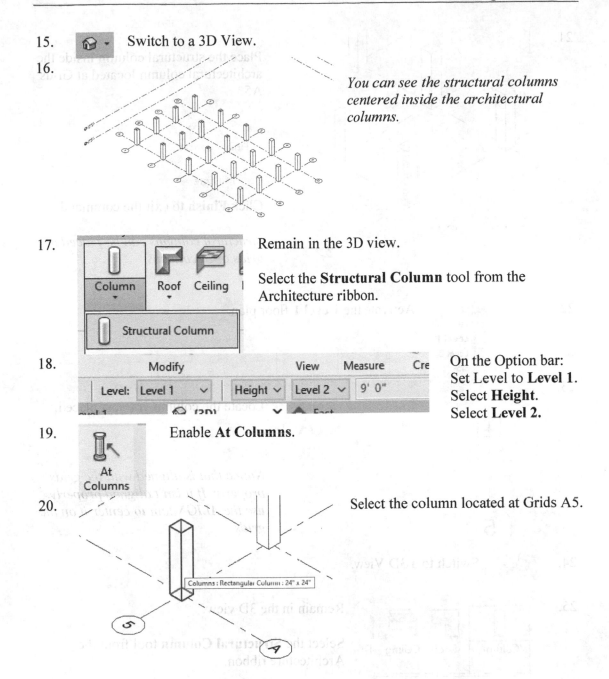 Switch to a 3D View.

16.

You can see the structural columns centered inside the architectural columns.

17. Remain in the 3D view.

Select the **Structural Column** tool from the Architecture ribbon.

18. On the Option bar:
Set Level to **Level 1**.
Select **Height**.
Select **Level 2**.

19. Enable **At Columns**.

20. Select the column located at Grids A5.

21.

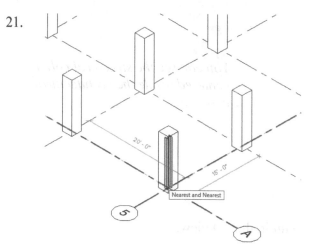

Place the structural column inside the architectural column located at Grids A5.

Click **Finish** to exit the command.

Structural columns can be placed at grids or at columns.

22.

Views (all)
   Floor Plans
      Level 1
      Level 2
      Site

Activate the **Level 1** floor plan.

23.

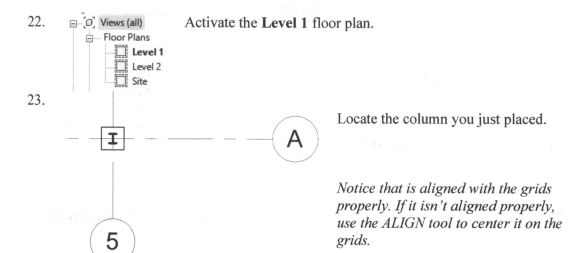

Locate the column you just placed.

Notice that is aligned with the grids properly. If it isn't aligned properly, use the ALIGN tool to center it on the grids.

24.   Switch to a 3D View.

25.

Remain in the 3D view.

Select the **Structural Column** tool from the Architecture ribbon.

26.

On the Option bar:
Set Level to **Level 1**.
Select **Height**.
Select **Level 2.**

27. 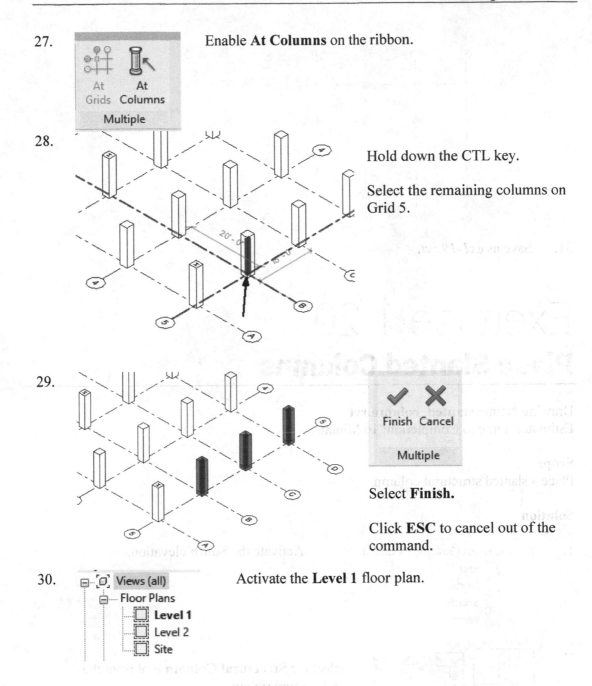 Enable **At Columns** on the ribbon.

28. Hold down the CTL key.

Select the remaining columns on Grid 5.

29. Select **Finish.**

Click **ESC** to cancel out of the command.

30. Activate the **Level 1** floor plan.

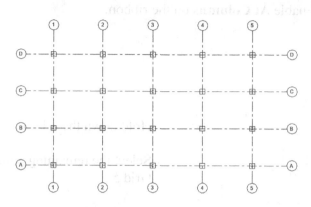

31.   Save as *ex1-19.rvt*.

Exercise 1-20
Place Slanted Columns

Drawing Name: **slanted_column.rvt**
Estimated Time to Completion: 10 Minutes

Scope
Place a slanted structural column

Solution

1.   Elevations (Building Elevation)   Activate the **South** elevation.
          East
          North
          South
          West

2.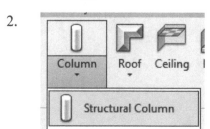

       Select the **Structural Column** tool from the
       Architecture ribbon.

3.

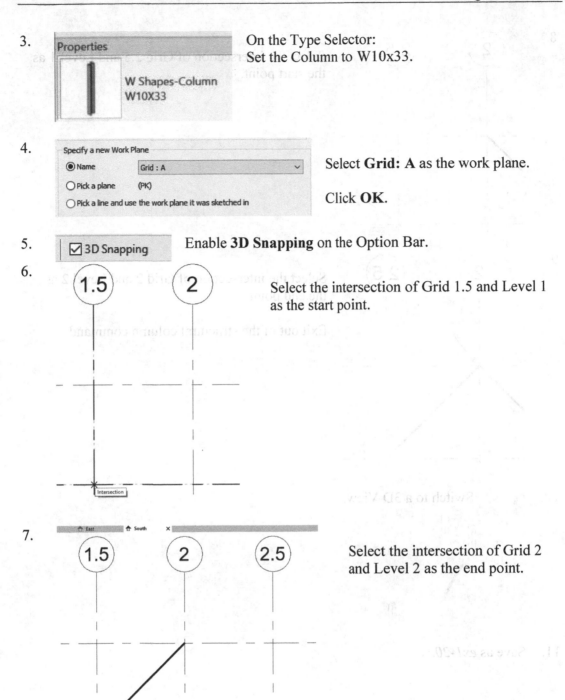

On the Type Selector:
Set the Column to W10x33.

4.

Select **Grid: A** as the work plane.

Click **OK**.

5. Enable **3D Snapping** on the Option Bar.

6. Select the intersection of Grid 1.5 and Level 1
as the start point.

7. Select the intersection of Grid 2
and Level 2 as the end point.

8.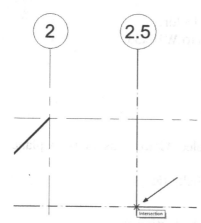

Select the intersection of Grid 2.5 and Level 1 as the start point.

9.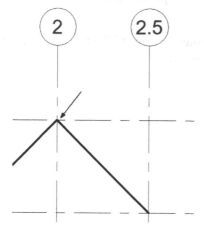

Select the intersection of Grid 2 and Level 2 as the end point.

Exit out of the structural column command.

10. 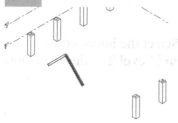 Switch to a 3D View.

11. Save as *ex1-20.rvt*.

Exercise 1-21
Columns and Materials

Drawing Name: **columns_materials.rvt**
Estimated Time to Completion: 20 Minutes

Scope
Explore how architectural and structural columns manage materials when
intersecting with a wall

Solution

1.   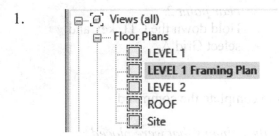   Activate the **LEVEL 1 Framing Plan** floor plan.

2.   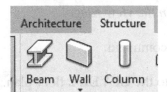   Activate the Structure ribbon.

Select the **Column** tool.

Note that the Structure ribbon only allows you to place structural columns while the Architecture ribbon allows you to place structural OR architectural columns.

3.   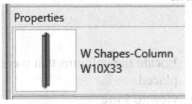   Select the **W10X33** column family using the Type Selector.

4.   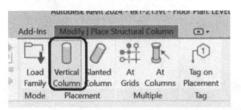   Enable **Vertical Column**.

5. On the Option bar:
Verify that **Height** is enabled.
LEVEL 2 is selected.
9'-0" is displayed as the default.

6. Enable **At Grids**.

If you enable At Grids before you check the Option bar, you will not be able to set the Height.

7. Use a crossing to select Grids 5 to Grids 1. *Click a point near point 1 and then click a point near point 2.*

8. Hold down the CTL key and select Grid A.

9. Click **Finish** to complete the command.

You don't see the columns that were placed because they are not in the view range.

Click ESC to exit the command.

10. Switch to the **LEVEL 2** floor plan.

11. Locate the columns that were placed.

12. Set the Detail Level of the view to **Fine**.

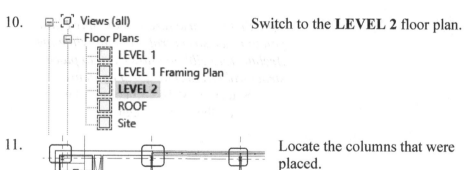

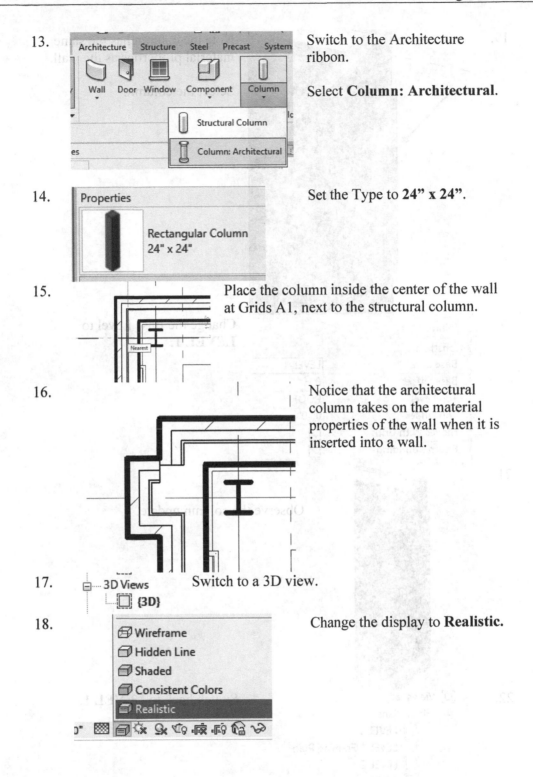

13. Switch to the Architecture ribbon.

Select **Column: Architectural**.

14. Set the Type to **24" x 24"**.

15. Place the column inside the center of the wall at Grids A1, next to the structural column.

16. Notice that the architectural column takes on the material properties of the wall when it is inserted into a wall.

17. Switch to a 3D view.

18. Change the display to **Realistic.**

19.

Note the column has the same material properties as the wall.

Select the column.

20.

Columns (1)	
Constraints	
Base Level	LEVEL 1
Base Offset	0' 0"
Top Level	ROOF
Top Offset	0' 0"
Moves With Grids	☑
Room Bounding	☑

Change the Base Level to **LEVEL 1**.

21.

Observe the column update.

22.

Switch back to **LEVEL 1**.

23.

Zoom into the structural column located at A3.

Enable **Thin Lines**.
This is located on the Quick Access Tool bar.

The structural column ignores the wall it is placed inside.

24.

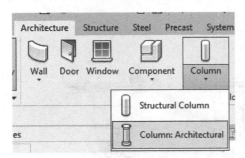

Switch to the Architecture ribbon.

Select **Column: Architectural**.

25.

Set the Type to **24" x 24"**.

26.

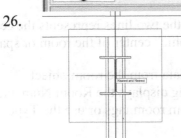

Place the column inside the center of the wall at Grids A3, on top of the structural column.

27.

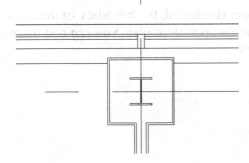

Note that the architectural column joins with the wall and adopts the material properties of the wall.

Save as *ex1-21.rvt*.

Rooms

A room is a subdivision of space within a building model, based on elements such as walls, floors, roofs, and ceilings.

Walls, floors, roofs, and ceilings are defined as room-bounding. Revit uses these room-bounding elements when computing the perimeter, area, and volume of a room.

You can turn on/off the Room Bounding parameter of many elements. You can also use room separation lines to further subdivide space where no room-bounding elements exist. When you add, move, or delete room-bounding elements, the room's dimensions update automatically.

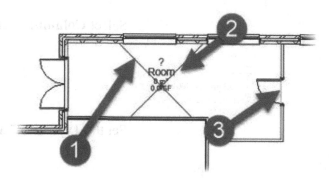

1: The Room Reference Crosshair- The intersection of the two lines represents the center of the room object. This is not necessarily the "geographic" center of the room or space.

2: Room Tag – This is a text annotation that takes the data from the room object paramenters and displays it in the tag. This particular tag displays the Room Number, Room Name, and Area. You can create your own custom room tags or use the Type Selector to display different room tag types.

3: Room Boundary Line – When the room is selected, the boundary of the room is displayed as a light blue line. If you have an open floor plan, you can add room boundary lines to designate different rooms/spaces.

Exercise 1-22

Rooms

Drawing Name: rooms_1.rvt
Estimated Time to Completion: 10 Minutes

Scope
Creating rooms.
Modify room tags.

Solution

1.
    Activate the **GROUND FLOOR – Exam Rooms** floor plan.

2.
    Activate the Architecture ribbon.

    Select the **Room** tool.

3.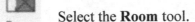
    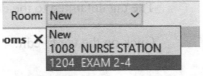
    From the Options bar:

    Select **1204 EXAM 2-4** from the drop-down list for rooms.

4.
    Verify that **Tag on Placement** is enabled on the ribbon.

5.

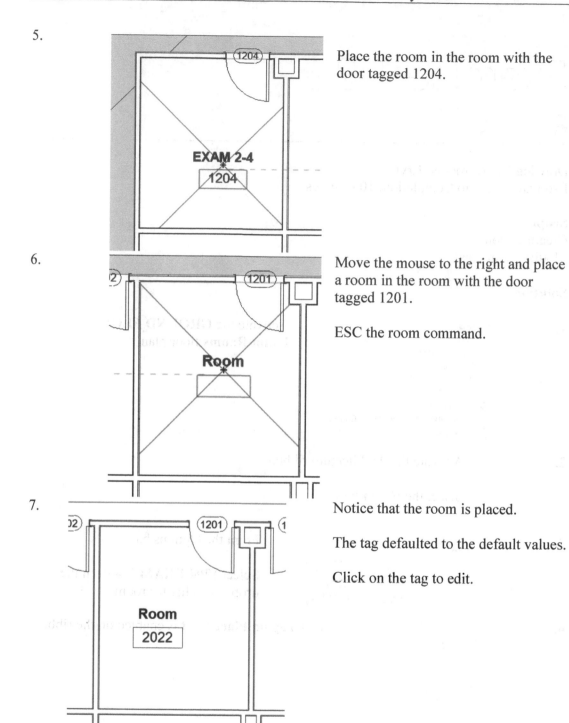

Place the room in the room with the door tagged 1204.

6.

Move the mouse to the right and place a room in the room with the door tagged 1201.

ESC the room command.

7.

Notice that the room is placed.

The tag defaulted to the default values.

Click on the tag to edit.

8.

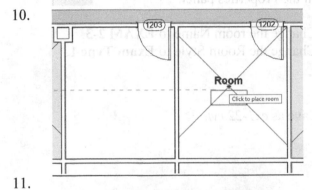

Change the tag to read:

EXAM 2-1
1201

9.

Activate the Architecture ribbon.

Select the **Room** tool.

10.

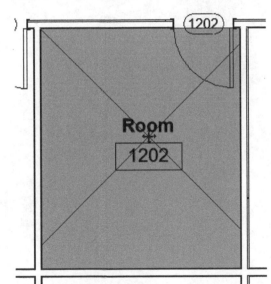

Place a room in the room with the door tagged 1202.

Place a room in the room with the door tagged 1203.

ESC the room command.

11.

Notice that the tag auto-incremented 1202.

Select the room (not the tag).

12.

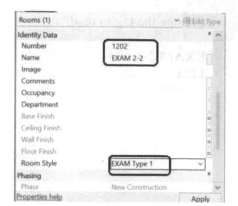

In the Properties panel:

Change the room Name to **EXAM 2-2**.
Change the Room Style to **Exam Type 1**.

13.

Select the room tagged 1203.

In the Properties panel:

Change the room Name to **EXAM 2-3**.
Change the Room Style to **Exam Type 1**.

14.

Save as *ex1-22.rvt*.

Exercise 1-23
Room Separators

Drawing Name: rooms_2.rvt
Estimated Time to Completion: 5 Minutes

Scope
Creating rooms using a room separator.
Modify room tags.

Solution

1.
```
Floor Plans
    Roof
    Working Second Floor
    SECOND FLOOR
    Working Ground Floor
    SITE PLAN
    GROUND FLOOR
        GROUND FLOOR - Staff
        GROUND FLOOR - Exam Rooms
```

Activate the **GROUND FLOOR –
Staff** floor plan.

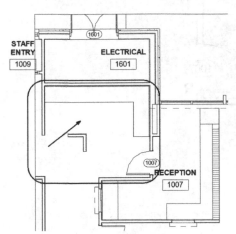

The area indicated is a nurse's station,
but it is missing necessary room-
bounding elements as it has openings
between the walls.

We need to use Room Separators to
define the room.

2. 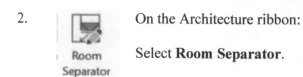 On the Architecture ribbon:

   Select **Room Separator**.

3. 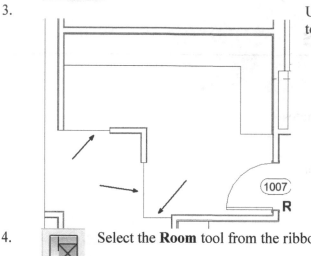 Using the Line tool, place three lines to define the room.

4. 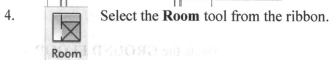 Select the **Room** tool from the ribbon.

5. 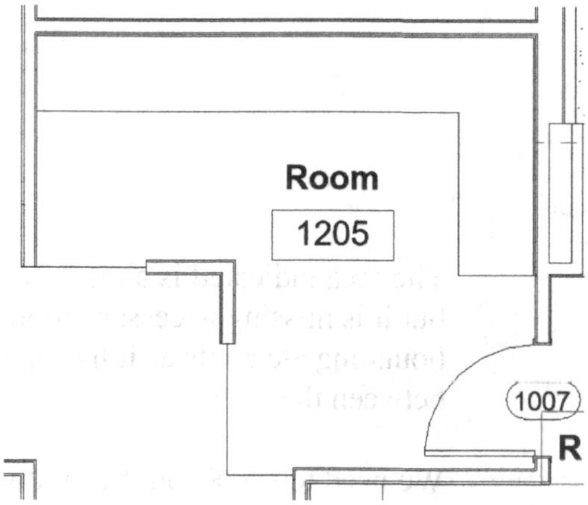 Place a room in the area defined by the walls and room separators.

Room

1205

6.

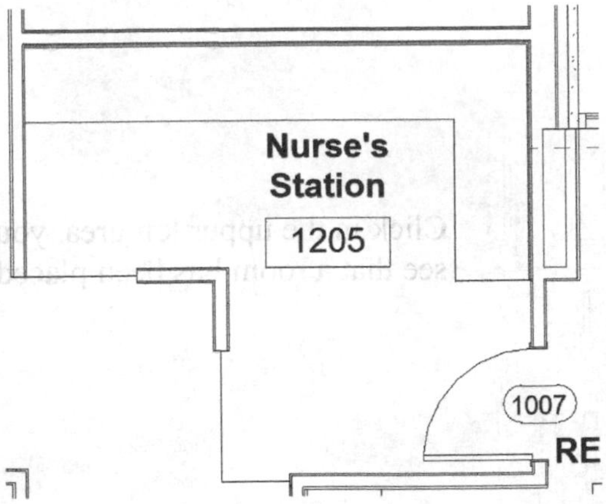

Edit the room tag to **Nurse's Station**.

Save as *ex1-23.rvt*.

Exercise 1-24
Defining an Element as Room Bounding

Drawing Name: rooms_3.rvt
Estimated Time to Completion: 5 Minutes

Scope
Changing a wall to room bounding

Solution

1.

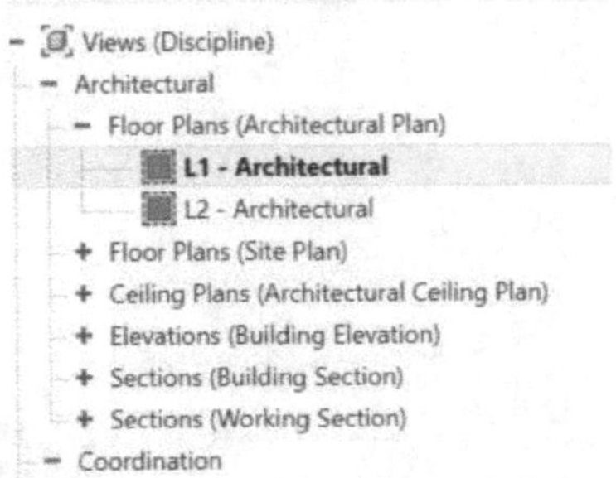

Activate the **L1 – Architectural** floor plan.

2.

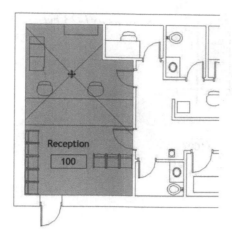

Click in the upper left area, you should see that a room has been placed.

3.

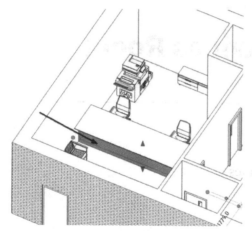

Switch to a 3D view.

You see there is a half wall separating the reception area and the administration area.

You want to divide the room using the half wall as a bounding element.

Select the half wall.

4.

Base Extension Distance	0.0
Top Constraint	Unconnected
Unconnected Height	958.0
Top Offset	0.0
Top is Attached	☐
Top Extension Distance	0.0
Room Bounding	☑
Related to Mass	☐
Cross-Section Definition	

Properties help

Enable **Room Bounding** on the Properties palette.

5.

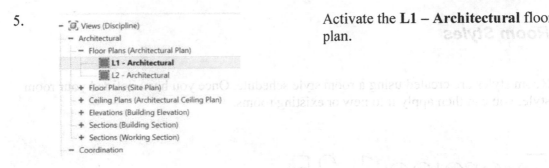

Activate the **L1 – Architectural** floor plan.

6.

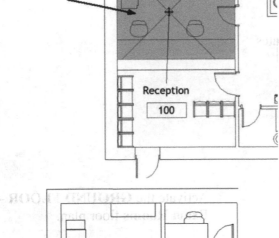

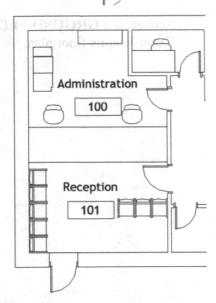

The room is now placed only in the administration area.

Move the room tag into the administration area and rename **Administration**.

You can select the room tag and disable Leader to delete the leader.

7.

Place a room in the reception area.

Change the room tag to **Reception**.

8.   Save as ex1-24.rvt.

Room Styles

Room styles are created using a room style schedule. Once you have defined your room style, you can then apply it to new or existing rooms.

Exercise 1-25

Room Styles

Drawing Name: room styles.rvt
Estimated Time to Completion: 15 Minutes

Scope
Creating a room style schedule.
Applying room styles to rooms.

Solution

1.

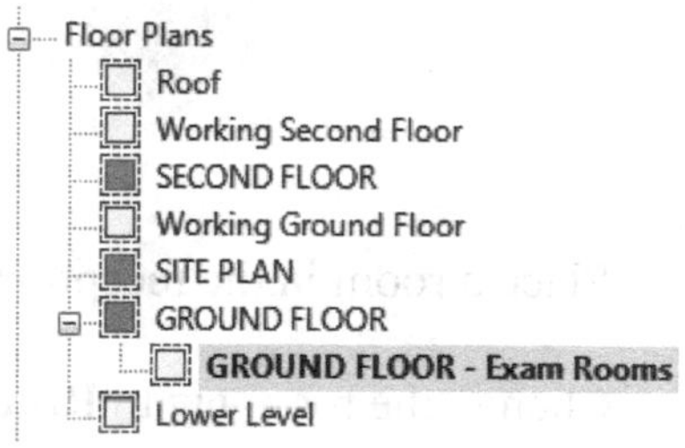

Floor Plans
    Roof
    Working Second Floor
    SECOND FLOOR
    Working Ground Floor
    SITE PLAN
    GROUND FLOOR
        GROUND FLOOR - Exam Rooms
    Lower Level

Activate the **GROUND FLOOR – Exam Rooms** floor plan.

2.

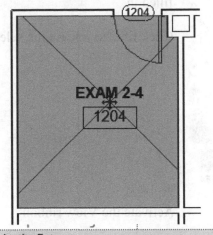

Select Exam 2-4 room.

3.

Identity Data	
Number	1204
Name	EXAM 2-4
Image	
Comments	
Occupancy	
Department	
Base Finish	VB-2
Ceiling Finish	ACT
Wall Finish	PNT-2
Floor Finish	VCT-2

Note that in the Identity Data the finishes have been filled in.

Click ESC to release the selection.

4.

Select EXAM 3-4.

5.

Note that the finish data has not been filled in.

Click ESC to release the selection.

Rooms (1)	⌄	Edit Type
Volume	Not Computed	
Computation Height	0' 0"	
Identity Data		≫
Number	1304	
Name	EXAM 3-4	
Image		
Comments		
Occupancy		
Department		
Base Finish		
Ceiling Finish		
Wall Finish		
Floor Finish		
Phasing		≫

6.

Activate the View ribbon.

Select **Schedules→Schedule/Quantities**.

7. Highlight **Rooms**.

⊞ Roofs
Rooms
RVT Links

8.

Enable **Schedule Keys**.
Change the Name to **ROOM STYLES**.

Note that the Key Name is automatically set to Room Style.

Click **OK**.

Name:

ROOM STYLES

○ Schedule building components
● Schedule keys

Key name:

Room Style

9.

Select the following fields in order:

- Key Name
- Ceiling Finish
- Floor Finish
- Wall Finish

Scheduled fields (in order):

Key Name
Ceiling Finish
Floor Finish
Wall Finish

10.

Select the Sorting/Grouping tab.

Sort by **Key Name**.

Fields | Sorting/Grouping | Formatting | Appearance

Sort by: Key Name ⌄ ● Ascending ○ Descending
☐ Header ☐ Footer: ☐ Blank line

11. Select the Appearance tab.

Enable **Grid Lines**.
Enable **Outline**.
Disable **Blank row before data**.

Click **OK**.

12. Use the **Insert Data Row** tool to add three data rows to the schedule.

Insert Data Row

13. 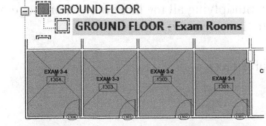 Fill in the data rows as shown.

14. GROUND FLOOR

GROUND FLOOR - Exam Rooms

Open the **GROUND FLOOR – Exam Rooms** floor plan view.

15. Hold down the CTL key and select the four exam rooms tagged 3-x.

16. 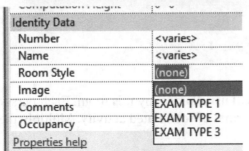 In the Properties panel:

Room Style is now available in the Identity Data.

Select **EXAM TYPE 3** to assign that room style to the selected rooms.

17.

Name	<varies>
Room Style	EXAM TYPE 3
Image	
Comments	
Occupancy	
Department	
Base Finish	
Ceiling Finish	ACT
Wall Finish	PNT-3
Floor Finish	VCT-3

Scroll down and you will see the finish values have been filled in based on the room style that was assigned.

18.

Save as *ex1-25.rvt*.

Volume Calculations

Revit can calculate the area and volume of rooms and display the information in schedules and tags. To see room volumes on the Properties palette, in tags, and in room schedules, enable volume computations.

Room volume is required to determine HVAC requirements.

Step 1

Determine the volume of a room in cubic feet. This is done by measuring the length, width and height of the room in feet and multiplying all the three dimensions together. Revit can automatically provide you with the volume for rooms.

Volume = Width X Length X Height (cubic feet)

Step 2
Multiply this volume by 6.

$C1$ = Volume X 6

Step 3
Estimate the number of people (N) that will usually occupy this room. Each person produces about 500 BTU/hr of heat for normal office-related activity. Multiply these two figures together.

$C2$ = N x 500 BTU/hr

Step 4

Add C1 and C2 together and you will get a very simplified cooling capacity needed for the room.

Estimated Cooling Capacity needed = C1 + C2 (BTU/hr)

Air Conditioning Calculations – Other Factors

Other factors to determine the sizing of the cooling capacity include the direction of your room. If the room is facing east or west, additional capacity is needed as it will be exposed to the morning and evening sun compared to a room that faces north or south. If the lighting of the room emits a lot of heat, additional capacity is needed. If electrical appliances that generate heat are used, additional capacity has to be factored in. The type of material of the room and windows is also an important consideration.

We can use the tools inside of Revit to make these calculations almost automatic.

Exercise 1-26
Volume Calculations of Rooms

Drawing Name: rooms_4.rvt
Estimated Time to Completion: 20 Minutes

Scope
Creating rooms.
Modify room tags.

Solution

1.  Activate the **GROUND FLOOR – Exam Rooms** floor plan.

2.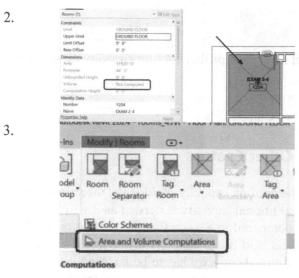

Click to select the room located in EXAM 2-4.

*In the Properties palette, note that the Volume is **Not Computed**.*

3.

Select **Area and Volume Computations**.

This is located on the expanded Room & Area panel.

4.

Enable **Areas and Volumes**.
Enable **At wall finish**.

Click **OK**.

5.

Click to select the room located in EXAM 2-4.

The volume of the room is now shown as cubic yards.

6.

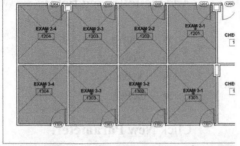

Switch to the Manage ribbon.

Click **Project Units**.

7.

Set the Volume units to **Cubic Feet (CF)**.

Click **OK**.

8.

Select all the exam rooms.

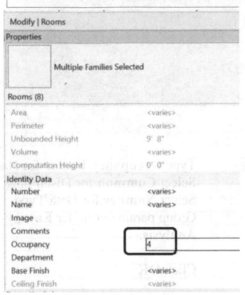

9.

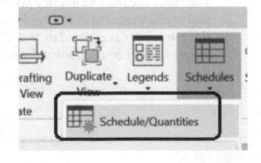

In the Properties palette:

Change the Occupancy to **4**.

10.

Switch to the View ribbon.

Click **Schedule/Quantities**.

11.

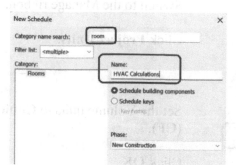

Type **room** in the search field.

Highlight **Rooms**.
Type **HVAC Calculations** in the Name field.

Click OK.

12.

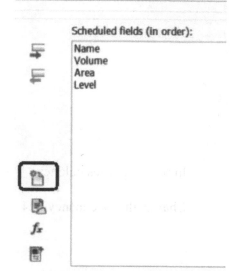

Add the following fields:

- Name
- Volume
- Area
- Level

Click **New Parameter**.

13.

Type **Occupancy Factor** for Name.
Select **Common** for Discipline.
Select **Number** for Data Type.
Group parameter under **Energy Analysis**.

Click **OK**.

You can't use the Occupancy parameter that is associated with Rooms because this parameter is defined as text. This means it can't be used for calculations.

14.

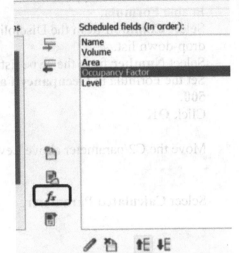

Move the Occupancy Factor parameter above Level.

Click **Calculated Parameter**.

15.

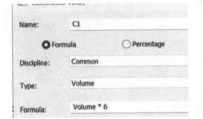

In the Name field, type **C1**.
Enable **Formula**.
Select **Common** from the Discipline drop-down list.
Select **Volume** from the Type list.
Set the Formula to **Volume *6**.
Click **OK**.

16.

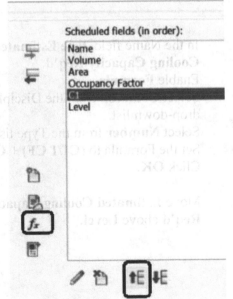

Move the C1 parameter above Level.

Select **Calculated Parameter**.

17.

Name: C2

○ Formula   ○ Percentage

Discipline: Common

Type: Number

Formula: Occupancy Factor * 500

In the Name field, type **C2**.
Enable **Formula**.
Select **Common** from the Discipline drop-down list.
Select **Number** from the Type list.
Set the Formula to **Occupancy Factor * 500**.
Click **OK**.

18.

Scheduled fields (in order):

Name
Volume
Area
Occupancy Factor
C1
C2
Level

Move the C2 parameter above Level.

Select **Calculated Parameter**.

19.

Name: Estimated Cooling Capacity Req'd

○ Formula   ○ Percentage

Discipline: Common

Type: Number

Formula: (C1 / 1 CF) + C2

In the Name field, type **Estimated Cooling Capacity Req'd**.
Enable **Formula**.
Select **Common** from the Discipline drop-down list.
Select **Number** from the Type list.
Set the Formula to **(C1/1 CF) + C2**.
Click **OK**.

20.

Scheduled fields (in order):

Name
Volume
Area
Occupancy Factor
C1
C2
Estimated Cooling Capacity Req'd
Level

Move **Estimated Cooling Capacity Req'd** above **Level.**

21.

Select the Filter tab.
Set Filter by: Name begins with EXAM
And:
Level equals GROUND FLOOR.
Click **OK**.

22.

Type 4 in the Occupancy Factor column.

Columns C2 and **Estimated Cooling Capacity Req'd** will automatically calculate.

23.

Save as *ex1-26.rvt*.

Defining Rooms in Linked Files

This information applies to linked Revit models and linked IFC models.

When you link models together, by default Revit does not recognize room-bounding elements in the linked model.

If you try to place a room between walls in the host model and walls (or other elements) in a linked model, Revit does not automatically recognize the room-bounding elements of the linked model. However, you can force Revit to use the room-bounding elements of a linked model.

Adding room separators in the host file does not affect the way rooms are defined in the linked file.

Exercise 1-27
Managing Room Boundaries in Linked Files

Drawing Name: rooms_host.rvt
Estimated Time to Completion: 15 Minutes

Scope
Manage room boundaries in a linked file.

Solution

1. 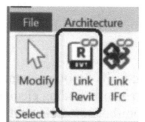   Open the Insert ribbon.

    Click **Link Revit**.

2. 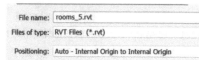   Locate *rooms_5.rvt* in the downloaded exercise files.
    Set the Positioning to **Auto-Internal Origin to Internal Origin**.

    Click **Open**.

3. 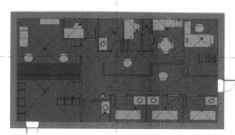   Click on the linked file.

    You will see that rooms have been placed.

    If you hover over the file, you do not see any rooms in the host file.

4.   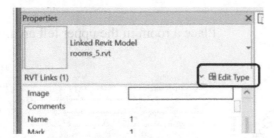   Click **Edit Type**.

5.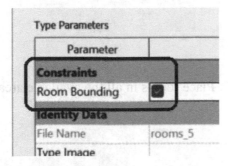

Enable **Room Bounding**.

Click **OK**.

Click **ESC** to release the selection.

6.   Open the Architecture ribbon.

Click the **Room** tool.

7.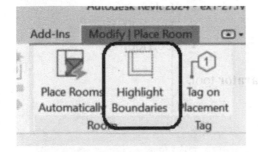

Disable Tag on Placement.

Enable **Highlight Boundaries**.

8.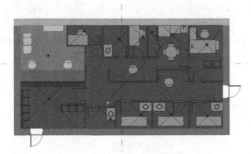

A room highlights and you also see the rooms placed in the linked file.

Click **Close**.

9.

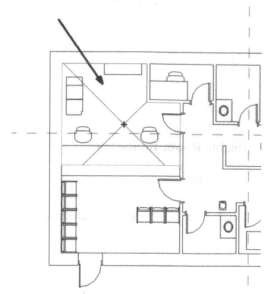

Place a room in the upper left area.

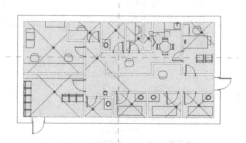

Place rooms in all the available areas.

10.

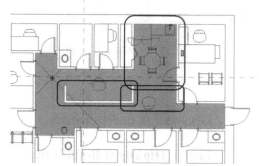

If you click the room assigned to the center of the building, you will see that there are actually four rooms in this area:

- Break Room
- Lab
- Nurse Station
- Corridor

11. Click the **Room Separator** tool.

12.

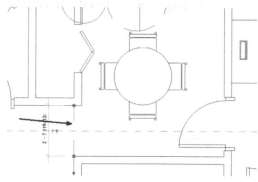

Place a line to create the break room.

13.

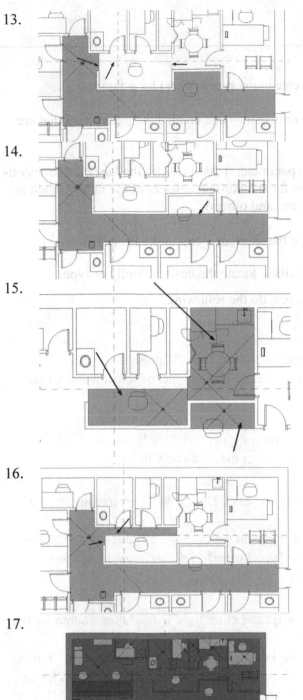

Add three lines to create a lab room area.

14.

Add a line to create a nurse station area.

15.

Add rooms to the three newly defined areas.

16.

Use the TRIM tool to adjust the boundary of the corridor area.

17.

Select the linked file.

Notice that the rooms defined in the linked files are not affected by the room separator lines placed in the host file.

Save as *ex1-27.rvt*.

Family Types

Using the Family Types tool, you can create many types (sizes) for a family.

To do this, you need to have labeled the dimensions and created the parameters that are going to vary.

Each family type has a set of properties (parameters) that includes the labeled dimensions and their values. You can also add values for standard parameters of the family (such as Material, Model, Manufacturer, Type Mark, and others).

To manage families and family types, use the shortcut menu in the Project Browser.

1.   In the Project Browser, under Families, locate the desired family or type.

2.   To manage families and family types, do the following:

If you want to...	then...
change type properties	right-click a type, and click Type Properties. As an alternative, double-click a type, or select it and Click Enter.
rename a family or type	right-click the family or type, and click Rename. As an alternative, select the type and Click F2.
add a type from an existing type	right-click a type, and click Duplicate. Enter a name for the type. The new type displays in the type list. Double-click the new type to open the Type Properties dialog, and change properties for the new type.
add a new type	right-click a family, and click New Type. Enter a name for the type. The new type displays in the list. Double-click the new type to open the Type Properties dialog, and define properties for the new type.
copy and paste a type into another project	right-click a type, and click Copy to Clipboard. Open the other project and, in the drawing area, Click Ctrl+V to paste it.

If you want to...	then...
reload a family	right-click a family, and click Reload. Navigate to the location of the updated family, select it, and click Open.
edit a family in the Family Editor	right-click a family, and click Edit. The family opens in the Family Editor.
delete a family or type	right-click a family or type, and click Delete from Project. In addition to deleting the family or type from the project, this function deletes instances of matching types that exist in the model. Note: You cannot delete the last type in a system family.

Exercise 1-28

Indentifying a Family

Drawing Name: **i_firestation_elem.rvt**
Estimated Time to Completion: 5 Minutes

Scope
Identify different elements and their families.

Solution

1. Open *i_firestation_elem.rvt*.

2. 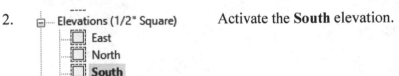 Activate the **South** elevation.

3. 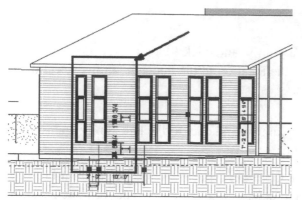 Select the second window from the left.

4. 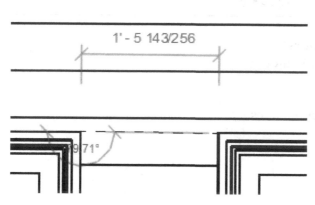 Select the **30″ W** from the window type list in the Properties pane.

5. Use the Measure tool from the Quick Access toolbar to check the distance between the outside edges of the two left windows.

6. Check to see if you got the same value.

1' - 5 143/256

9.71°

7. Close without saving.

Family Parameters

Family parameters define behaviors or Identity Data that apply across all types in that family. Different categories have different family parameters based on how Revit expects the component to be used. Family parameters can be Type or Instance. A Type parameter does not change regardless of where the element is placed. Type parameters are usually dimensions (such as length or height) or material. These parameters don't change unless a different type is selected. An Instance parameter is unique to each element. For example, the location mark of an element is changed depending on where it is placed. Door hardware can be different depending on the instance. You should be familiar with the concept of type and instance parameters and be able to identify whether or not a parameter is a type or instance parameter.

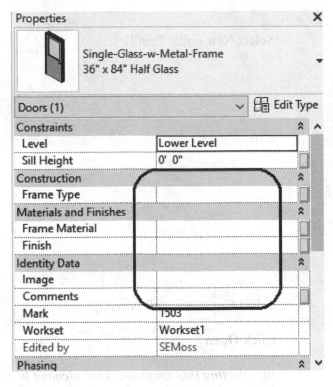

One way to determine whether a parameter is an instance parameter is to select the element and look in the Properties palette.

If you can enter or modify new data next to a parameter, it is an instance parameter.

If you have to select Edit Type to access and modify the parameter, then it is a type parameter.

Exercise 1-29
Create a Casework Family

Drawing Name: **new (using Casework wall-based.rft)**
Estimated Time to Completion: 45 Minutes

Scope
Create a new casework family.

Solution

1.

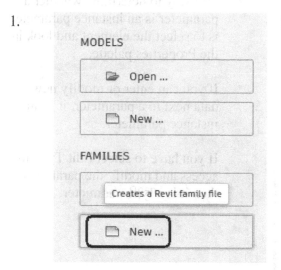

Select **New** under Families.

2.

Select *Casework wall-based.rft.*

Click **Open**.

By selecting this template, is the element to be hosted or non-hosted?

If it is hosted, what element will be the host?

3.

Switch to the **Right** elevation.

4.

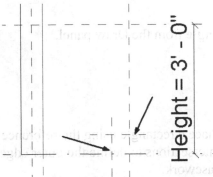

We want to add reference planes for the toe kick.

Select the **Reference Plane** tool from the ribbon.

5.

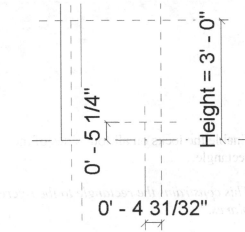

Draw a horizontal plane and a vertical plane.

6.

Select the **ALIGNED DIMENSION** tool from the Quick Access toolbar.

7.

Add a horizontal and vertical dimension between the reference planes.

Your values may be different depending on how you placed the reference planes.

8.

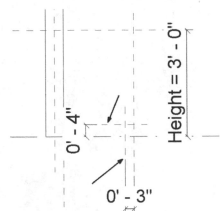

Change the horizontal dimension to 3" and the vertical dimension to 4".

Hint: *Select the reference plane and then the dimension in order to modify the dimension.*

9.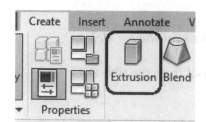

Switch to the Create ribbon.

Select the **Extrusion** tool.

10.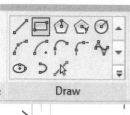

Select **Rectangle** from the Draw panel.

11.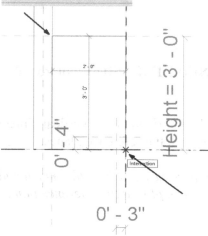

Place a rectangle using the reference plane intersections to define the outer edge of the casework.

12.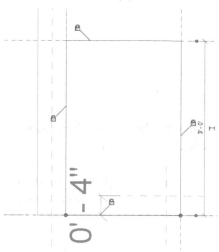

Enable the locks on all four sides of the rectangle.

This constrains the rectangle to the reference planes.

13. **Green check** to complete the extrusion.

14.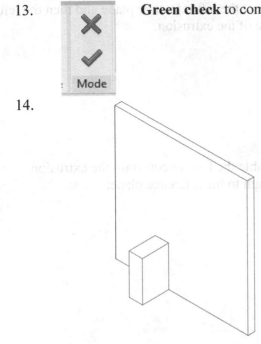

Switch to a 3D view.

Use the Viewcube to re-orient the view.

The casework is not very wide. The larger rectangle represents a wall. This wall does not come in when the casework is placed. It is there to help you test and check your model. It is part of the wall-based template.

15.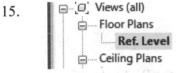

Switch to a **Ref. Level** view.

Views (all)
Floor Plans
Ref. Level
Ceiling Plans

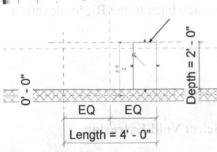

You can see the extrusion you created and how it is located relative to the existing reference planes.

There is currently no relationship between the length of the extrusion and the length parameter.

16. Select the extrusion.

Casework (1)		Edit Type
Constraints		
Extrusion End	1' 0"	
Extrusion Start	0' 0"	
Work Plane	Reference Plane : Right	

On the Properties palette, you can see the Extrusion End is set to **1'-0"**.

Click ESC to release the selection.

17. Select the **ALIGN** tool from the ribbon.

18.

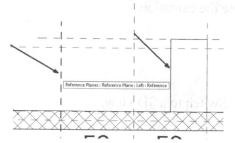

Select the left reference plane and then the left side of the extrusion.

19.

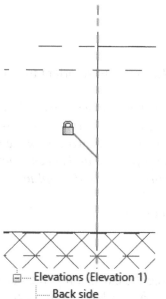

Enable the lock to constrain the extrusion length to the reference plane.

20.

Switch back to the **Right** elevation.

21.

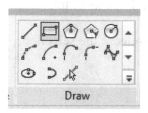

Switch back to the Create ribbon.

Select **Void Extrusion**.

22.

Select **Rectangle** from the Draw panel.

23.

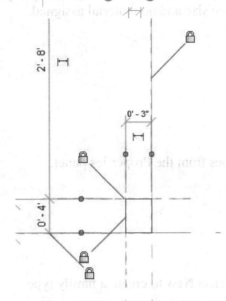

Place a rectangle using the reference plane intersections to define the outer edge of the toe kick.

24.

Enable the locks on all four sides of the rectangle.

This constrains the rectangle to the reference planes.

25.

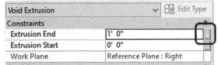

In the Properties palette:

The Extrusion End is currently set to 1'-0".

Select the small button at the right end of the field.

26.

<none>
Default Elevation
Depth
Height
Length
Width

Highlight **Length**.

This associates the extrusion end value with the Length parameter.

This is similar to using the ALIGN command and locking to a reference plane.

Click **OK**.

27.  **Green check** to complete the extrusion.

28. 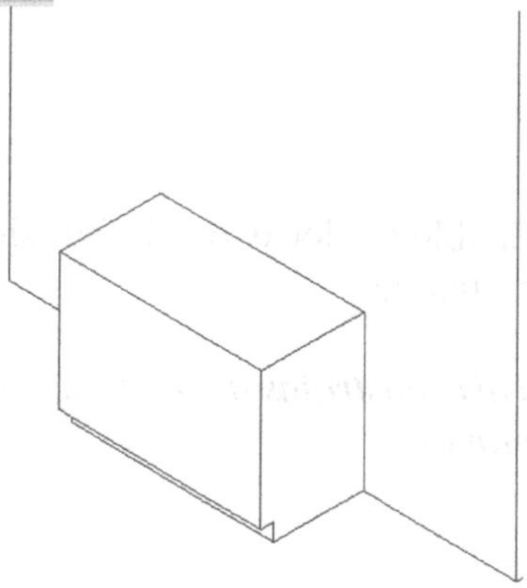 Switch to a 3D view.

Right now, the casework is a single type with one size and no material assigned.

29. 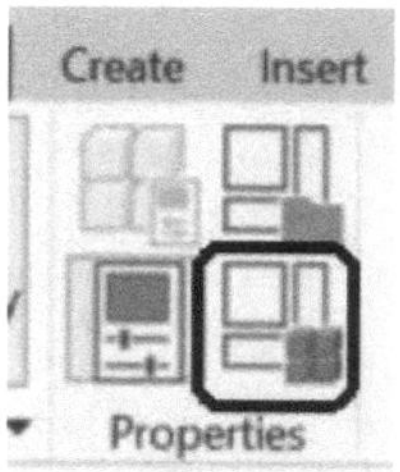 Select **Family Types** from the Properties panel.

30. 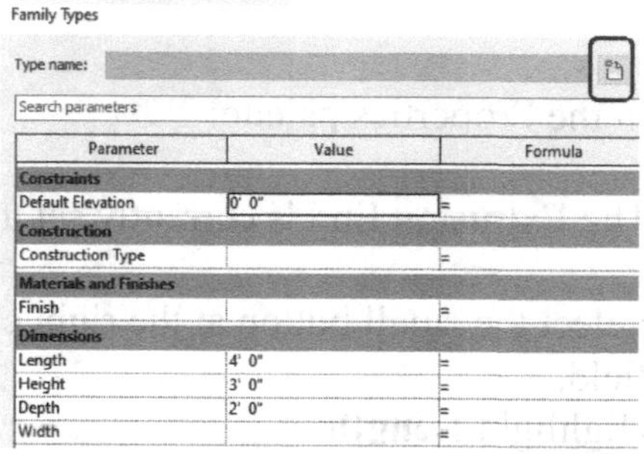 Select **New** to create a family type.

31. 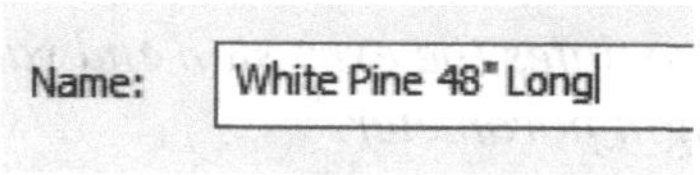 Type **White Pine 48" Long**.

Click **OK**.

32.  Type **White Pine** in the **Finish** field.

Click **Apply**.

The Finish value is simply text, so it will not actually control the appearance of the extrusion.

33.  Select **New** to create a family type.

34.

Name: | White Pine 36" Long|

Type **White Pine 36" Long**.

Click **OK**.

35.

Dimensions	
Length	3' 0"
Height	3' 0"
Depth	2' 0"
Width	

Change the Length to **3'-0"**

Click **Apply.**

The casework should adjust.

Click **OK** to close the dialog.

36.

File name: | Cambridge_Book_Shelf.rfa
Files of type: | All Supported Files (*.rvt, *.rfa, *.rte)

Open *Cambridge_Book_Shelf.rfa* from the downloaded exercise files.

37.

Transfer
s Project Standards

Switch back to the casework family you were working on.

Go to the Manage ribbon.

Click **Transfer Project Standards**.

38.

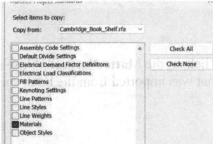

Select items to copy:
Copy from: | Cambridge_Book_Shelf.rfa
☐ Assembly Code Settings
☐ Default Divide Settings
☐ Electrical Demand Factor Definitions
☐ Electrical Load Classifications
☐ Fill Patterns
☐ Keynoting Settings
☐ Line Patterns
☐ Line Styles
☐ Line Weights
☑ Materials
☐ Object Styles
Check All
Check None

Set Copy from: to *Cambridge_Book_Shelf.rfa.*

Click **Check None**.

Enable **Materials**.

Click **OK**.

39.

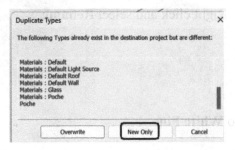

Duplicate Types   ✕

The following Types already exist in the destination project but are different:

Materials : Default
Materials : Default Light Source
Materials : Default Roof
Materials : Default Wall
Materials : Glass
Materials : Poche
Poche

Overwrite | New Only | Cancel

Click **New Only**.

40.

Select the extrusion.

EQ | EQ

Length = 3' - 0"

41.

Casework (1)		Edit Type
Constraints		
Extrusion End	3' 0"	
Extrusion Start	0' 0"	
Work Plane	Reference Plane : Right	
Graphics		
Visible	☑	
Visibility/Graphics Overrides	Edit...	
Materials and Finishes		
Material	<By Category>	
Identity Data		
Subcategory	<None>	

Click in the **Material** field to assign a material.

This will control the appearance of the casework.

42.

Search

Project Materials: All ▼ ▾

Name

BMCD2AR3\Solid Materials \Matte\Textured\Black

BMCD2AR3\Solid Materials \Matte\Normal\White

BMCD2AR3\Solid Materials \Flat Colors\Smooth\White

Click on the **Matte\Normal\White** material that was imported from the book shelf family.

43.

BMCD2AR3\Solid Materials \Matte\Textured\Black

BMCD2AR3\Solid Material

BMCD2AR3\Solid Material

Analytical Wall Surface

Analytical Slab Surface

Edit
Duplicate Material and Assets
Duplicate Using Shared Assets
Rename
Delete
Add to ▸

Right click and select **Rename**.

44.

Name

White Pine

Rename to **White Pine**.

Click **OK**.

45.

Visibility/Graphics Overrides	Edit...
Materials and Finishes	
Material	White Pine
Identity Data	
Subcategory	<None>

White Pine is now displayed as the material for the extrusion.

Click **ESC** to release the selection.

46.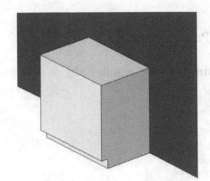

Switch to a 3D view.

Change the display to **Realistic**.

Save as *casework_2.rfa*.

Exercise 1-30
Create a New Family Type

Drawing Name: **family_type.rvt**
Estimated Time to Completion: 5 Minutes

Scope
Create a new family type.

Solution

1. 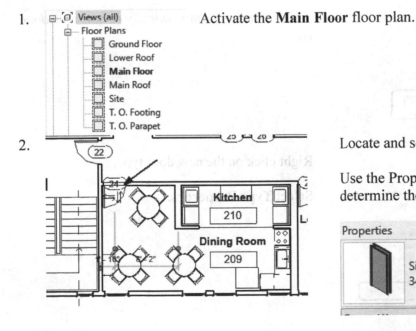 Activate the **Main Floor** floor plan.

2. Locate and select Door 24.

Use the Properties panel to determine the door type.

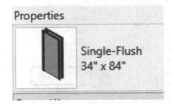

3.

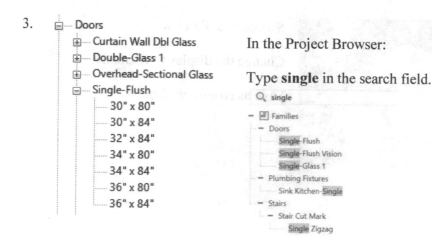

In the Project Browser:

Type **single** in the search field.

Locate the **Single-Flush** door family.

Expand to see the different door types available.

4.

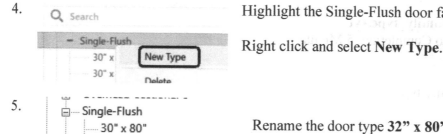

Highlight the Single-Flush door family.

Right click and select **New Type**.

5.

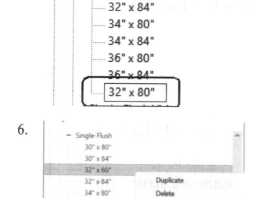

Rename the door type **32" x 80"**.

Notice that Revit automatically re-sorts the door types.

6.

Right click on the new door type.

Select **Type Properties**.

7.

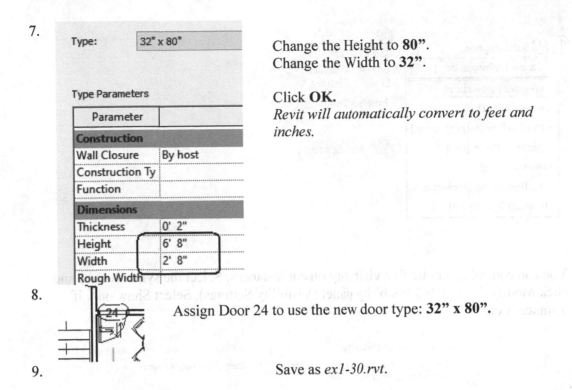

Change the Height to **80"**.
Change the Width to **32"**.

Click **OK.**
Revit will automatically convert to feet and inches.

8.

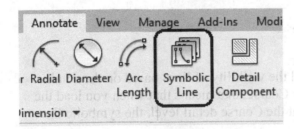

Assign Door 24 to use the new door type: **32" x 80".**

9.

Save as *ex1-30.rvt.*

Symbolic Lines

The **Symbolic Line** tool is available when you are creating or editing a family. It is located on the Annotate ribbon→Detail panel. This tool allows you to sketch lines that are meant for symbolic purposes only. Symbolic lines are not part of the actual geometry of the family. Symbolic lines are visible parallel to the view in which you sketch them.

For Example: You might sketch symbolic lines in an elevation view to represent a door swing.

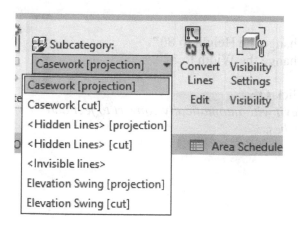

You can control symbolic line visibility on cut instances. Select the symbolic line and click Modify |Lines tab→Visibility panel (Visibility Settings). Select Show only if instance is cut.

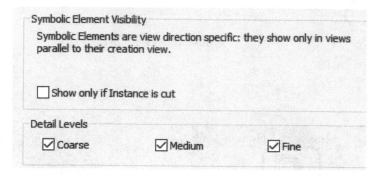

In the displayed dialog, you can also control the visibility of lines based on the detail level of the view. For example, if you select Coarse, that means that when you load the family into a project and place it in a view at the Coarse detail level, the symbolic lines are visible.

Exercise 1-31
Using Symbolic Lines in Families

Drawing Name: **casework_3.rfa, test_casework.rvt**
Estimated Time to Completion: 20 Minutes

Scope
Add symbolic lines to a family.
Control the visibility of family elements.
Load a family into a project.
Change the display of a loadable family.

Solution

1.   Views (all)
      Floor Plans
        Ref. Level

                 Open *casework_3.rfa*.
                 Open the **Ref. Level.**

2.   Reference Line   Reference Plane   Datum

                 Select the **Reference Plane** tool from the Create ribbon.

3.   Draw a horizontal reference plane through the casework.

      EQ   EQ

4.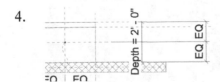

                 Use the ALIGNED DIMENSION to center the reference plane on the casework.

                 Hint: Place a segmented dimension and set the dimensions equal. Make sure you select reference planes. Use the TAB key to cycle through the selections if necessary.

5. 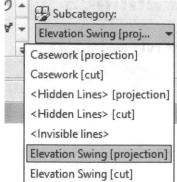   Switch to the Annotate ribbon.

Select **Symbolic Line**.

6.   Select **Elevation Swing [projection]** to assign the Subcategory.

7. 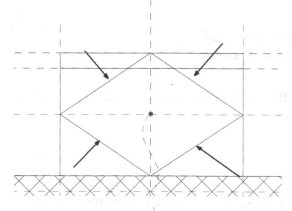   Place the four lines as shown using the midpoints.

8. 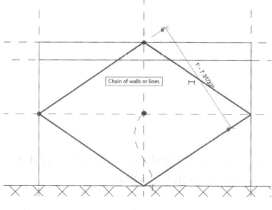   Select one line and then click the TAB key.
This will select all four lines as they are connected.
Left click to accept the selection.

9. Select **Visibility Settings**.

10. Disable **Coarse**.

Click **OK**.

This means the symbolic lines will only be visible if the Detail Level is set to Medium or Fine.

11. Click in the window to release the selection.

12. Switch to the Manage ribbon.

Click **Object Styles**.

13.

Category	Line Weight		Line Color	Line Pattern
	Projection	Cut		
Casework	1	3	■ Black	
<Hidden Lines>	1	1	▨ Cyan	Dash
Elevation Swing	1	1	■ Blue	Double dash
Walls	2	2	■ Black	
<Hidden Lines>	2	2	■ Black	Dash
Common Edges	2	2	■ Black	

Change the Hidden Lines to **Cyan**. Change the Elevation Swing to **Blue** and **Double dash**.

Click **OK**.

14. Save as *casework_4.rfa*.

Open *test_casework.rvt*.

15. 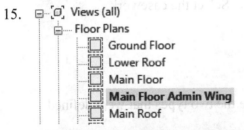 Open the **Main Floor Admin Wing floor** plan.

16. 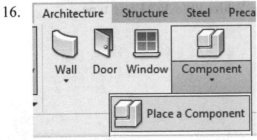 Select **Place a Component** from the Architecture ribbon.

17. 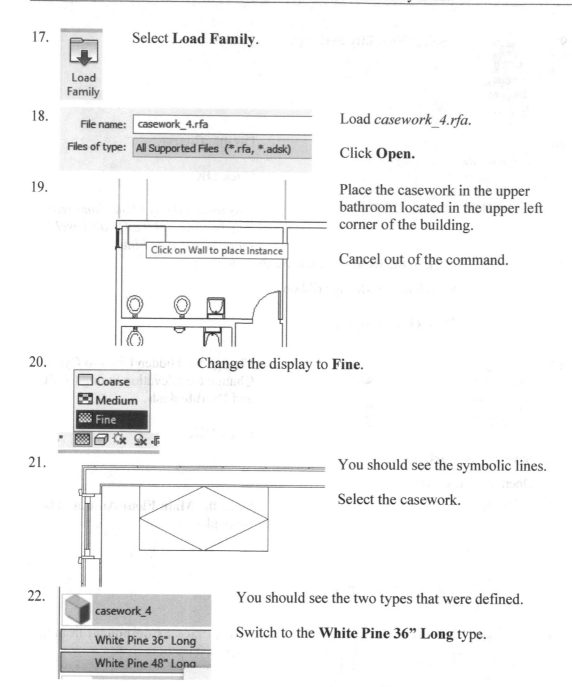 Select **Load Family**.

18. Load *casework_4.rfa*.

File name: casework_4.rfa

Files of type: All Supported Files (*.rfa, *.adsk)

Click **Open.**

19. Place the casework in the upper bathroom located in the upper left corner of the building.

Click on Wall to place Instance

Cancel out of the command.

20. Change the display to **Fine**.

☐ Coarse
☒ Medium
▨ Fine

21. You should see the symbolic lines.

Select the casework.

22. You should see the two types that were defined.

casework_4

White Pine 36" Long

White Pine 48" Long

Switch to the **White Pine 36" Long** type.

23.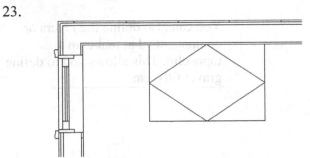

The casework should adjust.

Do the symbolic lines display with the correct color?

Do you need to modify the Object Styles in the project in order for them to display correctly?

24. Save as *ex1-31.rvt*.

Toposolids

To create a toposolid, you can either draw a sketch boundary, similar to creating a floor, or convert an imported CAD file, or use a CSV file that contains the XYZ points used to generate the toposolid.

When modifying the elevation of a point located on a toposolid, the elevation can be defined relative to the connected level, the survey point, the project base point, or the internal origin.

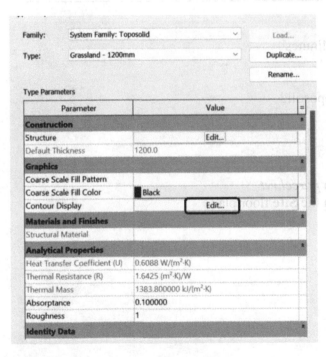

The contour display for a toposolid is controlled in the type properties.

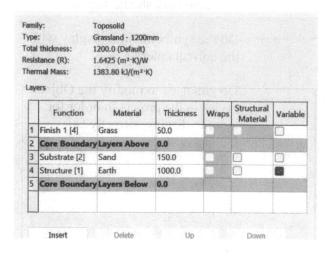

Family: Toposolid
Type: Grassland - 1200mm
Total thickness: 1200.0 (Default)
Resistance (R): 1.6425 (m²·K)/W
Thermal Mass: 1383.80 kJ/(m²·K)

Layers

	Function	Material	Thickness	Wraps	Structural Material	Variable
1	Finish 1 [4]	Grass	50.0	☐	☐	☐
2	Core Boundary	Layers Above	0.0			
3	Substrate [2]	Sand	150.0	☐	☐	☐
4	Structure [1]	Earth	1000.0	☐	☐	☑
5	Core Boundary	Layers Below	0.0			

Insert   Delete   Up   Down

You can also define the layers or components that make up a toposolid. This allows you to define gravel fills, etc.

Exercise 1-32
Create a Toposolid Using Points

Drawing Name: parcel.rvt
Estimated Time to Completion: 10 Minutes

Scope
Create and place points to define a toposolid
Assign point elevations.

Solution

1.
- Floor Plans
  - Level 1
  - Level 2
  - **Site**

Open *parcel.rvt*
Open the **Site** floor plan

2.
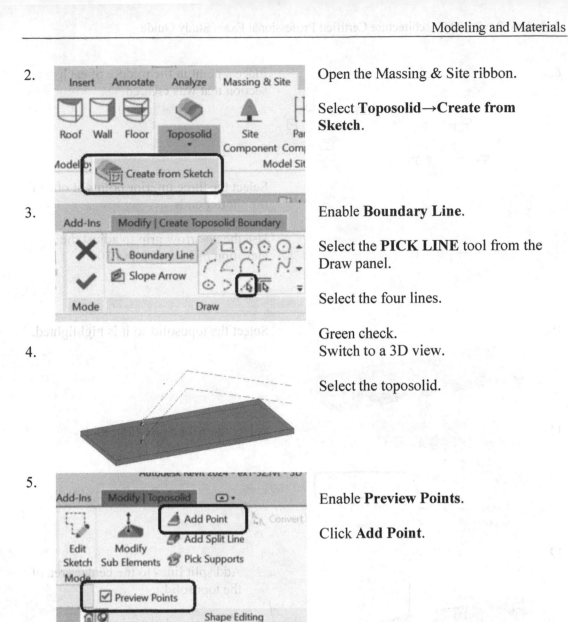

Open the Massing & Site ribbon.

Select **Toposolid→Create from Sketch**.

3.

Enable **Boundary Line**.

Select the **PICK LINE** tool from the Draw panel.

Select the four lines.

Green check.
Switch to a 3D view.

4.

Select the toposolid.

5.

Enable **Preview Points**.

Click **Add Point**.

6.

Add a point to the middle of each side on the top face.

7.

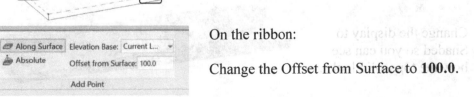

On the ribbon:

Change the Offset from Surface to **100.0**.

8. 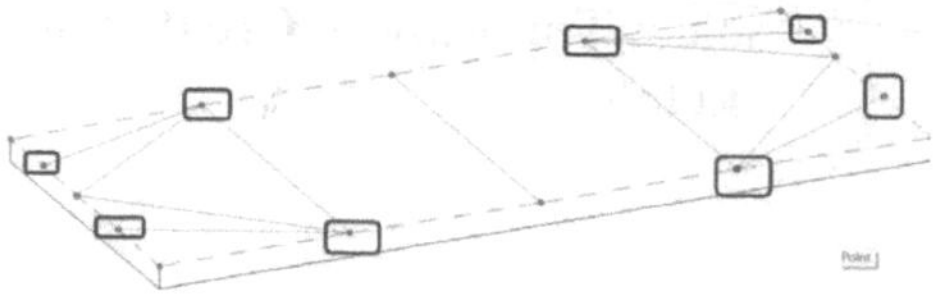   Place a point in the middle of each section that was created.

9. 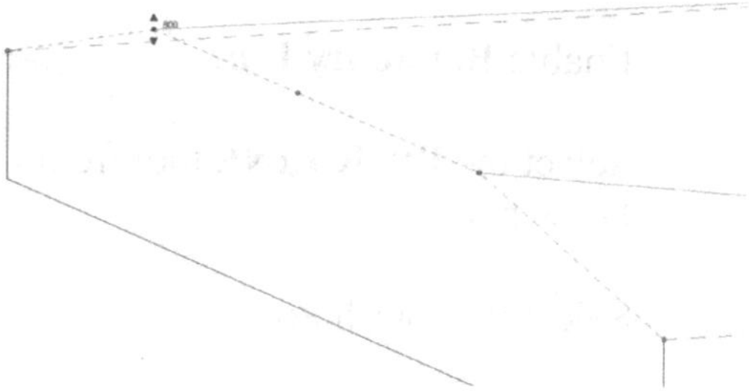   Select the three interior points at one of the shorter ends.

Use the up arrow grip to adjust the elevation of each point to 800.

10. 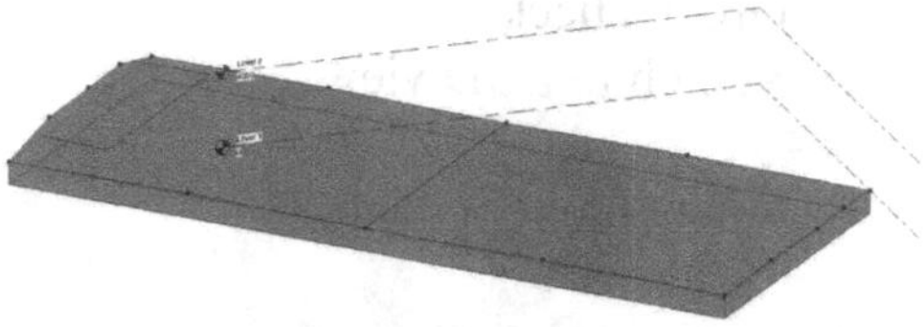 Select the toposolid so it is highlighted.

11. 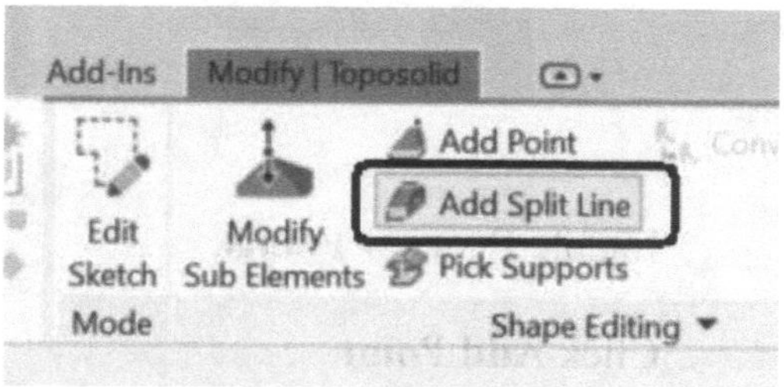 Select the **Add Split Line** tool.

12. 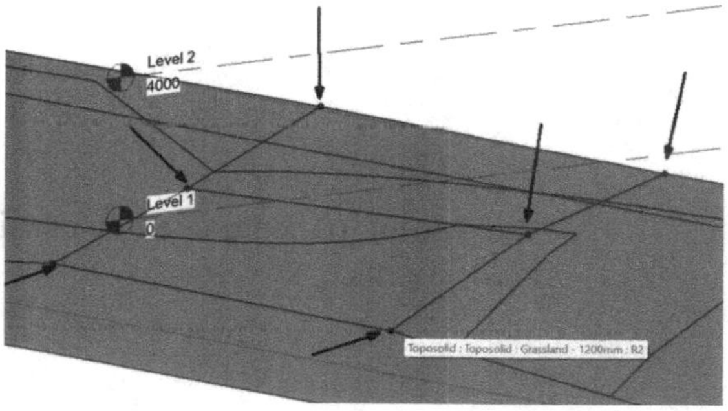 Add split lines to the center area of the toposolid.

Adjust the elevation of the endpoints of the split lines.

13. Change the display to Shaded so you can see how the toposolid looks.

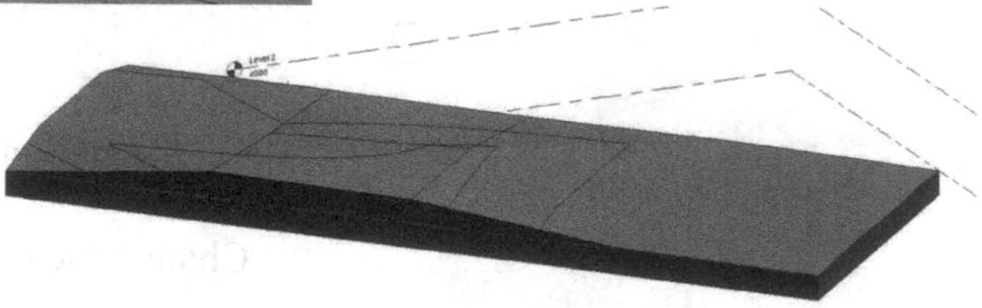

14. Save as *1-32.rvt*.

Exercise 1-33
Create a Toposolid Using a CSV File

Drawing Name: *new, topo-points.csv*
Estimated Time to Completion: 5 Minutes

Scope
Use a CSV file to create a toposolid

Solution

1.  Go to the Application menu.
Select **New→Project**.

2. Select the *default.rte* template under English-Imperial.

   Click **Open**.

3. Click **OK**.

4. Go to the Massing & Site ribbon.

   Select **Toposolid→Create from Import**.

5. Click **Create from CSV** on the ribbon.

6. Locate the *topo-points.csv* file in the downloaded exercises.

7. One unit in the file equals one:

Feet

OK   Cancel

Click **Open**.
Set the units to **Feet**.
Click **OK**.

8.

Ignore the warning.

Switch to a 3D view so you can inspect the toposolid that was created.

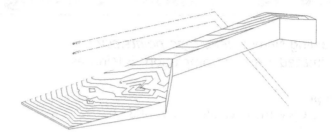

9. Save as *ex1-33.rvt*.

Exercise 1-34
Create a Toposolid Using a DWG File

Drawing Name: *new, parcel.dwg*
Estimated Time to Completion: 10 Minutes

Scope
Use a DWG file to create a toposolid

Solution

1. File Architecture Structure Steel Precast Systems Inse

Creates a Revit file.

New ▸ Project
Creates a Revit project file.

Go to the Application menu.
Select **New→Project**.

2. English-Imperial

Name
Commercial-Default.rte
Construction-Default.rte
default.rte
Default-Multi-discipline.rte
Electrical-Default.rte

Select the *default.rte* template under English-Imperial.

Click **Open**.

3. Click **OK**.

4. Go to the **Insert** ribbon.
   Click **Import CAD**.

5. Locate the *parcel.dwg* file in the downloaded exercises.

   Set Colors to **Preserve**.
   Set Positioning to **Auto – Center to Center**.
   Set Import units to **Auto-Detect**.

   Click **Open**.

6. Click on the imported file.
   Select **Query** from the ribbon.

7. Select the top line indicated.

8. Notice the layer assigned to the line.

   Click **OK**.

9. Select the blue line.

10. Notice the layer assigned to the line.

    Click **OK**.

11. Go to the Massing & Site ribbon.

    Select **Toposolid→Create from Import**.

12.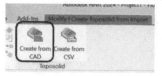

Click **Create from CAD** on the ribbon.

Click on the imported CAD file.

13.

Enable the following layers:

- 0
- Parcels
- Railroads
- Roads
- Streams
- Waterbodies.

Click **OK**.

14.

Switch to a 3D view so
you can inspect the
toposolid that was
created.

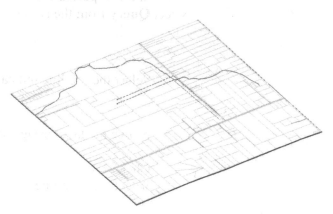

15.

Save as *ex1-34.rvt*.

Exercise 1-35
Add a Subdivision to a Toposolid

Drawing Name: *topo_1.dwg*
Estimated Time to Completion: 5 Minutes

Scope
Add a subdivision to a toposolid
Apply a material to a subdivision

Solution

1. — Floor Plans   Open the **Site** floor plan.
    ☐ Level 1
    ☐ Level 2
    ☐ **Site**

2. Sub-Divide   Select the toposolid.

    Toposolid Shaping   Click **Sub-Divide** on the ribbon.

3. Boundary Line   Select the **LINE** tool from the DRAW panel.
    Slope Arrow
    Draw

4.   Draw four lines using the midpoint OSNAPS to place the lines.

    Green check.

5. | Toposolid (1) | |
    | --- | --- |
    | Sub-Division | |
    | Material | Concrete, Cast-in-Place, Gray |
    | Sub-divide Height | 304.8 |
    | Inherit Contours | ☑ |

    Assign the material: **Concrete, Cast-in-Place, Gray** to the sub-division.

    Click to release the selection.

6.   Switch to a 3D View.

      Save as *ex1-35.rvt*.

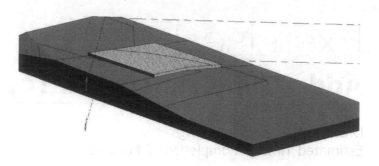

Groups

You can group elements in a project or family and then place that group many times in a project or family.

Grouping elements is useful when you need to create entities that represent repeating layouts or are common to many building projects, such as hotel rooms, apartments, or repeating floors.

With each instance of a group that you place, there is associativity among them. For example, you create a group with a bed, walls, and window and then place multiple instances of the group in your project. If you modify a wall in one group, it changes for all instances of that group, simplifying the modification process.

You can create model groups, detail groups, and attached detail groups. Attached detail groups contain view-specific elements, such as annotations, along with model elements. You cannot create a group that includes both model and detail elements.

Exercise 1-36
Model Groups

Drawing Name: **model groups.rvt**
Estimated Time to Completion: 15 Minutes

Scope
Create a model group.
Place a model group.
Edit a model group.

Solution

1.

Activate the **GROUND FLOOR – Exam Rooms** floor plan.

2.

Exam room 2-4 has a patient bed, a chair, a desk, a rolling chair and some casework.

We are going to create a model group of all the furniture in the exam room.

Select all the items in the room except for the casework.

Remember to hold down the CTL key to select more than one element.

3.

Select the **Create Group** tool from the ribbon.

4.

Name: Exam Room

☐ Open in Group Editor

OK   Cancel

Type **Exam Room** for the name.

Click **OK**.

5.

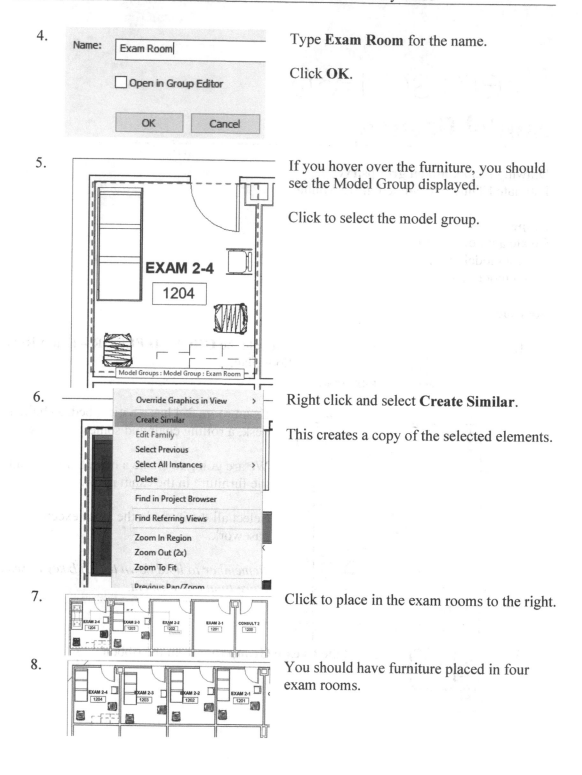

If you hover over the furniture, you should see the Model Group displayed.

Click to select the model group.

EXAM 2-4

1204

Model Groups : Model Group : Exam Room

6.

Override Graphics in View   >

Create Similar

Edit Family

Select Previous

Select All Instances   >

Delete

Find in Project Browser

Find Referring Views

Zoom In Region

Zoom Out (2x)

Zoom To Fit

Previous Pan/Zoom

Right click and select **Create Similar**.

This creates a copy of the selected elements.

7.

EXAM 2-4 1204 EXAM 2-3 1203 EXAM 2-2 1202 EXAM 2-1 1201 CONSULT 2 1200

Click to place in the exam rooms to the right.

8.

EXAM 2-4 1204 EXAM 2-3 1203 EXAM 2-2 1202 EXAM 2-1 1201

You should have furniture placed in four exam rooms.

9.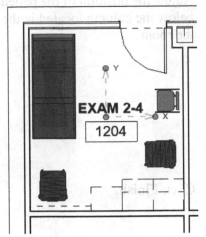

Select the model group located in Exam 2-4.

10.

Select **Edit Group** on the ribbon.

11.

Select **Add.**

Notice that elements that are already part of the model group are grayed out.

12.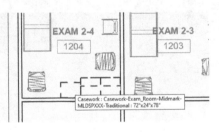

Select the casework in Exam 2-4.

Right click and select **Cancel**.

13. 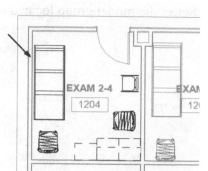 Adjust the position of the patient bed so it is below the column located in the upper right of the room.

14. 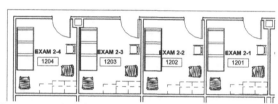 Click **Finish**.

15.  Notice that all the model groups update.

Adjust the position of the model groups so they fit properly in each exam room.

16. Save as *ex1-36.rvt*

Practice Exam

1. Select the answer which is NOT an example of bidirectional associativity:

    A. Flip a section line and all views update.
    B. Draw a wall in plan view and it appears in all other views.
    C. Change an element type in a schedule and the change is displayed in the floor plan view as well.
    D. Flip a door orientation so the door swing is on the exterior of the building.

2. Select the answer which is NOT an example of a parametric relationship:

    A. A floor is attached to enclosing walls. When a wall moves, the floor updates so it remains connected to the walls.
    B. A series of windows are placed along a wall using an EQ dimension. The length of the wall is modified, and the windows remain equally spaced.
    C. A door is placed in a wall. The wall is moved, and the door remains constrained in the wall.
    D. A shared parameter file is loaded to the server.

3. Which tab does NOT appear on Revit's ribbon?

    A. Architecture
    B. Basics
    C. Insert
    D. View

4. Which item does NOT appear in the Project Browser?

    A. Families
    B. Groups
    C. Callouts
    D. Notes

5. Which is the most recently saved backup file?

    A. office.0001
    B. office.0002
    C. office.0003
    D. office.0004

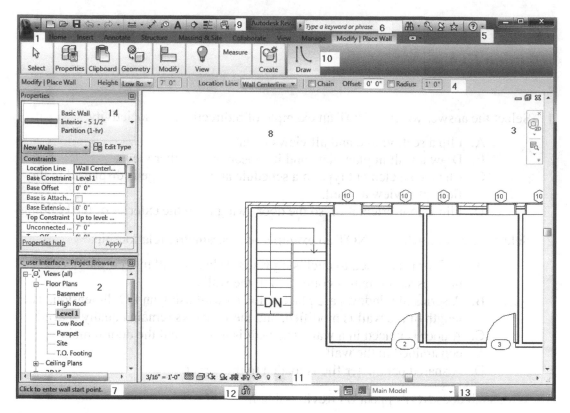

6. Match the numbers with their names.

View Control Bar	InfoCenter
Project Browser	Status Bar
Navigation Bar	Properties Pane
Options Bar	Application Menu
Design Options	Drawing Area
Help	Quick Access Toolbar
Ribbon	Worksets

7. Which of the following can NOT be defined prior to placing a wall?

    A. Unconnected Height
    B. Base Constraint
    C. Location Line
    D. Profile
    E. Top Offset

8. Identify the stacked wall.

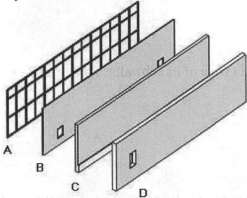

9. Walls are system families. Which name is NOT a wall family?

    A. BASIC
    B. STACKED
    C. CURTAIN
    D. COMPLICATED

10. Select the TWO that are wall type properties:

    A. COARSE FILL PATTERN
    B. LOCATION LINE
    C. TOP CONSTRAINT
    D. FUNCTION
    E. BASE CONSTRAINT

11. Use this key to cycle through selections:

    A. TAB
    B. CTRL
    C. SHIFT
    D. ALT

12. The construction of a stacked wall is defined by different wall _____.

    A. Types
    B. Layers
    C. Regions
    D. Instances

13. To change the structure of a basic wall you must modify its:

    A. Type Parameters
    B. Instance Parameters
    C. Structural Usage
    D. Function

14. If a stacked wall is based on Level 1 but one of its subwalls is on Level 7, the base level for the subwall is Level _____.

    A. 7
    B. 1
    C. Unconnected
    D. Variable

15. A designer wants to adjust the horizontal rail indicated in the image. Where should the designer click in the Type Properties dialog?

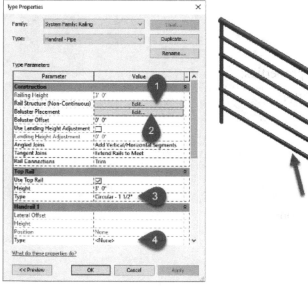
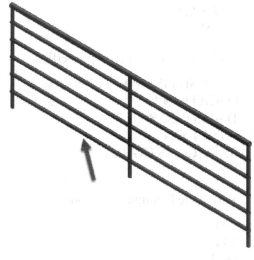

16. Which of the following methods is used to join a wall to a roof and have the wall shape conform to the underside of the roof?

    A. Select the wall, click Attach Top/Base on the ribbon, and then select the roof.
    B. Edit the profile of the wall, select the top boundary, click Attach Top/Base, and then select the roof.
    C. Use the Wall by Face tool and create a wall based on the underside face of the roof.
    D. Select the roof, click Attach Top/Base, and then select the wall.

17. A designer is working with a room schedule. The designer only wants to see rooms that are offices in the schedule, and not restrooms, break rooms, etc. How should this be accomplished?

    A. Delete the non-office room rows in the schedule.
    B. Only add the office rooms to the list of fields.
    C. Sort the schedule to only show office rooms.
    D. Add a filter to filter out the non-office rooms.

18. Which of the following best describes the hierarchy of elements in a Revit project?

    A. Roofs > Walls > Floors > Ceilings
    B. Category > Family > Type > Instance
    C. Category > Type > Family > Instance
    D. Family > Category > Instance > Type

19. Which of the following two methods can be used to cut a horizontal or vertical projection from a wall? (Select 2)

    A. Open the wall in the Family Editor and add a reveal profile to the wall type.
    B. Use the Wall: Reveal tool to add a vertical or horizontal reveal in a wall.
    C. Activate the Cut Geometry tool, select a reveal profile, and place it in a wall.
    D. Click Opening By Face, select a wall, and then sketch the projection to cut.
    E. Modify the wall type and add a reveal profile to the wall structure.

20. Which tool is used to create a wall that follows the floor as shown in the image?

A. Attach Top/Base
B. Edit Profile
C. Edit Workplane
D. Pick New Host

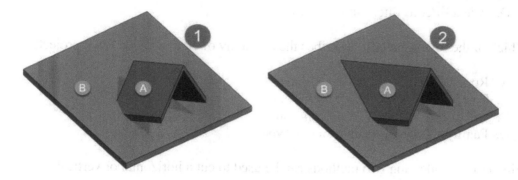

21. In the image, the rear edge of Roof A is extended to meet the slope of Roof B. How is this accomplished?

A. Use Trim/Extend Single Element tool to extend Roof A tp Roof B.
B. Use the Roof Joins tool to extend the rear edge to Roof A to the face of Roof B.
C. Edit the boundary sketch of Roof A to overlap Roof B, then use Attach Roof.
D. Use the Join Geometry tool to join the two roofs together.

22. An architectural designer has duplicated a material in the Material Browser and modified the appearance asset used by the material. Several other materials have had their appearance assets inherit the same changes. How should the designer have duplicated the material to prevent this from happening?

    A. Duplicate as New Material Asset
    B. Duplicate Material and Assets
    C. Duplicate as Independent
    D. Duplicate Using Shared Assets

23. To place a door in a curtain wall, which steps should be followed:

    A. Create a wall opening in the curtain wall. Then place a curtain wall door type into the opening using the Door tool.
    B. Add mullions to the curtain wall to create a door-sized panel. Then use the Door tool on the Architecture ribbon and place a Curtain Wall Door type to replace the panel.
    C. Click the Door tool on the Architecture ribbon. Select a Curtain Wall Door type using the Type Selector. Place the door in the curtain wall.
    D. Add Curtain Grid Lines to the curtain wall to create a door-sized panel. Select the panel and change it to a Curtain Wall Door family using the Type Selector.

24. A railing is hosted by a sloped wall. A post needs to be placed in the railing in Positions 1 and 2. How can this be done?

    A)   Split the railing into three different railings.
    B)   Edit the railing type and in the baluster placement settings set the Corner Posts At option to Each Segment End.
    C)   Edit the railing type and in the baluster placement settings set the Corner Posts At option to Angles Greater than 0 degrees.
    D)   Edit the railing path and split the sketch line at position 1 and 2 to create separate segments.

Answers:
1) D; 2) D; 3) B; 4) D; 5) D; 6) 1- Application Menu, 2- Project Browser, 3- Navigation Bar, 4- Options Bar, 5- Help, 6- InfoCenter, 7- Status Bar, 8- Drawing Area, 9- Quick Access Toolbar, 10- Ribbon, 11- View Control Bar,12- Worksets, 13- Design Options; 7) D ; 8) C; 9) D; 10) A & D; 11) D; 12) A; 13) A; 14) B; 15) 1; 16) A; 17) D; 18) B; 19) B & E; 20) B; 21) B; 22) B; 23) D; 24) D

Notes:

Documentation

This lesson addresses the following exam questions:

- Create and control model views, element visibility and graphics
- Configure and manage annotative elements
- Use revision control
- Create and manage legends
- Create detail and drafting views
- Create custom tags
- Create and modify detail groups

View Templates

A view template is a collection of view properties, such as view scale, discipline, detail level, and visibility settings.

Use view templates to apply standard settings to views. View templates can help to ensure adherence to office standards and achieve consistency across construction document sets.

Before creating view templates, first think about how you use views. For each type of view (floor plan, elevation, section, 3D view, and so on), what styles do you use? For example, an architect may use many styles of floor plan views, such as power and signal, partition, demolition, furniture, and enlarged.

You can create a view template for each style to control settings for the visibility/graphics overrides of categories, view scales, detail levels, graphic display options, and more.

Exercise 2-1
Create a View Template

Drawing Name: **view_templates.rvt**
Estimated Time to Completion: 30 Minutes

Scope
Apply a wall tag.
Create a view template.
Create a view filter.
Apply view settings to a view.

Solution

1.   Views (all)
         Floor Plans
            Level 1
            Level 2
            Level 3
            Site   Activate Level 1.

2.   Annotate   Activate the **Annotate** ribbon.

3.   Select **Tag All**.

         Tag
       y   All

4.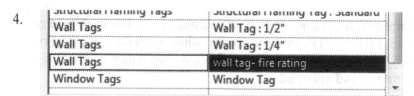
                                               Highlight the **Wall tag - fire rating** as the tag to be used and click **OK**.

The wall tag - fire rating is a custom family. It was pre-loaded into this exercise but is included with the exercise files on the publisher's website for your use.

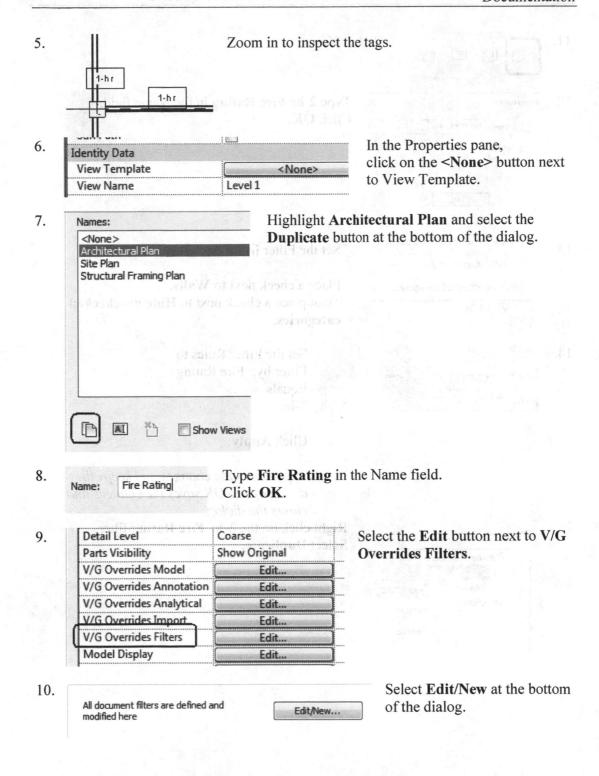

5.   Zoom in to inspect the tags.

6.   In the Properties pane, click on the **<None>** button next to View Template.

Identity Data	
View Template	<None>
View Name	Level 1

7.   Highlight **Architectural Plan** and select the **Duplicate** button at the bottom of the dialog.

Names:
<None>
Architectural Plan
Site Plan
Structural Framing Plan

☐ Show Views

8.   Type **Fire Rating** in the Name field. Click **OK**.

Name: Fire Rating

9.   Select the **Edit** button next to **V/G Overrides Filters**.

Detail Level	Coarse
Parts Visibility	Show Original
V/G Overrides Model	Edit...
V/G Overrides Annotation	Edit...
V/G Overrides Analytical	Edit...
V/G Overrides Import	Edit...
V/G Overrides Filters	Edit...
Model Display	Edit...

10.   Select **Edit/New** at the bottom of the dialog.

All document filters are defined and modified here   Edit/New...

11.  Select **New.**

12. 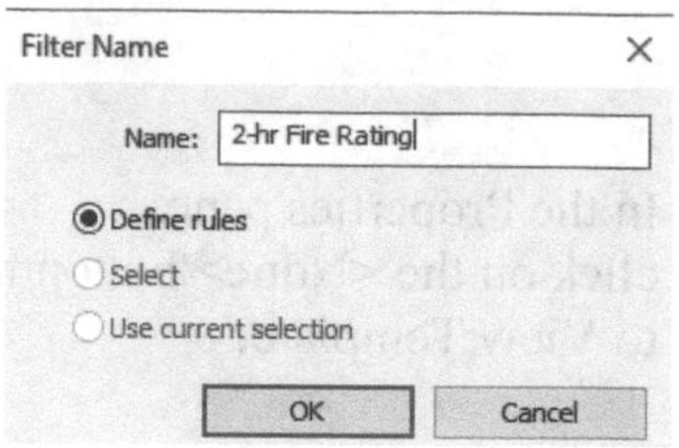 Type **2-hr Fire Rating** in the Name field.
Click **OK.**

13. 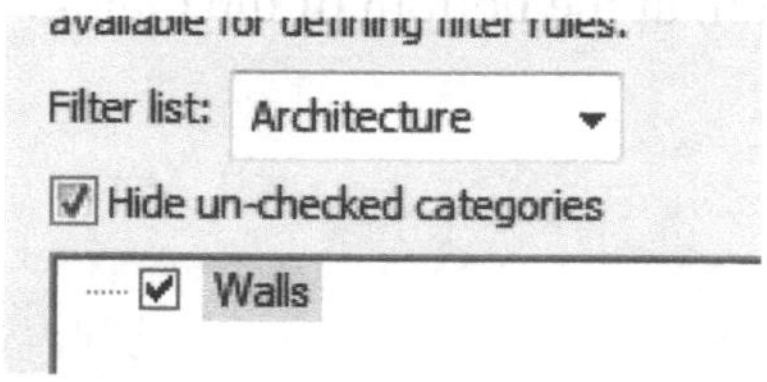 Set the Filter list to Architecture.

Place a check next to **Walls.**
Then place a check next to **Hide un-checked categories.**

14. 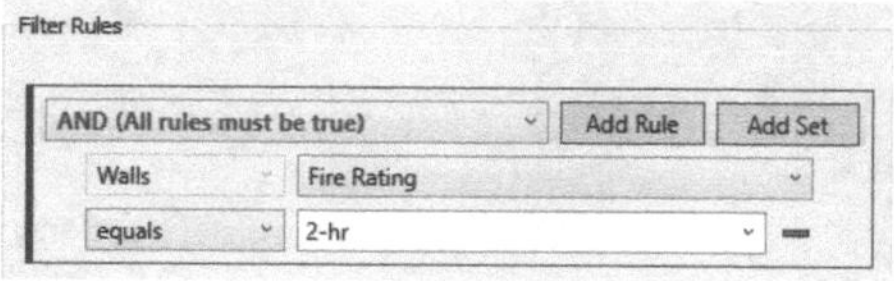 Set the Filter Rules to
Filter by: Fire Rating
Equals
2-hr.

Click **Apply.**

Apply saves the changes and keeps the dialog open. OK saves the changes and closes the dialog.

15. 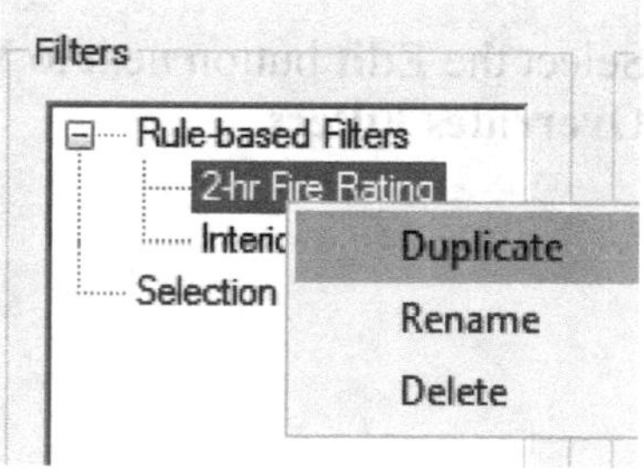 Right click on the **2-hr Fire Rating** filter.
Select **Duplicate.**

16.

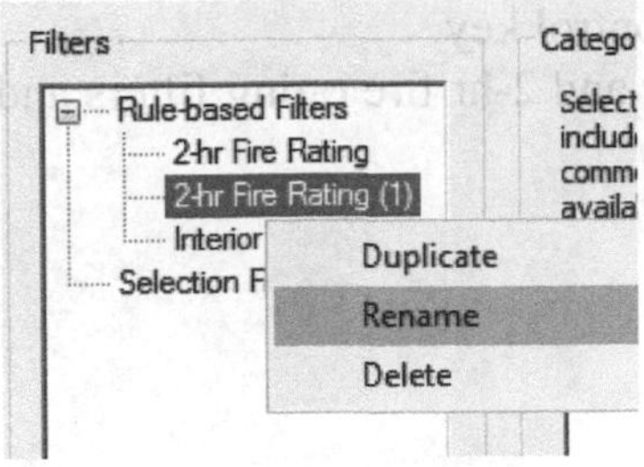

Highlight the copied filter.
Right click and select **Rename**.

17.

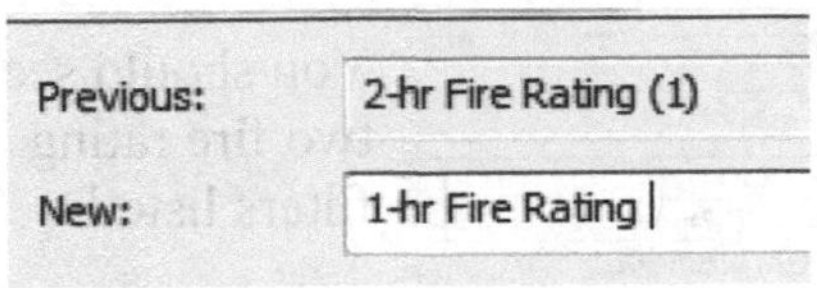

Rename to **1-hr Fire Rating**.
Click **OK**.

18.

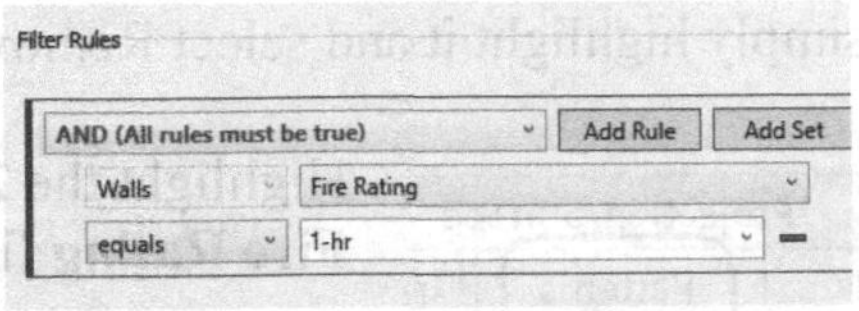

Set the Filter Rules to
Filter by: Fire Rating
Equals
1-hr.

Click **Apply**.
Click **OK**.

19.

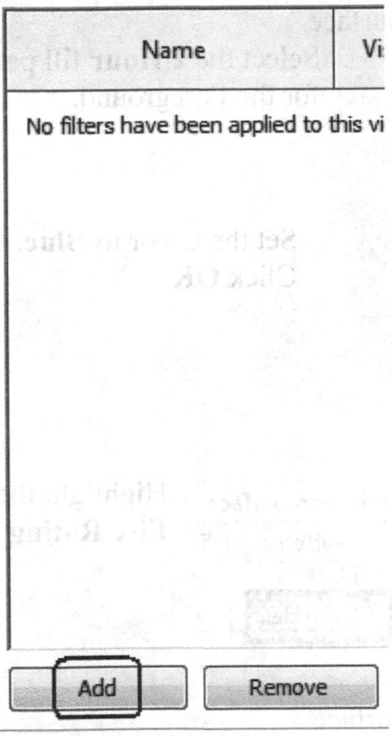

Select the **Add** button at the bottom of the Filters tab.

20.

Select one or more filters to insert.

Rule-based Filters
  1-hr Fire Rating
  2-hr Fire Rating
  Interior
Selection Filters

Hold down the Control key.
Highlight the 1-hr and 2-hr fire rating filters and click **OK**.

21.

Name	Visibility	Projection/Surface			Cut		Halftone
		Lines	Patterns	Transparen...	Lines	Patterns	
2-hr Fire Rating	☑	Override...	Override...	Override...	Override...	Override...	☐
1-hr Fire Rating	☑						☐

You should see the two fire rating filters listed.

If the interior filter was accidentally added, simply highlight it and select Remove to delete it.

22.

Name	Visibility	Projection/Surface	
		Lines	Patterns
2-hr Fire Rating	☑	Override...	Override...
1-hr Fire Rating	☑		

Highlight the **2-hr Fire Rating** filter.

23. Select the **Pattern Override** under Projection/Surface.

24.

Pattern Overrides
Foreground ☑ Visible
Pattern: 2 Hour

Select the **2 Hour** fill pattern for the Foreground.

25.

Pattern Overrides
Foreground ☑ Visible
Pattern: 2 Hour
Color: Blue

Set the Color to **Blue**.
Click **OK**.

26.

Name	Visibility	Projection/Surface	
		Lines	Patterns
2-hr Fire Rating	☑		
1-hr Fire Rating	☑	Override...	Override...

Highlight the **1-hr Fire Rating** filter.

27. Select the **Pattern Override** under Projection/Surface.

28. Set the fill pattern for the Foreground to **1 Hour** and the Color to **Magenta**. Click **OK**.

Pattern Overrides

Foreground ☑ Visible

Pattern: 1 Hour [IIIIIIIIIIIIIIIIIIII] ∨ [...]

Color: [▮] Magenta

29. Apply the same settings to the Cut Overrides. Click **OK**.

Name	Visibility	Projection/Surface			Cut	
		Lines	Patterns	Transparency	Lines	Patterns
2-hr Fire Rating	☑		IIIIIIIIIIII			IIIIIIIIIIII
1-hr Fire Rating	☑	Override...	IIIIIIIIIIII	Override...	Override...	IIIIIIIIIIII Hidden

30. Highlight the **Fire Rating** View Template.

Click **OK**.

Names:

<None>
Architectural Plan
Fire Rating
Site Plan
Structural Framing Plan

31. The view updates.

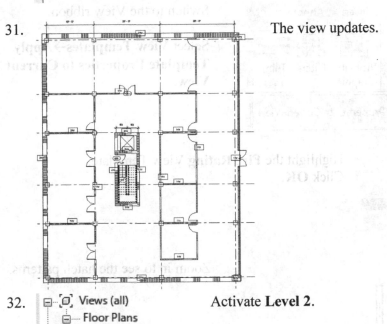

32. Activate **Level 2**.

☐ [□] Views (all)
    ☐ Floor Plans
        [□] Level 1
        [□] **Level 2**
        [□] Level 3

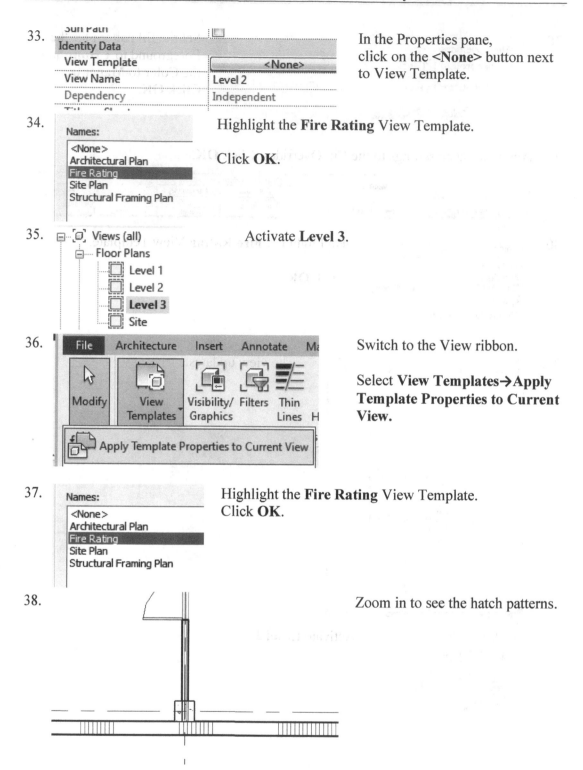

33. In the Properties pane, click on the **<None>** button next to View Template.

34. Highlight the **Fire Rating** View Template.

    Click **OK**.

35. Activate **Level 3**.

36. Switch to the View ribbon.

    Select **View Templates→Apply Template Properties to Current View.**

37. Highlight the **Fire Rating** View Template.
    Click **OK**.

38. Zoom in to see the hatch patterns.

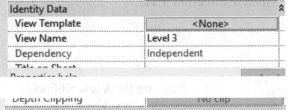

Note that the name of the view template is not displayed in the Properties pane.

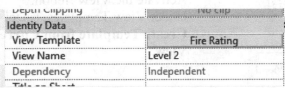

If you switch to the Level 1 or Level 2 floor plan views, notice that the template has been assigned to the view.

39. Close without saving.

The 1-hr, 2-hr, and 3-hr hatch patterns are custom fill patterns provided inside this exercise file. The *.pat files are included with the exercise files on the publisher's website for your use.

Exercise 2-2

Apply a View Template to a Sheet

Drawing Name: **view templates2.rvt**
Estimated Time to Completion: 10 Minutes

Scope
Use a view template to set all the views on a sheet to the same view scale.
View templates are used to standardize project views. In large offices, different people are working on the same project. By using a view template, everybody's sheets and views will be consistent.

Solution

1.   Sheets (all)
          A101 - Sections

    Open the **A101 – Sections** sheet.

2.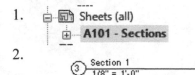

    Zoom into each title bar and see the scale for the view.

3. 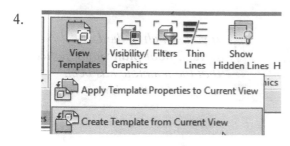 Open the **Section 1** view.

4. 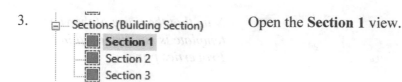 Activate the View ribbon.
Under View Templates, select
Create Template from Current View.

5. 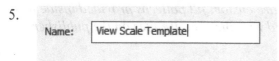 Name the new template
View Scale Template
Click **OK**.

6. Uncheck all the boxes EXCEPT View Scale.
Click **OK**.

7. Open the **A101 – Sections** sheet.

8. 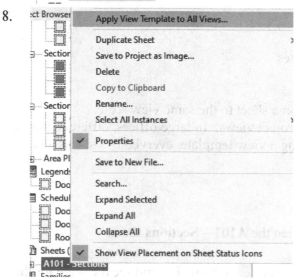 Highlight the **A101** sheet in the browser.
Right click and select **Apply View Template to All Views.**

9.

Names:
Architectural Elevation
Architectural Section
Site Section
Structural Framing Elevation
Structural Section
View Scale Template

Select the **View Scale Template**.
Click **OK**.

10. All the views on the sheet adjusted. Re-position the views.

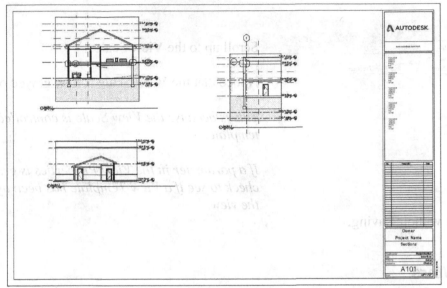

11.

└── Sections (Building Section)
      Section 1
      Section 2
      Section 3

Open **Section 2** view.

12.

Graphics	
View Scale	1/8" = 1'-0"
Scale Value 1:	96
Display Model	Normal

Note that the view scale has been modified for the view.

13.

Identity Data	
View Template	<None>
View Name	Section 2
Dependency	Independent
Title on Sheet	

Scroll down the Property panel.

The View Template was not assigned.

Click **<None>.**

14.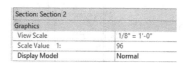

Select the **View Scale Template**.

Click **OK**.

Scroll up to the View Scale.

Notice that the View Scale is now grayed out.

This is because the View Scale is controlled by the template.

If a parameter in the View Properties is grayed out, check to see if a View Template has been assigned to the view.

15. Close without saving.

View Filters

View filters allow you to override the visibility or graphic display (color, hatch pattern, transparency, line type, etc.) of elements. You can create two types of view filters: rule-based or selection-based.

You can apply more than one filter to a view. The order in which they are listed on the Filter tab determines their priority. The filter at the top of the list takes precedence.

View filters can be copied from one project to another using the Transfer Project Standards tool.

Exercise 2-3

Create and Apply a View Filter

Drawing Name: **graphic overrides.rvt**
Estimated Time to Completion: 90 Minutes

Scope

Apply several different display overrides to the same wall.
Remove the display overrides to see which overrides have priority.

Solution

1.

In the Project Browser, there are several views that show different overrides.

Open the **Level 1- 2000 Remodel New Construction** floor plan.

Views (all)
Floor Plans
Level 1-2000 Remodel Demo
Level 1-2000 Remodel New Construction
Level 1-As Built
Level 1-Override by Category
Level 1-Override By Element
Level 1-Override Filter

2.

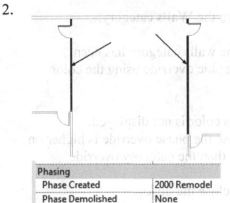

Note the two walls that are magenta.

Select one of the vertical magenta walls.

Phasing	
Phase Created	2000 Remodel
Phase Demolished	None

The walls are assigned the 2000 Remodel phase.

3.

Switch to the Manage ribbon.

Click **Phases**.

Phases

Phasing

4.

Click the **Graphic Overrides** tab.

Note that New construction has been assigned the color Magenta.

Click **OK**.

5.

Open the **Level 1 – Override by Category** floor plan view.

6.

The two walls are displayed as magenta. This is the Phases graphic override.

Select the two walls.

7.

Phasing	
Phase Created	2000 Remodel
Phase Demolished	None

vg

Type **VG** to open the Visibility Graphics dialog. On the Model Categories tab:

8.

Scroll down to the **Walls** category.

Notice that the walls category has been assigned a graphic override using the color yellow.

However, this color is not displayed.
That is because the phase override is higher on the hierarchy than the category override.

Click **OK** to close the dialog.

9.

Phasing	
Phase Created	As-Built
Phase Demolished	None

Select the two magenta walls and change the Phase Created to **As-Built**.

Click to release the selection.

The walls color changes to the new phase assignment.

10. Switch to the Manage ribbon.

    Click **Phases**.

11. 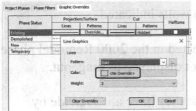 Click the **Graphic Overrides** tab.

    Set the Color for the Existing phase to **No Override** in both of the Lines columns.

    Click **OK**.

12. 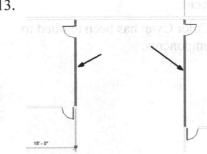 *The walls now appear as yellow.*

    Yellow is the color assigned to the walls category.

13.  Select the two walls indicated and assign them to the 2000 Remodel phase.

14. 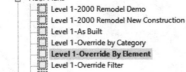 Open the **Level 1 – Override by Element** floor plan view.

15.

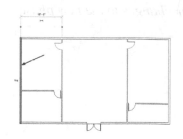

Select one of the outer walls.

Notice it was created in the As-Built phase.

However, the color is not yellow. The Visibility Graphics override is view-specific.

Release the selection.

16.

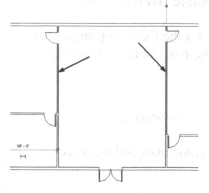

Select the two walls indicated.

They were created in the 2000 Remodel phase, but they are not displayed as magenta.

This is because they have an Override by Element assigned.

Element graphic overrides supercede phase graphic overrides.

17.

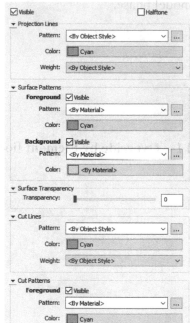

With the two walls selected:
Right click and select **Override Graphics in View→By Element.**

18.

Notice that the Color Cyan has been applied to different wall components.

Click **OK**.

19.

Floor Plans
- Level 1-2000 Remodel Demo
- Level 1-2000 Remodel New Construction
- Level 1-As Built
- Level 1-Override by Category
- Level 1-Override By Element
- **Level 1-Override Filter**
- Level 2
- Site

Open the **Level 1 – Override Filter** floor plan view.

20.

The outer walls are assigned the As-Built Phase. The indicated walls are assigned the 2000 Remodel Phase but have a graphic override by element applied.

Type **VG** to bring up the dialog box.

21.

Select the **Filters** tab.

There is a filter called walls that has assigned the color red to some walls.

22.

All document filters are defined and modified here

Edit/New...

Click **Edit/New**.

23.

Filter Rules

AND (All rules must be true) | Add Rule | Add Set
Walls | Phase Created
equals | 2000 Remodel

The filter changes the color for walls assigned to the **2000 Remodel** phase.

Click **OK**.

24.

- Vertical Circulation
- Walls
  - <Hidden Lines>
  - Common Edges
  - Non-Core Layers

Go to the Model Categories tab.

Scroll down to Walls.
No Graphic overrides have been applied.

Click **OK** to close the dialog.
Select the two walls.

25.

16' - 0"

Click to select, TAB for alternates, CTRL adds, SHIFT unselects.

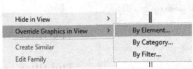

Hide in View >
Override Graphics in View > | By Element...
Create Similar | By Category...
Edit Family | By Filter...

Right click and select **Override Graphics in View→By Element.**

26.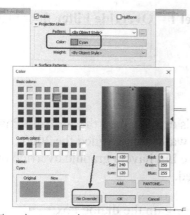

Click in each color box.

Click **No Override**.

Click **OK**.

27.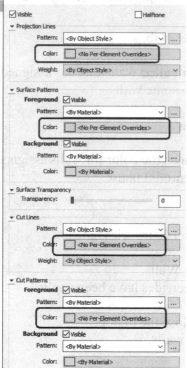

The box should show that no graphic overrides are active.

Click **OK**.

Release the selection.

28.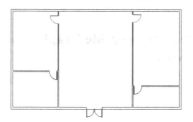

The filter override is now applied.

The filter graphic override has a higher place on the hierarchy than the phase graphic override.

Save as *ex2-3.rvt*.

Element Visibility Hierarchy

Most users are unaware that Revit applies a hierarchy to how elements are displayed and their visibility.

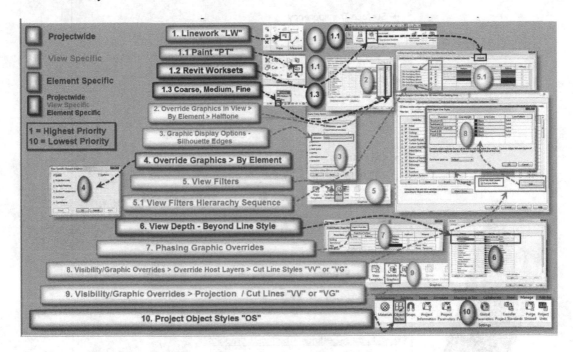

Understanding the Effect of Multiple Overrides

A wall has the following overrides:

- Visibility/Graphic Override for the Walls category for the view

- Override Graphics in View→By Element for the wall instance

- Phasing graphic override

- View filter overriding the graphics for walls over a specific thickness

With the four overrides defined, the element override on the wall instance is visible because it is the highest in the hierarchy:

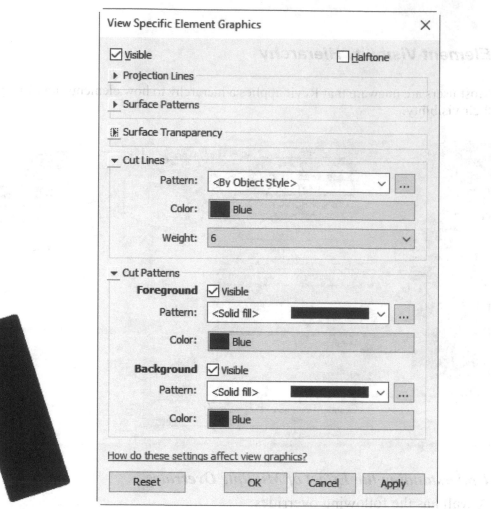

Remove the element override and the view filter is visible:

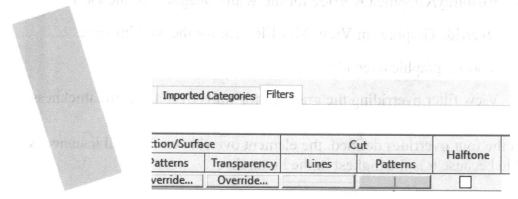

Remove the view filter and the phasing graphic override is visible:

Project Phases	Phase Filters	Graphic Overrides

Phase Status	Projection/Surface		Cut		
	Lines	Patterns	Lines	Patterns	
Existing	———————		———————	Hidden	Hidden
Demolished	- - - - - - - - -		- - - - - - - - -	Hidden	Hidden
New	━━━━━━━		━━━━━━━		
Temporary	················		··················	///////	///////

Change the phase filter of the view and the Visibility/Graphic override for the Walls category is visible:

Hidden Elements

Temporarily hiding or isolating elements or element categories may be useful when you want to see or edit only a few elements of a certain category in a view.

The Hide tool hides the selected elements in the view, and the Isolate tool shows the selected elements and hides all other elements in the view. The tool affects only the active view in the drawing area.

Element visibility reverts back to its original state when you close the project, unless you make the changes permanent. Temporary Hide/Isolate also does not affect printing.

Exercise 2-4
Reveal Hidden Elements

Drawing Name: **i_visibility.rvt**
Estimated Time to Completion: 5 Minutes

Scope
Turn on the display of hidden elements

Solution

1. 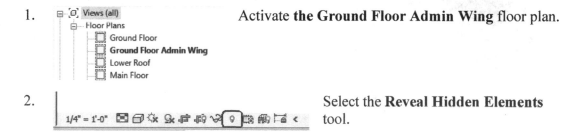   Activate **the Ground Floor Admin Wing** floor plan.

2.   Select the **Reveal Hidden Elements** tool.

3. Items highlighted in magenta are hidden. Window around the two tables while holding down the CONTROL key to select them.

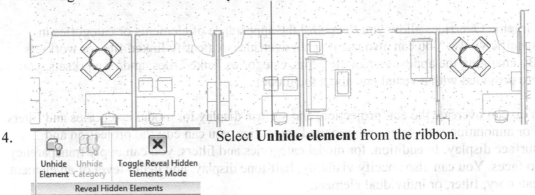

4.  Select **Unhide element** from the ribbon.

The tables will no longer be displayed as magenta (hidden elements).

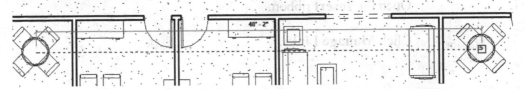

5. Select the **Close Hidden Elements** tool.

6. The view will be restored.

7. Select the **Measure** tool from the Quick Access toolbar.

8. Determine the distance between the center of the two tables.
Did you get 48' 2"?

If you didn't get that measurement, check that you selected the midpoint or center of the two tables.

9. Close without saving.

Graphic Overrides

Graphic Overrides allow you to control the appearance of elements or categories in specific views. You can also use graphic overrides to identify linked files or worksets. By changing the appearance of elements, categories, linked files, and/or worksets, you are provided with a visual cue about the model.

You can override the cut, projection, and surface display for model categories and filters. For annotation categories and imported categories, you can edit the projection and surface display. In addition, for model categories and filters, you can apply transparency to faces. You can also specify visibility, half-tone display, and detail level of an element category, filter, or individual element.

Exercise 2-5

Graphic Overrides of Linked Files

Drawing Name: **graphic overrides_2.rvt**
Estimated Time to Completion: 15 Minutes

Scope
Manage Links
Apply Graphic Overrides to a linked file

Solution

1.   Open the **Insert** ribbon.

   Click **Manage Links**.

2. 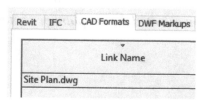   Switch to the CAD Formats tab.

   You should see the Site Plan.dwg file listed.

   Highlight the file name.

3. Click **Reload From...**

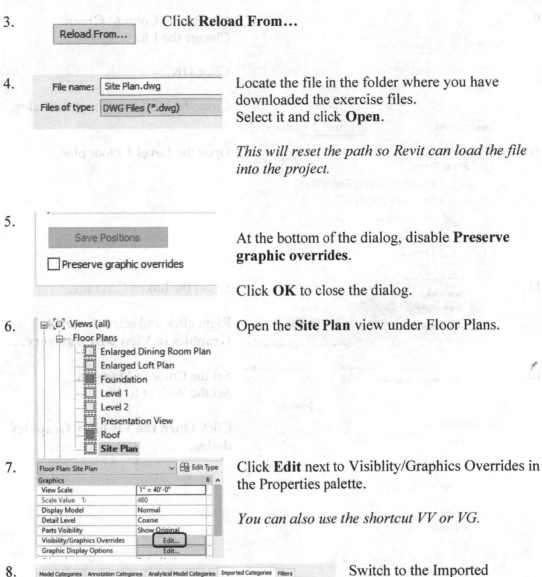

4. Locate the file in the folder where you have downloaded the exercise files.
Select it and click **Open**.

This will reset the path so Revit can load the file into the project.

5. At the bottom of the dialog, disable **Preserve graphic overrides**.

Click **OK** to close the dialog.

6. Open the **Site Plan** view under Floor Plans.

7. Click **Edit** next to Visiblity/Graphics Overrides in the Properties palette.

You can also use the shortcut VV or VG.

8. Switch to the Imported Categories tab.

Highlight the *Site Plan.dwg*.

Click **Override** in the Lines column.

9.

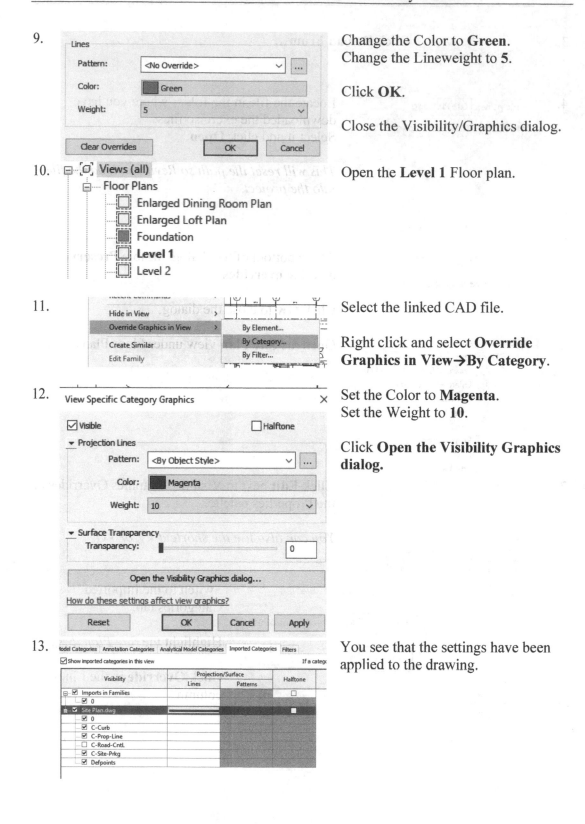

Change the Color to **Green**.
Change the Lineweight to **5**.

Click **OK**.

Close the Visibility/Graphics dialog.

10.

Open the **Level 1** Floor plan.

11.

Select the linked CAD file.

Right click and select **Override Graphics in View→By Category**.

12.

Set the Color to **Magenta**.
Set the Weight to **10**.

Click **Open the Visibility Graphics dialog.**

13.

You see that the settings have been applied to the drawing.

14.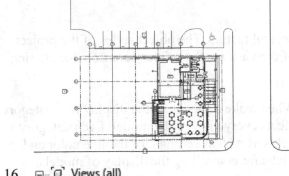

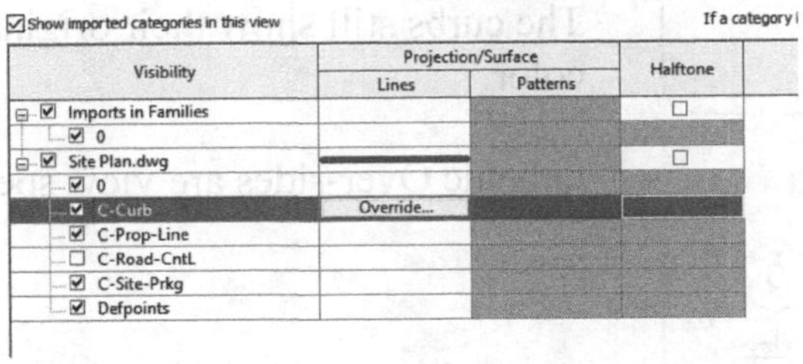

Highlight the **C-Curb** layer.

Click **Override** in the Lines column.

15.

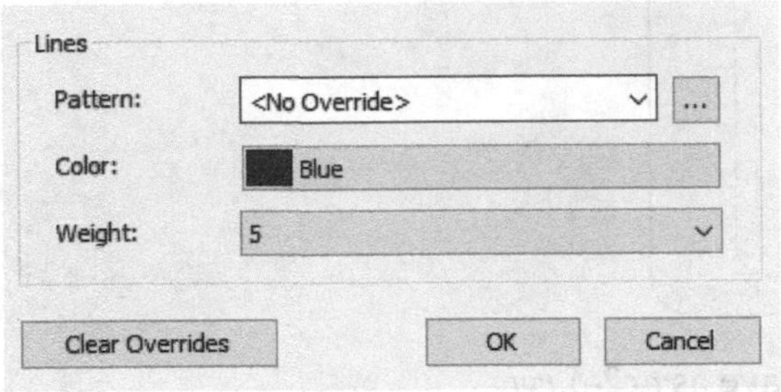

Set the Color to **Blue**.
Set the Line Weight to **5**.

Click **OK**.

Close the dialog.

The color of the curb changes to blue.

16. □⋯ [Ō] Views (all)
    □⋯ Floor Plans
        ⬜ Enlarged Dining Room Plan
        ⬜ Enlarged Loft Plan
        ⬛ Foundation
        ⬜ Level 1
        ⬜ Level 2
        ⬜ Presentation View
        ⬛ Roof
        ⬜ **Site Plan**

Open the **Site Plan** view.

17.

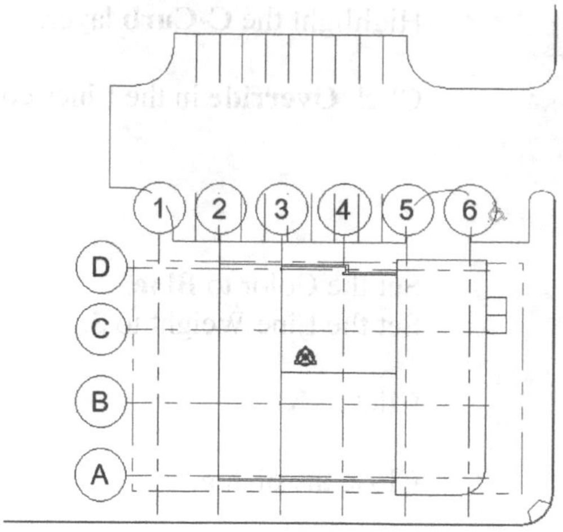

The curbs still show their original color.

Graphic Over-rides are view specific.

18.   Save as *ex2-5.rvt*.

Object Styles

The display of elements is controlled by several factors. The object styles of the project, and the visibility and graphics overrides of the view are the primary controls for altering the display of objects.

Object Styles are **global** model settings. When make a change to the display of a category in the Object Styles dialog, that change affects every view of the model. Each category has a value for projection and cut line weights as well as settings for the line color and line pattern to be used. There are separate tabs for controlling the display of models, annotation, and imported objects.

Exercise 2-6
Object Styles

Drawing Name: **object styles.rvt**
Estimated Time to Completion: 10 Minutes

Scope
Change the object styles for an element.

Solution

1.    Open the **Insert** ribbon.

     Click **Manage Links**.

2. 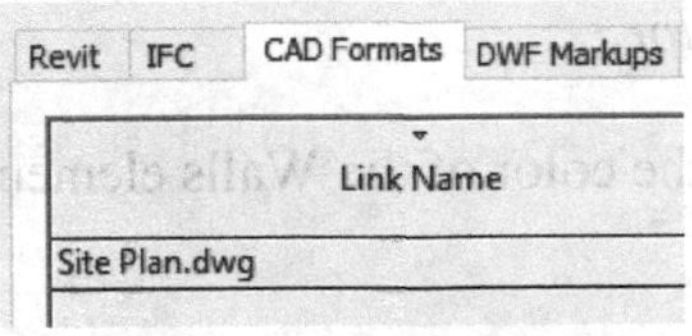   Switch to the CAD Formats tab.

     You should see the Site Plan.dwg file listed.

     Highlight the file name.

3.    Click **Reload From…**

4. 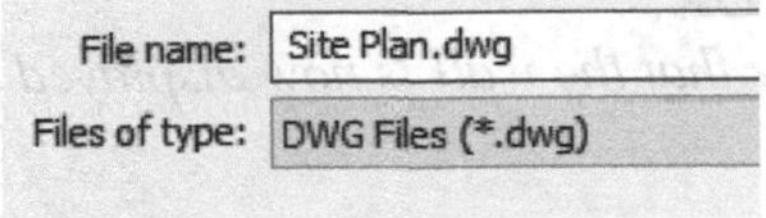   Locate the file in the folder where you have
     downloaded the exercise files.
     Select it and click **Open**.

     *This will reset the path so Revit can load the file
     into the project.*

5.    Open the Site Plan floor plan.

     Notice that the curbs are red.

6.

Open Level 1 floor plan.

Notice that the curbs are blue.

7.

Open the Manage ribbon.

Click **Object Styles**.

8.

Open the **Imported Objects** tab.

Notice that the C-Curb layer is assigned a red color – the color that is displayed in the Site Plan view.

9.

Open the Model Objects tab.
Scroll down to Walls.
Notice that they are currently assigned the Black color.

10.

Change the color of the Walls element to **Blue**.

Change the color of the Common Edges element to **Blue**.

Click **OK**.
Notice that the wall is now displayed as blue.

11.

Switch to the Site Plan floor plan.

12.

Notice that the wall is now displayed as blue.

Object Styles settings are global.

Save as *ex2-6.rvt*.

View Scale

The view scale controls the scale of the view as it appears on the drawing sheet.

You can assign a different scale to each view in a project. You can also create custom view scales.

Many view types in Revit contain a "View Scale" property- such as Floor Plans, Ceiling Plans, Sections, Elevations, Callouts, and Drafting Views. The "View Scale" parameter allows you to set the scale at which each particular view will be printed out. You can also modify the view scale to ensure that a view fits on a sheet.

Each view has its own "View Scale" property.
- You can change the scale of a view at any time- using the View Control Bar or View Properties.
- Revit maintains annotations (tags, dimensions, etc.) at their actual printed size, regardless of the scale of the view.

If you assign a template to a view, the view scale may be defined by the template.

Exercise 2-7
Change the View Scale

Drawing Name: **view_scale.rvt**
Estimated Time to Completion: 10 Minutes

Scope
Changing a view scale.
Adding a custom view scale.

Solution

1. 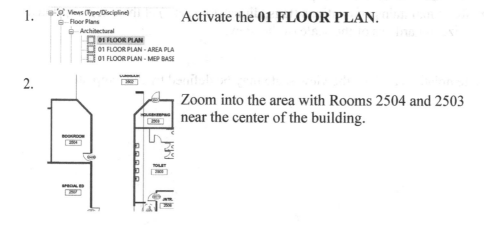 Activate the **01 FLOOR PLAN**.

2. Zoom into the area with Rooms 2504 and 2503 near the center of the building.

3.

Look down at the bottom left-hand corner of the active view window. Locate the View Control Bar.

The first button on the View Control Bar is the View Scale.

You can see that this view is set to Scale: **1:96**.

Click on the View Scale button.

A pop-up displays a list of all the available scales.

At the top of the list is Custom. This is used to define a scale that is not available in the default list.

4.

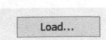

Change the view scale to **1:100**.

Did you notice how the room tags adjusted size in relation to the view scale?

5.

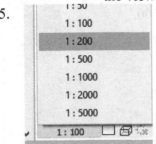

Change the view scale to **1:200**.

Did the room tags get bigger or smaller?

In fact, the room tags remained the same size. The model elements increased in scale.

6.

In the browser, locate the Sheets category.

Highlight and right click to select **New Sheet**.

7.

Click **Load** to load a new titleblock.

Load...

8.  Browse to the *Titleblocks* folder under English\US.

9. File name: A2 metric.rfa
   Files of type: All Supported Files (*.rfa, *.adsk)

   Select *A2 metric.rfa.*
   Click **Open**.
   Click **OK** to create a new sheet.

10. 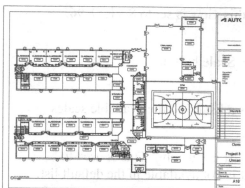 Drag and drop the 01 FLOOR PLAN onto the sheet.

11.

Viewports (1)	
Graphics	
View Scale	1 : 200
Scale Value 1:	1 : 25
Display Model	1 : 50
Detail Level	1 : 100
Parts Visibility	1 : 200
Detail Number	1 : 500
Rotation on Sheet	1 : 1000
	1 : 2000

Select the view on the sheet.

On the Properties pane:

Set the View Scale to 1:500.

Click **Apply.**

12. The view updates.

13.

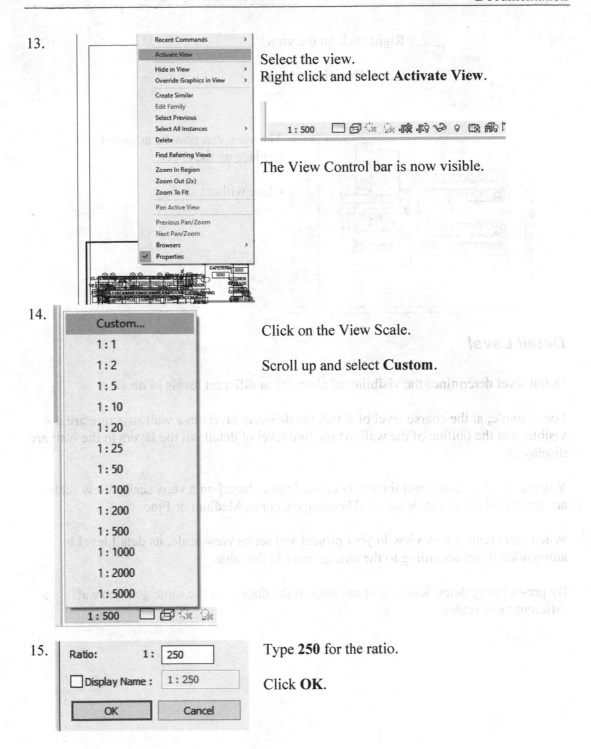

Select the view.
Right click and select **Activate View**.

The View Control bar is now visible.

14.

Click on the View Scale.

Scroll up and select **Custom**.

15.

Type **250** for the ratio.

Click **OK**.

16. 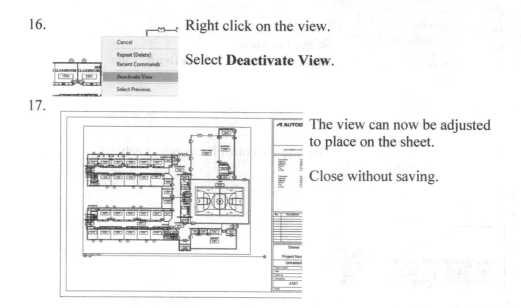 Right click on the view.

Select **Deactivate View**.

17.

The view can now be adjusted to place on the sheet.

Close without saving.

Detail Level

Detail level determines the visibility of elements at different levels of detail.

For example, at the coarse level of detail, the different layers in a wall structure are not visible, just the outline of the wall. At the fine level of detail, all the layers in the wall are displayed.

You can set the detail level for newly created views based on a view scale. View scales are organized under the detail level headings: Coarse, Medium or Fine.

When you create a new view in your project and set its view scale, its detail level is automatically set according to the arrangement in the table.

By pre-defining detail levels, you can control the display of the same geometry at different view scales.

Exercise 2-8

Change the Detail Level of a View

Drawing Name: **detail_level.rvt**
Estimated Time to Completion: 15 Minutes

Scope
Defining Detail Levels for View Scales

Solution

1. ─ Floor Plans
   ⊟── Architectural
       ──── **01 FLOOR PLAN**

   Activate the **01 Floor Plan**.

2. 1 : 250 ☐ ▱

   On the View Control bar:
   The View Scale is set to **1:250**.
   The Detail Level is set to **Coarse**.
   The Display Style is set to **Hidden Line**.

3.

   You should be able to identify the different icons associated with Detail Level and Display Style for the certification exam. Zoom into the lower left area of the floor plan – where Rooms 2010 and 2008 are located.

4.

   Switch to the Manage ribbon.

   Select **Additional Settings** on the ribbon.

5. Detail Level

   Select **Detail Level** from the drop-down list.

6.

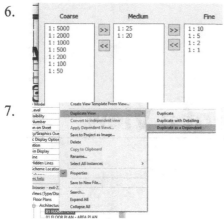

Note that each View Scale has been assigned a Detail Level setting.

Click **OK**.

7.

Highlight the **01 FLOOR PLAN**.

Right click and select **Duplicate View→Duplicate as a Dependent.**

8.

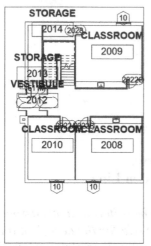

Crop the view using the crop region so only Rooms 2008, 2009, and 2010 are displayed.

Change the view scale to **1:10**.

The Detail Level is set to **Coarse**.
The View Display is set to **Hidden Line**.

You may recall on the Detail Level settings, the Detail Level for a View Scale of 1:10 should be Fine.

9.

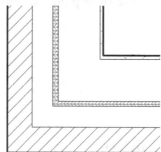

Set the Detail Level to **Fine.**

Inspect the wall.

10.

Change the View Scale to **1:500**.
Change the Display to **Coarse**.

Notice how the wall display changes.

11. Close without saving.

Segmented Views

Split a section or elevation view to permit viewing otherwise obscured parts of the view.

This function allows you to vary a section view or an elevation view to show disparate parts of the model without having to create a different view. For example, you may find that landscaping obscures the parts of the model that you would like to see in an elevation view. Splitting the elevation allows you to work around these obstacles.

Exercise 2-9

Segmented Views

Drawing Name: **segmented views.rvt**
Estimated Time to Completion: 10 Minutes

Scope
Modify a section view into segments.

Solution

1. Sheets (all)
    A101 - Segmented Elevation

    Activate the **A101- Segmented Elevation** sheet.

    There are two views on the sheet. One is the floor plan and the other is a section view defined in the floor plan.

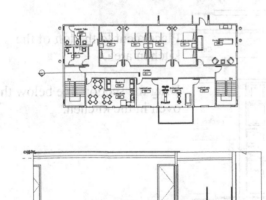

2. 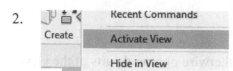 Right click on the **Main Floor** floor plan view. Select **Activate View**.

This is the top view on the sheet.

3. 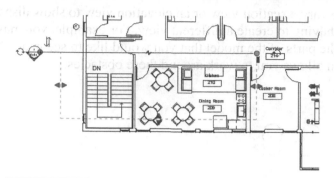 Select the section line.

4. 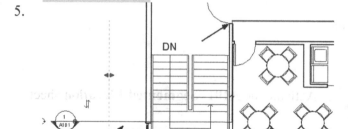 Select **Split Segment** on the ribbon.

5. 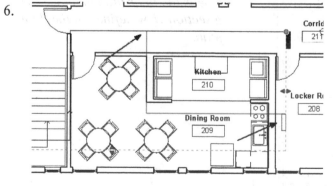 Place a cut to the right of the stairs.
Drag the section line segment below the stairs.

6. Place a cut to the left of the kitchen area.

Drag the section line below the oven in the kitchen.

7. 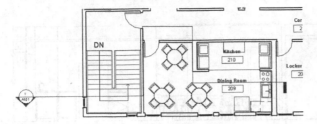 Right click and cancel out the command.
The new section line is now segmented.

8. 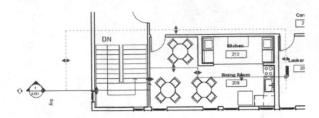 Reverse the direction of the section line so the arrow is pointed up.

9. Adjust the segments of the section line so the section lines are as shown.

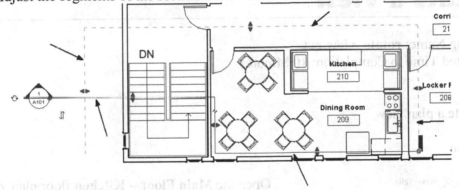

10. 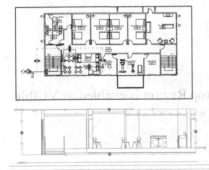 You can see how the new segmented view appears in the lower elevation view.

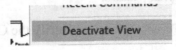

Right click and select **Deactivate View**.

Note how the section view has updated.

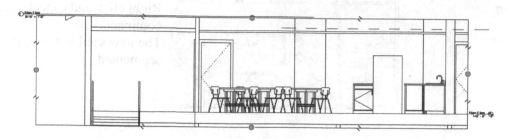

11. Close without saving.

Exercise 2-10
Rotate a View

Drawing Name: **rotate_view.rvt**
Estimated Time to Completion: 10 Minutes

Scope
To rotate a plan view

Solution

1. 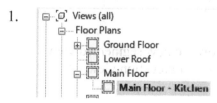 Open the **Main Floor – Kitchen** floor plan view.

2.

Extents	
Crop View	☑
Crop Region Visible	☑
Annotation Crop	☑
View Range	

Verify the Crop Region is enabled as **Visible**.

3. Select the crop region outline.

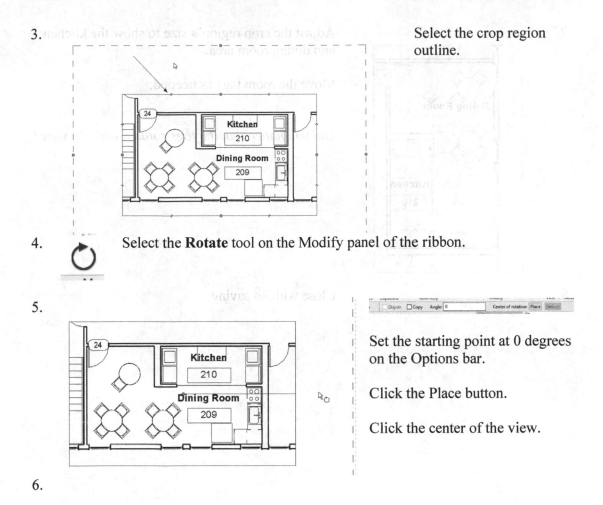

4. Select the **Rotate** tool on the Modify panel of the ribbon.

5. Set the starting point at 0 degrees on the Options bar.

    Click the Place button.

    Click the center of the view.

6. Rotate the view 90 degrees.

7.

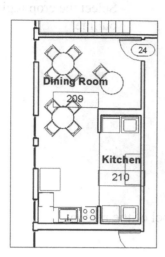

Adjust the crop region's size to show the kitchen and dining room area.

Move the room tags as needed.

Did you notice that the tags rotated with the view?

8.

Close without saving.

Duplicating Views

In Revit you can duplicate a view such as a floor plan or section. Once a view is placed on a sheet, it is "consumed" and cannot be re-used. So, the best way to place a different version of the same view is to duplicate views. For example, you might have one version of a floor plan that displays the fire ratings of the walls. You might have one version of the floor plan that displays a color scheme for area or department use. You might have a version of the floor plan that displays how furniture will be placed. You might want to create a view of the floor plan that focuses on a specific section of the floorplan, like the lavatories or stairs.

There are three methods for copying a view. The three types are as follows:

- Duplicate
- Duplicate with detailing
- Duplicate as Dependent

Duplicate

This is the most common way to duplicate a view. This option:

- Copies visibility settings
- No detailing
- Independent from original view

This creates multiple copies of the same view, such as a floor plan, but use them for different reasons such as fire escape plans, room area plans & dimensions. Each view will be of the same floor plan, but each view will have different annotations, tags, and displays.

Duplicate with Detailing

This is similar to duplicate; however, as the name suggests it will also copy any detailing such as dimensions and text. The new duplicated view is still independent from the original view.

- Copies visibility settings
- Copies detailing
- Independent from original view

This can be useful if you want to preserve any tags or notes that have been added to the view. For example, you want to use the same view but at different scales or color schemes.

Duplicate as Dependent

Duplicating as a dependent will create identical copies of the original view, the new views are also tied to the original view as child objects. What this means is if you add a dimension in the original view, all dependent views will also have the new dimension.

- Shares visibility settings
- Copies detailing
- Is synchronised to the original view
-

This can be extremely useful if you have a large floor plan and you want to make several views which are cropped to specific regions.

Exercise 2-11

Duplicating Views

Drawing Name: **duplicating_views.rvt**
Estimated Time to Completion: 15 Minutes

Scope
Duplicate view with Detailing
Duplicate view as Dependent
Duplicate view
Understand the difference between the different duplicating views options

Solution

1. 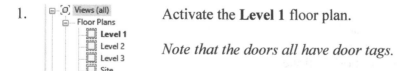   Activate the **Level 1** floor plan.

   Note that the doors all have door tags.

2.

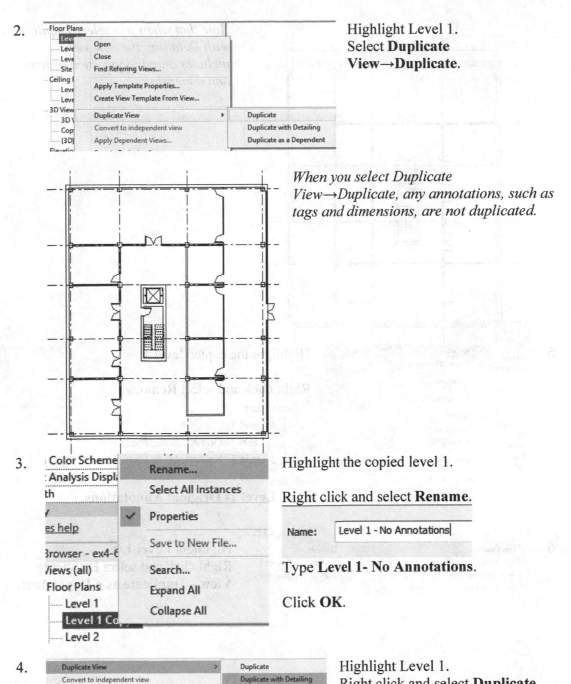

Highlight Level 1.
Select **Duplicate
View→Duplicate**.

*When you select Duplicate
View→Duplicate, any annotations, such as
tags and dimensions, are not duplicated.*

3.

Highlight the copied level 1.

Right click and select **Rename**.

Type **Level 1- No Annotations**.

Click **OK**.

4.

Highlight Level 1.
Right click and select **Duplicate
View→Duplicate with Detailing**.

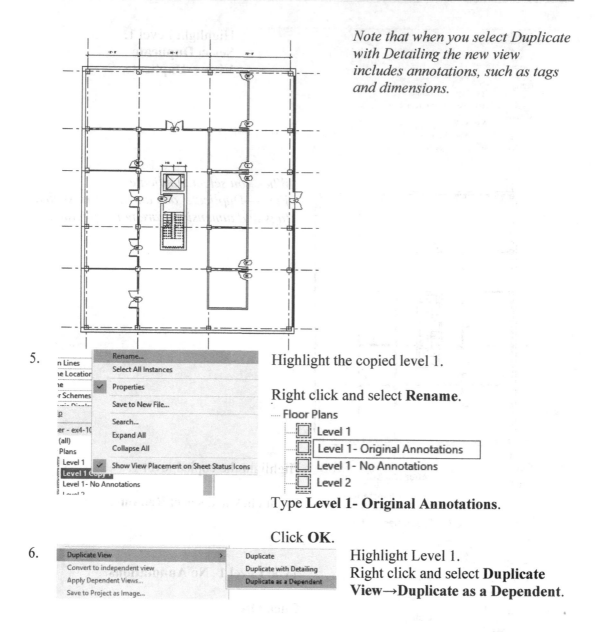

Note that when you select Duplicate with Detailing the new view includes annotations, such as tags and dimensions.

5.

Highlight the copied level 1.

Right click and select **Rename**.

---- Floor Plans
    ☐ Level 1
    ☐ Level 1- Original Annotations
    ☐ Level 1- No Annotations
    ☐ Level 2

Type **Level 1- Original Annotations**.

Click **OK**.

6.

Highlight Level 1.
Right click and select **Duplicate View→Duplicate as a Dependent**.

7.

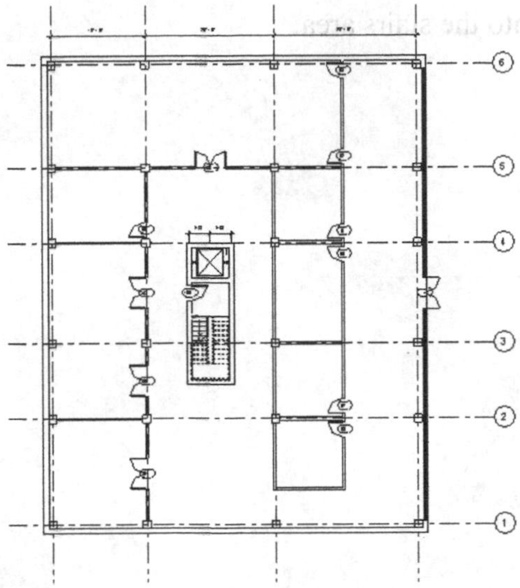

Notice that the duplicated view includes the annotations.

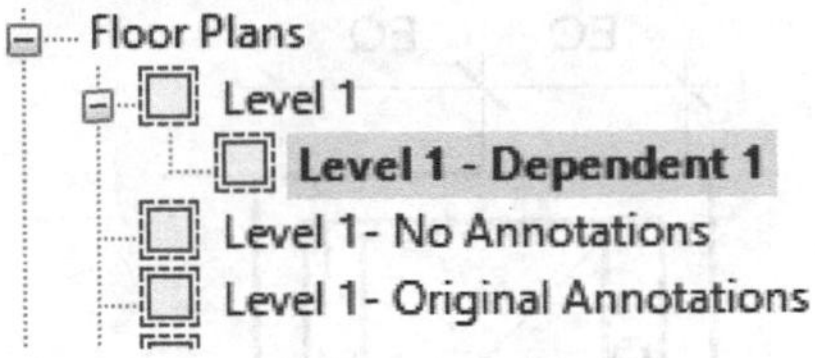

In the Project Browser, the dependent view is listed underneath the parent view.

8.

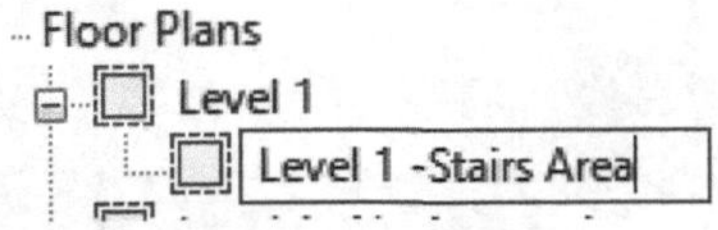

Rename the dependent view **Level 1- Stairs area**.

9.

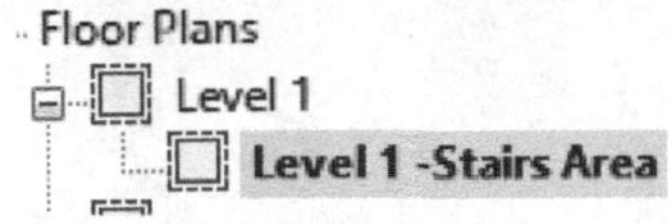

Activate **Level 1- Stairs area**.

10.

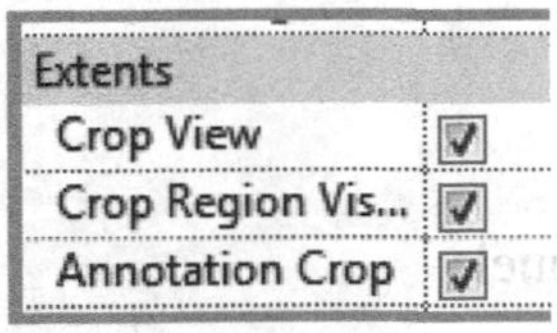

In the Properties pane,
Enable **Crop View**.
Enable **Crop Region Visible**.

11.

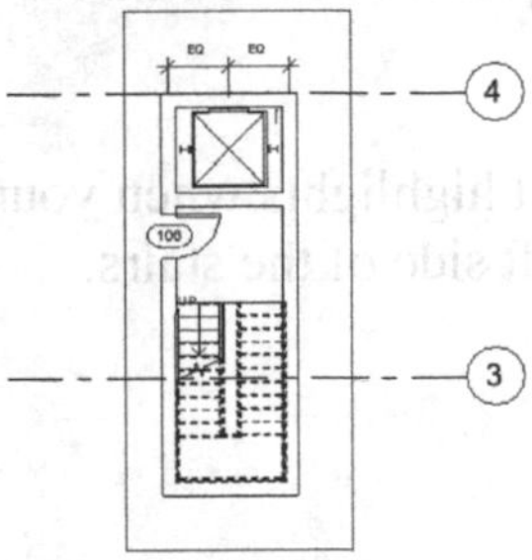

Adjust the crop region to focus the view on the stairs area.

12.

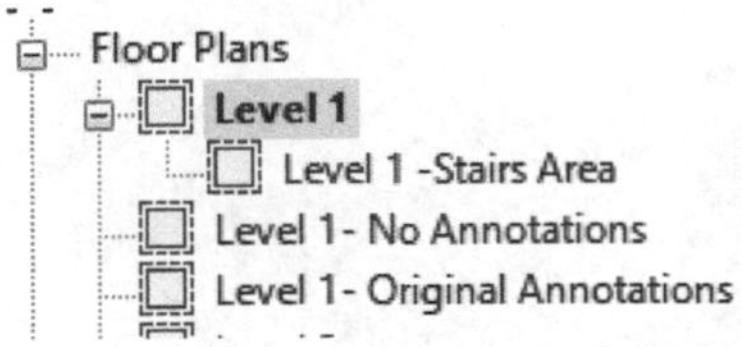

Activate **Level 1**.

13.

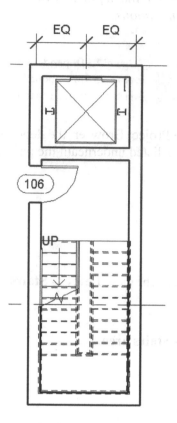

Zoom into the stairs area.

14. Annotate | Activate the **Annotate** ribbon.

15. Select the **Tread Number** tool on the Tag panel.

Tread Number

Tread Numbers can only be applied to Component-based stairs.

16.

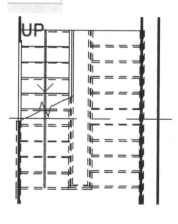

Select the middle line that highlights when your mouse hovers over the left side of the stairs.

17. 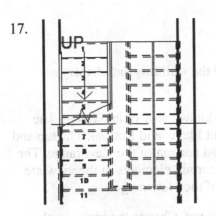 Select the middle line on the right side of the stairs. Right click and select CANCEL to exit the command.

18. 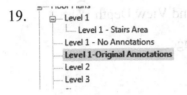 Activate **Level 1- Original Annotations**.

Notice that the new annotations - the tread numbers - are not visible in this view.

19. 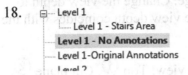 Activate **Level 1 - No Annotations**.

Notice that the new annotations - the tread numbers - are not visible in this view.

20. 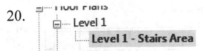 Activate the **Level 1- Stairs area** dependent view.

Notice that any annotations added to the parent view are added to the dependent view.

21. Close without saving.

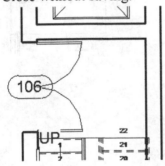

Extra: Change the door labeled 106 to a double flush door. Which Level 1 views display the new door type?

View Range

The view range is a set of horizontal planes that control the visibility and display of objects in a plan view.

Every plan view has a property called **view range**, also known as a visible range. The horizontal planes that define the view range are Top, Cut Plane, and Bottom. The top and bottom clip planes represent the topmost and bottommost portion of the view range. The cut plane is a plane that determines the height at which certain elements in the view are shown as cut. These 3 planes define the primary range of the view range.

View depth is an additional plane beyond the primary range. Change the view depth to show elements below the bottom clip plane. By default, the view depth coincides with the bottom clip plane.

The following elevation shows the view range ⑦ of a plan view: Top ①, Cut plane ②, Bottom ③, Offset (from bottom) ④, Primary Range ⑤, and View Depth ⑥.

The plan view on the right shows the result for this view range.

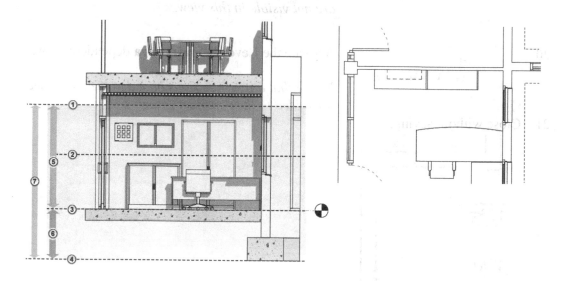

You will have at least one question on the Professional exam where you need to answer a question regarding one of the settings in the dialog.

Exercise 2-12

View Range

Drawing Name: **view_range.rvt**
Estimated Time to Completion: 5 Minutes

Scope
Determine the view depth of a view.

Solution

1. Floor Plans
   - Ground Floor
   - Lower Roof
   - Main Floor
   - Main Roof
   - Site
   - T.O. Footing
   - T.O. Parapet

   Activate the **Site** view.

2. | Extents | |
   |---|---|
   | Crop View | ☐ |
   | Crop Region Visible | ☐ |
   | Annotation Crop | ☐ |
   | View Range | Edit... |
   | Associated Level | Ground Floor |
   | Scope Box | None |
   | Depth Clipping | No clip |

   In the Properties pane:
   Scroll down to **View Range** located under the Extents category.
   Select the **Edit** button.

3. `<< Show`

   Select the **Show** button located at the bottom left of the dialog.

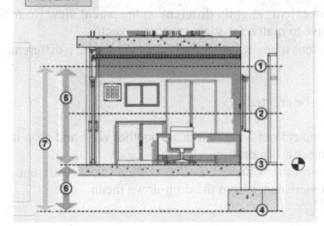

A representation of how view range settings work is displayed to assist you in defining the view range of a view.

Each number represents a field in the dialog box.

The elevation shows the view range ⑦ of a plan view.

4.

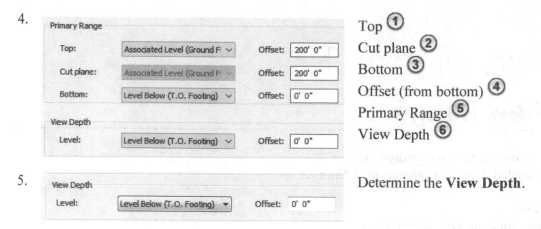

Top ①
Cut plane ②
Bottom ③
Offset (from bottom) ④
Primary Range ⑤
View Depth ⑥

5.

Determine the **View Depth**.

6. Click **OK**.

7. Close without saving.

Call-outs

The standard 'Callout' command places a rectangle defining the callout view extents. This will create a new view, and place the view reference rectangle on the active view. For reasons listed below, this method is not preferred.

- You cannot change the reference.
- You cannot move the callout to a different parent view.
- You cannot change the callout view family between Plan/Detail (Plans only).
- The callout rectangle can be stretched or rotated in the parent view – this will make the same change of extents or rotation to the callout view itself, which is not always desirable.
- You cannot have the callout extents slightly different on the parent view (to make callouts readable) - they have to match the view cropping exactly.
- If you copy and paste a Callout it creates an entirely new view with a different reference.

Reference Other View Callouts can be created by:

Create a view to be referenced by the callout – e.g. duplicate another view and crop it
Make sure the view is cropped (unless it is a drafting view).
On the view to place the callout, select the 'Callout' command, then select 'Reference Other View', and select a relevant view name from the drop-down menu.

This method has many advantages:

- Callouts can be moved to another view
- Callout references can be changed to refer to different view
- References will update if the view/sheet number is changed
- Callouts can be copied to another parent view

It also has a few disadvantages:
It is possible for the user to select the wrong view in the list - there is no automatic check for this.

If you want to have "Sim" showing on a callout for similar details on multiple callouts, that is a Type property of the view being referred to, not a property of the callout symbol.

That means it is all or nothing – i.e. all instances of the callout must have "Sim" or not. It also changes the view type so it may move in your project browser, depending on your browser organization scheme.

Exercise 2-13
Create a Call-out View

Drawing Name: **callouts.rvt**
Estimated Time to Completion: 10 Minutes

Scope
Place a callout to a drafting view

Solution

1.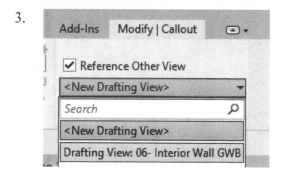

   Open the **Section 3** view under Sections (Building Section).

2. Switch to the View ribbon.

   Select the **Callout→Rectangle** tool.

3. On the ribbon:

   Enable **Reference Other View**.

   In the drop-down: select **Drafting View: 06-Interior Wall GWB**.

4.

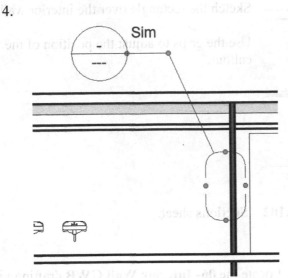

Sketch the rectangle over the interior wall.

Use the grips to adjust the position of the callout.

5.

Sections (Building Section)
   Architectural
      Section 1
      Section 2
      Section 3
      Section 4

Open the **Section 4** view under Sections (Building Section).

6.

Collaborate   View

Section   Callout

Rectangle

Switch to the View ribbon.

Select the **Callout→Rectangle** tool.

7.

Add-Ins   Modify | Callout

☑ Reference Other View

<New Drafting View>

Search   🔍

<New Drafting View>

Drafting View: 06- Interior Wall GWB

On the ribbon:

Enable **Reference Other View**.

In the drop-down: select **Drafting View: 06- Interior Wall GWB**.

8. Sketch the rectangle over the interior wall.

Use the grips to adjust the position of the callout.

9. Open the **A102 – Sections** sheet.

10. Locate the **06- Interior Wall GWB** drafting view in the Project Browser.

11. Place the drafting view on the sheet.

12. Notice that the callouts in the section views update with the view and sheet number.

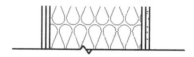

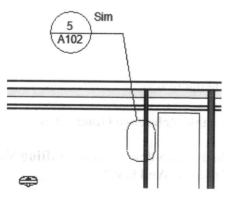

13. Save as *ex2-13.rvt*.

Scope Boxes

Scope boxes are created in plan views but are visible in other view categories.

To create a Scope Box, you must be in either a Floor Plan View or in a Reflected Ceiling Plan. However, once a scope box is created, it is going to be visible in the other view categories: sections, callouts, elevations and 3D views. In elevations and sections, the scope box is only going to be visible if it intersects the cut line. You can adjust the extents of the scope box in all view categories.

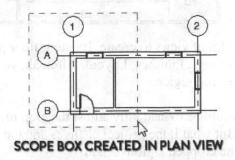

SCOPE BOX CREATED IN PLAN VIEW

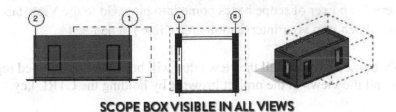

SCOPE BOX VISIBLE IN ALL VIEWS

Consider this office building renovation project. The area affected is in the middle of the building. You want the views to be cropped to fit the red rectangle.

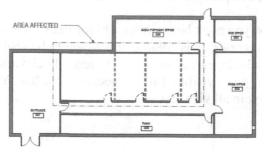

The thing is: you have a lot of views to create. Existing floor plan. Demolished floor plan. New floor plan. Ceilings. Finishes. Layout. All in all, you'll have about 10 views that need the exact same crop region.

An archaic workflow would be to manually adjust the crop region of each view. That would probably work. But what if the project changes and the area affected gets bigger? You have to adjust all the cropped regions again?

That's where the power of scope boxes come into play. Go to the View tab and create a Scope Box. Match it to your intervention area. Give it a name.

Now, apply the scope box to all the views that will be using this cropped region. To save time, select all the views in the project browser by holding the CTRL key.

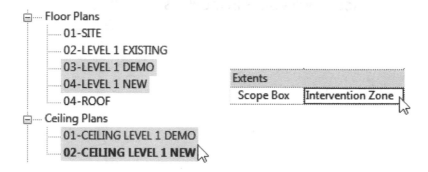

Look at all these views, sharing the exact same crop region. Adjusting the new Scope Box will affect all these views.

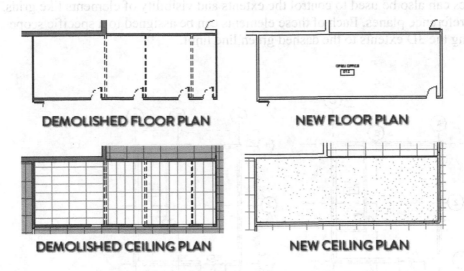

DEMOLISHED FLOOR PLAN   NEW FLOOR PLAN

DEMOLISHED CEILING PLAN   NEW CEILING PLAN

ALL PLANS SHARE THE SAME CROP REGION

Even though Scope Boxes can only be applied in plan view, they are 3D objects. That means they have an assigned height.

Name: Scope Box 1   Height: 12000

The Height is assigned in the Option bar.

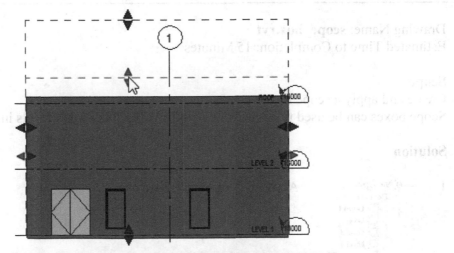

DRAG BLUE ARROWS TO ADJUST SCOPE BOX HEIGHT

Scope boxes can also be used to control the extents and visibility of elements like grids, levels and reference planes. Each of these elements can be assigned to a specific scope box, limiting the 3D extents to the dashed green line limit.

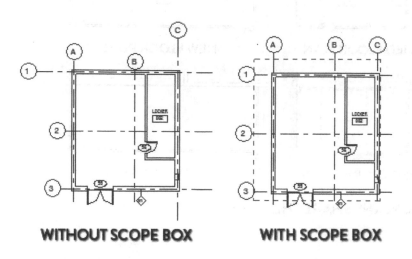

WITHOUT SCOPE BOX   **WITH SCOPE BOX**

Exercise 2-14

Create a Scope Box

Drawing Name: **scope_box.rvt**
Estimated Time to Completion: 15 Minutes

Scope
Create and apply a scope box.
Scope boxes can be used to control the visibility of grid lines and levels in views.

Solution

1. Views (all)
    Floor Plans
        Level 1
        Level 2
        Level 3
        Level 4

Activate the **Level 1** floor plan.

2.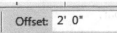
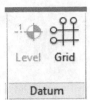

Activate the **Architecture** ribbon.
Select the **Grid** tool from the Datum panel.

3. Offset: 2' 0"

Set the Offset to **2′ 0″** on the Options bar.

4.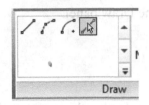

Select the **Pick Lines** tool from the Draw panel.

5.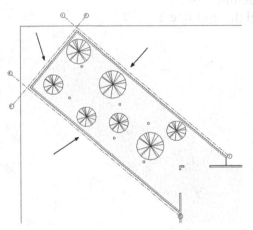

Place three grid lines using the exterior side of the walls to offset.

6.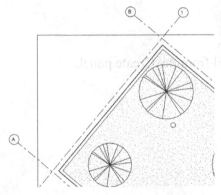

Re-label the grid bubbles so that the two long grid lines are A and B and the short grid line is 1.

7.

Select the **Grid** tool from the Datum panel.

8. Set the Offset to **180' 0"** on the Options bar.

9. Select the **Pick Lines** tool from the Draw panel.

10.

Place the lower grid line by selecting the upper wall and offsetting 180'.
Re-label the grid line **2**.

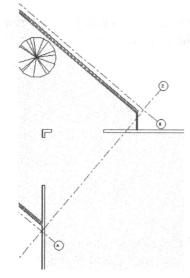

11.

Activate the **View** ribbon.
Select the **Scope Box** tool from the Create panel.

12.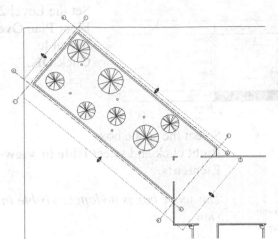

Place the scope box.

Use the Rotate icon on the corner to rotate the scope box into position, so it is aligned with the grid lines.

Use the blue grips to control the size of the scope box.

13.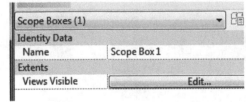

Select the grid line labeled **B**.

14.

Grids (1)	
Identity Data	
Name	B
Extents	
Scope Box	Scope Box 1

In the Properties pane:
Set the Scope Box to Scope Box 1; the scope box which was just placed.

15. Repeat for the other three grid lines.
You can use the Control key to select more than one grid line and set the Properties of all three at the same time.

16.

Scope Boxes (1)		▼
Identity Data		
Name	Scope Box 1	
Extents		
Views Visible	Edit...	

Select the Scope Box.
Select **Edit** next to Views Visible in the Properties pane.

17.

View Type	View Name	Automatic visibility	Override
3D View	{3D}	Visible	None
Ceiling Plan	Level 1	Visible	None
Ceiling Plan	Level 2	Visible	None
Elevation	South Elevation	Invisible	None
Floor Plan	Level 1	Visible	None
Floor Plan	Level 2	Visible	Invisible
Floor Plan	Site	Visible	None
Floor Plan	Level 3	Visible	None

Click on the column headers to change the sort order.

Set the Level 2
Floor Plan Override
Invisible.
Click **OK**.

18.

Cancel

Repeat [Scope Box]

Recent Commands ▶

Hide in View ▶   Elements

Override Graphics in View ▶   Category

Create Similar   By Filter...

Select the scope box.
Right click and select **Hide in View→
Elements**.

*The scope box is no longer visible in the
view.*

19. Floor Plans
.... Level 1
.... **Level 2**
.... Level 3

Activate Level 2.

The grid lines and scope box are not visible.

20. Close the file without saving.

Tip: To make the hide/isolate mode permanent to the view: on the View Control bar,
click the glasses icon and then click Apply Hide/Isolate to the view.

Exercise 2-15

Use a Scope Box to Crop Multiple Views

Drawing Name: **scope_multiple.rvt**
Estimated Time to Completion: 5 Minutes

Scope
Create and apply a scope box to crop multiple views.

Solution

1. Activate the **Level 1** floor plan.

2. Activate the **View** ribbon.
   Select the **Scope Box** tool from the Create panel.

3. Place the scope box so it encloses the building.

   Use the blue grips to control the size of the scope box.

4. In the Properties panel:

   Change the Name of the Scope Box to **Building Model.**

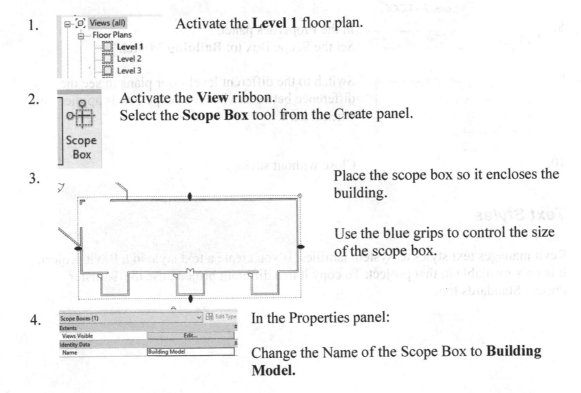

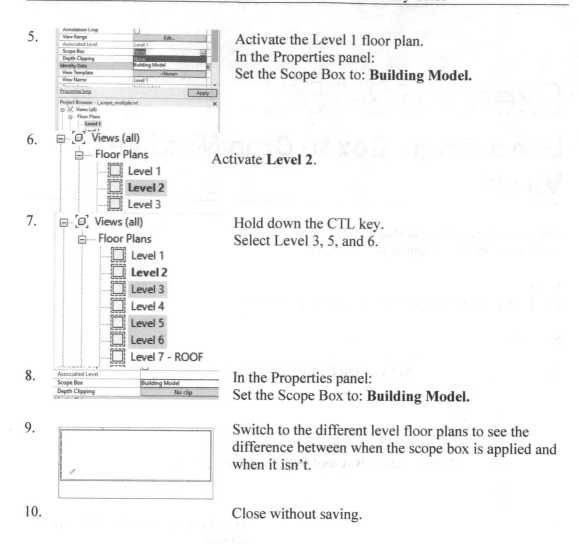

5. Activate the Level 1 floor plan.
In the Properties panel:
Set the Scope Box to: **Building Model.**

6. Activate **Level 2**.

7. Hold down the CTL key.
Select Level 3, 5, and 6.

8. In the Properties panel:
Set the Scope Box to: **Building Model.**

9. Switch to the different level floor plans to see the difference between when the scope box is applied and when it isn't.

10. Close without saving.

Text Styles

Revit manages text styles as system families. If you create a text style in a Revit project, it is only available in that project. To copy it to a different project, use the Transfer Project Standards tool.

To create a new text family, select an existing text element, and use the Edit Type→Duplicate to define a new text style. You can select any Windows available font for the text style.

If you import an AutoCAD file into Revit that uses *.shx fonts, Revit automatically maps *shx fonts to a similar font style. If you are unhappy with the results, select the text and apply a new text style.

Exercise 2-16

Create a Text Style and Leader

Drawing Name: **detail groups.rvt**
Estimated Time to Completion: 10 Minutes

Scope
Create a text family
Add text with a leader

Solution

1.
   - Working Ground Floor
   - SITE PLAN
   - GROUND FLOOR
     - GROUND FLOOR -Exam Rooms

   Activate the **GROUND FLOOR – Exam Rooms** floor plan.

2.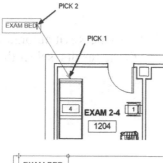

Switch to the Annotate ribbon.

Select the **Text** tool on the Text panel.

3.

On the ribbon:

Enable the two straight leader option.

4.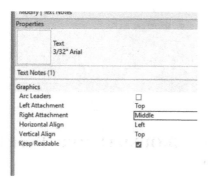

Select a point at the top of the exam bed as shown in the image.

Select a point outside of the view.

Type **EXAM BED**.

Click to enter.

5. Adjust the position of the text so it is visible in the view.

You may need to enlarge the cropped area, adjust the text and then restore the cropped area.

6.

Select the text.

On the Properties palette:

Set the Right Attachment to **Middle**.

How does the text change?

Select **Edit Type**.

7.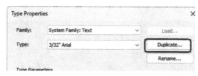

Click **Duplicate**.

8.

Name: 3/32" Tahoma

Rename to **3/32" Tahoma**.

Click **OK**.

9.

Rename...

Type Parameters	
Parameter	**Value**
Graphics	
Color	▦ Red
Line Weight	1
Background	Opaque
Show Border	☐
Leader/Border Offset	5/64"
Leader Arrowhead	Arrow Filled 30 Degree
Text	
Text Font	Tahoma
Text Size	3/32"
Tab Size	1/2"
Bold	☐
Italic	☐
Underline	☐
Width Factor	1.000000

Change the Color to **Red**.
Change the Leader Arrowhead to **Arrow Filled 30 Degree**.
Change the Font to **Tahoma**.

Click **OK**.

10.

EXAM BED

Release the selection.

Save as *ex2-16.rvt*.

Keynotes

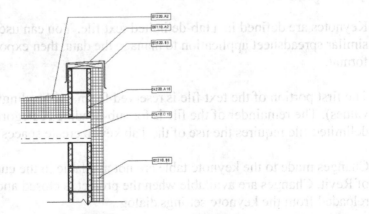

A keynote parameter is available for all model elements (including detail components)

and materials. You can tag each of these elements using a keynote tag family. The keynote value is derived from a separate text file that contains a list of keynotes. Reference Keynotes, or simply "keynotes", identify materials and products on the drawings and connect the drawing to the specifications. Some companies prefer to use their own keynote content. You can configure Revit to reference a specific keynote data set or modify the existing keynote data file to comply with your company's standards.

An element is individually identified with a note having a numeric identifier in a sequential list of items, with leaders pointing toward its position in the view. Keynote values are defined as a type parameter for each family. The keynote itself reads this parameter to resolve the label for the keynote. Keynotes typically reference a standard specification system used in the building industry. For example you might use keynotes that reference the 16 Divisions established by 1995 Construction Specification Institute (CSI) Master format system, or to one of the 50 divisions included in the 2004 version of the system.

For a better coordination between drawings and specifications, the keynotes use an alpha numeric suffix following the section number and separated by a decimal point, which references the specific sections within the project manual.

The keynotes text should be generic and use the same terminology as in the specifications. It should not detail the item's characteristics, because these can be changed during the project and consequently all the keynotes would need to be changed.

If there are different family types of the same building element, they should have different keynotes, associating each of them with the corresponding designation in the specifications. You can assign keynotes to materials or to elements, so that they are automatically tagged with the correct keynote.

Keynotes are defined in a tab-delimited text file. You can use Microsoft® Excel or a similar spreadsheet application to manage the data, then export it to a tab-delimited file format.

The first portion of the text file is reserved for major headings/categories (the parent values). The remainder of the file is for sub-headings/categories (the child values). A tab-delimited file requires the use of the Tab key to create spaces between data entries.

Changes made to the keynote table are not available in the current project session of Revit. Changes are available when the project is closed and reopened, or the file is reloaded from the keynote settings dialog.

Exercise 2-17
Configure Keynote Settings

Drawing Name: **keynotes_1.rvt**
Estimated Time to Completion: 5 Minutes

Scope
Configure keynote settings.

.
Solution

1. 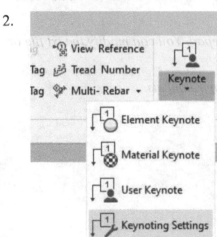 Activate the **GROUND FLOOR – Exam Rooms** floor plan.

2. Activate the **Annotate** ribbon.

    Select **Keynote → Keynoting Settings**.

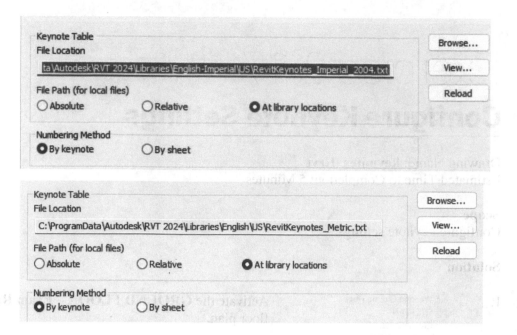

3.   *This shows the path for the keynote database.*

This is a txt file that can be edited using Notepad. You can modify the txt file or select a different txt file for keynote use.

The file can be placed on a server in team environments.

Click **OK**.

Exercise 2-18

Insert Keynotes

Solution

Drawing Name: insert_keynotes.rvt [m_keynotes.rvt]
Estimated Time: 30 minutes

This exercise reinforces the following skills:

- Elevation View
- Cropping a View
- Keynote Tags
- Sheets
- Views

1. Open *insert_keynotes.rvt. [m_keynotes.rvt]*

2. Activate the **Level 1 - Lobby Detail** view.

3. Type **VV** to bring up the Visibility/Graphics dialog.

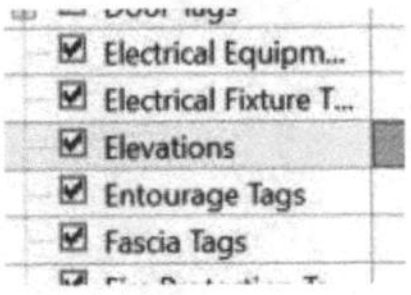

   Activate the **Annotation Categories** tab.

   Enable the visibility of Elevations.

   Click **OK**.

4. Activate the View ribbon.

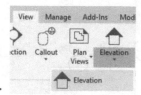

Select **Elevation→ Elevation**.

5. 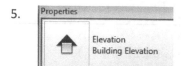 Check in the Properties palette.

You should see that you are placing a Building Elevation.

6. Place an elevation marker as shown in front of the exterior front door.
Right click and select **Cancel** to exit the command.

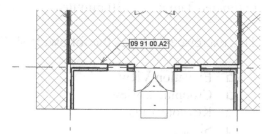

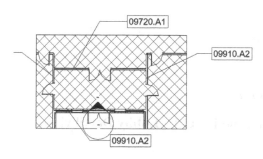

7. Click on the triangle part of the elevation marker to adjust the depth of the elevation view.

Adjust the boundaries of the elevation view to contain the lobby room.

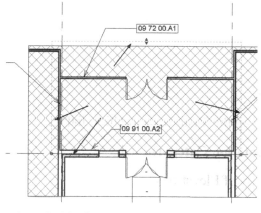

8. Locate the elevation in the Project Browser.

Rename to **South - Lobby**.

⊟···· Elevations (Building Elevation)
    ▢ East
    ▢ South - Lobby
    ▢ North
    ▢ South
    ▢ West

9. 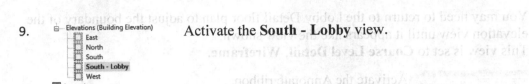 Activate the **South - Lobby** view.

10. Adjust the crop region to show the entire lobby including floor and ceiling.

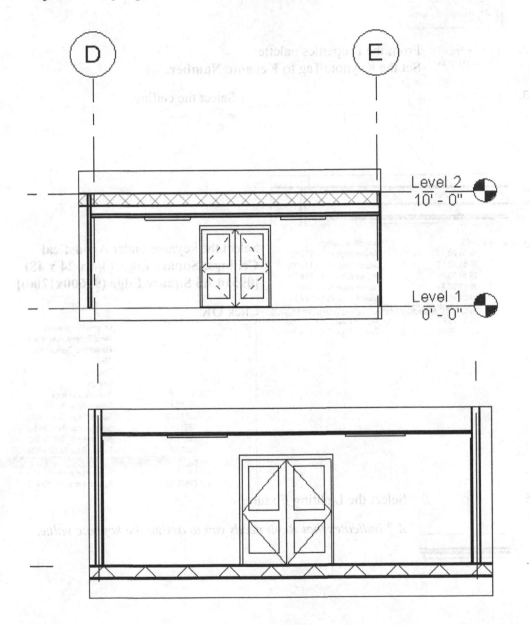

You may need to return to the Lobby Detail floor plan to adjust the boundary of the elevation view until it appears like the view shown.
This view is set to **Coarse Level Detail**, **Wireframe**.

11. 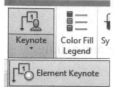 Activate the Annotate ribbon.

Select **Keynote→Element Keynote**.

12. Keynote Tag
    Keynote Number

    From the Properties palette:
    Set the Keynote Tag to **Keynote Number**.

13. 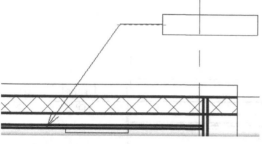 Select the ceiling.

14.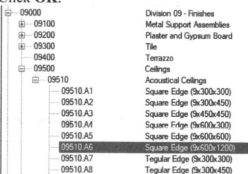

09 51 00	Acoustical Ceilings
09 51 00.A1	Square Edge (3/4 x 12 x 12)
09 51 00.A2	Square Edge (3/4 x 12 x 18)
09 51 00.A3	Square Edge (3/4 x 18 x 18)
09 51 00.A4	Square Edge (3/4 x 24 x 12)
09 51 00.A5	Square Edge (3/4 x 24 x 24)
09 51 00.A6	Square Edge (3/4 x 24 x 48)
09 51 00.A7	Tegular Edge (3/4 x 12 x 12)
09 51 00.A8	Tegular Edge (3/4 x 12 x 18)

Select the keynote under **Acoustical Ceilings: Square Edge (3/4 x 24 x 48) [09510.A6 Square Edge (9x600x1200)]**.

Click **OK**.

09000	Division 09 - Finishes
09100	Metal Support Assemblies
09200	Plaster and Gypsum Board
09300	Tile
09400	Terrazzo
09500	Ceilings
09510	Acoustical Ceilings
09510.A1	Square Edge (9x300x300)
09510.A2	Square Edge (9x300x450)
09510.A3	Square Edge (9x450x450)
09510.A4	Square Edge (9x600x300)
09510.A5	Square Edge (9x600x600)
09510.A6	Square Edge (9x600x1200)
09510.A7	Tegular Edge (9x300x300)
09510.A8	Tegular Edge (9x300x450)

15. **?** Select the Lighting Fixture.

  A ? indicates that Revit needs you to assign the keynote value.

16.

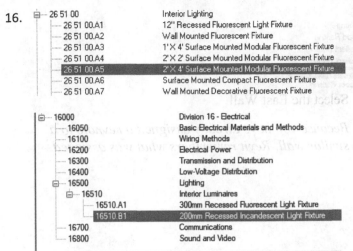

	26 51 00	Interior Lighting
	26 51 00.A1	12" Recessed Fluorescent Light Fixture
	26 51 00.A2	Wall Mounted Fluorescent Fixture
	26 51 00.A3	1' X 4' Surface Mounted Modular Fluorescent Fixture
	26 51 00.A4	2' X 2' Surface Mounted Modular Fluorescent Fixture
	26 51 00.A5	2' X 4' Surface Mounted Modular Fluorescent Fixture
	26 51 00.A6	Surface Mounted Compact Fluorescent Fixture
	26 51 00.A7	Wall Mounted Decorative Fluorescent Fixture

	16000	Division 16 - Electrical
	16050	Basic Electrical Materials and Methods
	16100	Wiring Methods
	16200	Electrical Power
	16300	Transmission and Distribution
	16400	Low-Voltage Distribution
	16500	Lighting
	16510	Interior Luminaires
	16510.A1	300mm Recessed Fluorescent Light Fixture
	16510.B1	200mm Recessed Incandescent Light Fixture
	16700	Communications
	16800	Sound and Video

Locate the **2′ x 4′ Surface Mounted Modular Fluorescent Fixture [16510.B1 200MM Recessed Incandescent Light Fixture]**.

Click **OK**.

17.

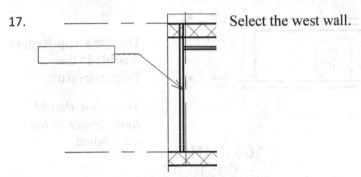

Select the west wall.

18.

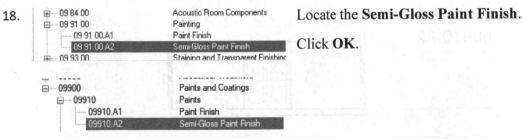

	09 84 00	Acoustic Room Components
	09 91 00	Painting
	09 91 00.A1	Paint Finish
	09 91 00.A2	Semi-Gloss Paint Finish
	09 93 00	Staining and Transparent Finishing

	09900	Paints and Coatings
	09910	Paints
	09910.A1	Paint Finish
	09910.A2	Semi-Gloss Paint Finish

Locate the **Semi-Gloss Paint Finish**.

Click **OK**.

19.

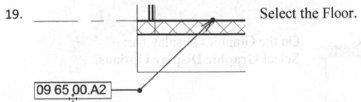

09 65 00.A2

Select the Floor.

20.

	09 84 00	Wood Flooring
	09 65 00	Resilient Flooring
	09 65 00.A1	Resilient Flooring
	09 65 00.A2	Vinyl Composition Tile
	09 65 00.A3	Rubber Flooring

Locate the **Vinyl Composition Tile**.

Click **OK**.

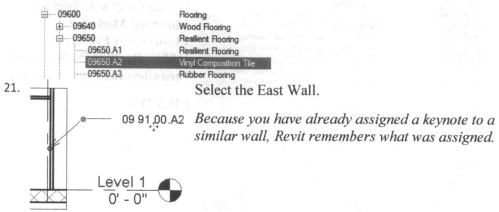

21. Select the East Wall.

09 91 00.A2 *Because you have already assigned a keynote to a similar wall, Revit remembers what was assigned.*

22.

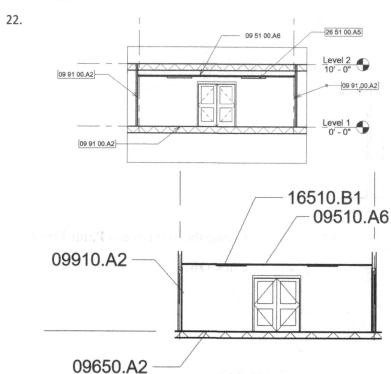

Disable **Crop Region Visible** in the Properties pane.

Your view should look similar to the one shown.

23.

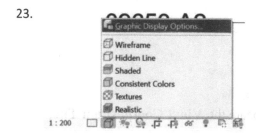

On the Graphics Display bar:
Select **Graphic Display Options**.

24.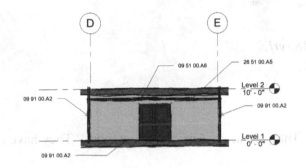

Set the Style to **Shaded**.
Enable **Show Edges**.

25.

Enable **Sketchy Lines**.
Set the Jitter to **6**.
Set the Extension to **6**.

Click **Apply** to preview.

26.

If you are happy with the
result, click **OK**.

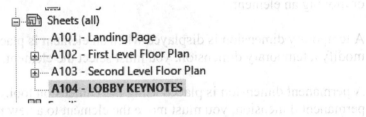

27.

Activate the **A103 - LOBBY KEYNOTES
[A104 LOBBY KEYNOTES]** sheet.

28.

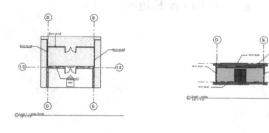

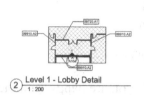

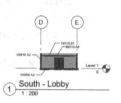

Add the **Level 1 - Lobby Detail** view to the sheet.

Add the **South Lobby Elevation** view to the sheet.

29. Save as *ex2-18.rvt [m_ex2-18.rvt]*.

Dimension Styles

Dimensions are system families. They are in the Annotation category. They have type and instance properties.

Revit has three types of dimensions: listening, temporary and permanent.

A listening dimension is the dimension that is displayed as you are drawing, modifying, or moving an element.

A temporary dimension is displayed when an element is placed or selected. In order to modify a temporary dimension, you must select the element.

A permanent dimension is placed using the Dimension tool. In order to modify a permanent dimension, you must move the element to a new position. The permanent dimension will automatically update. To reposition an element, you can modify the temporary dimension or move the element using listening dimensions.

When you enter dimension values using feet and inches, you do not have to enter the units. You can separate the feet and inches values with a space and Revit will fill in the units. If a single unit is entered, for example '10', Revit assumes that value is 10 feet, not 10 inches.

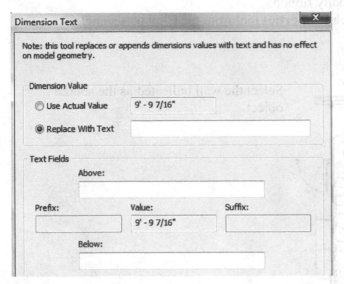

Dimension Text

Note: this tool replaces or appends dimensions values with text and has no effect on model geometry.

Dimension Value

○ Use Actual Value   9' - 9 7/16"

● Replace With Text

Text Fields

Above:

Prefix: Value: Suffix:

9' - 9 7/16"

Below:

To override a dimension value, select the permanent dimension and enable Replace with Text and put in the desired text.

You can also add additional notes using the Text Fields for Above, Prefix, Suffix, or Below.

Exercise 2-19

Create Dimensions

Drawing Name: **i_dimensions.rvt**
Estimated Time to Completion: 20 Minutes

Scope
Match Properties
Placing dimensions

Solution

1.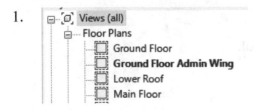

Activate the **Ground Floor Admin Wing** floor plan.

2. 	Activate the Modify ribbon.
	Select the **Match Properties** tool from the Clipboard panel.

3. 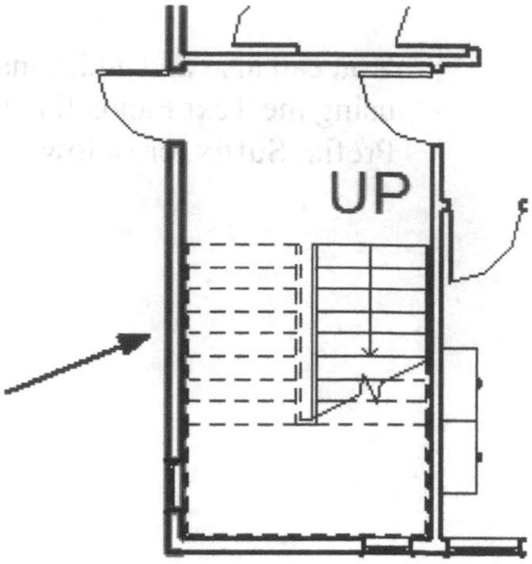	Select the wall indicated as the source object.

4. 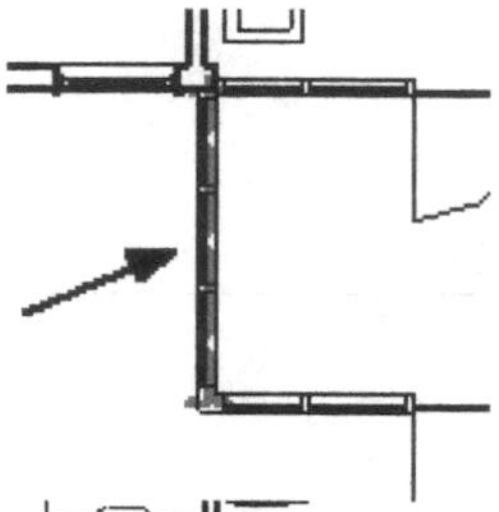	Select the wall indicated as the target object.

5. 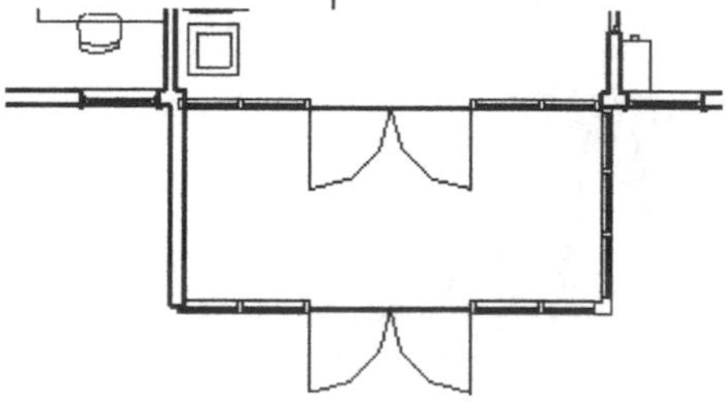	Note that the curtain wall changed to an Exterior-Siding wall.

	Cancel out of the Match Properties command.

6. 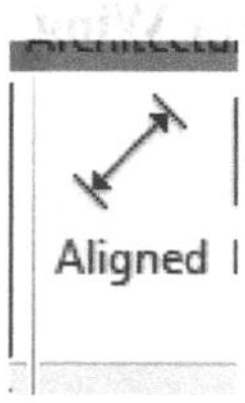	Activate the Annotate ribbon.
	Select the **Aligned Dimension** tool.

7.	On the Options bar, set the dimensions to select the **Wall faces**.
	Enable Pick: **Entire Walls**.

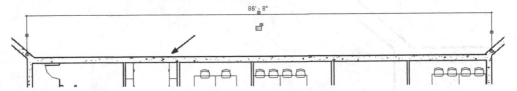

8. Pick the wall indicated. Move the mouse above the selected wall to place the dimension.

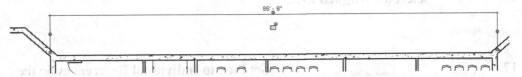

9. Note the entire wall is selected and the dimension is located at the faces of the walls.
    Cancel or escape to end the Dimension command.

10. Select the dimension.
    Note that there are several grips available. The grips are the small blue bubbles.

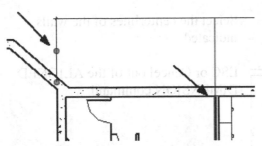

11. 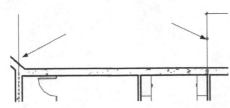 Select the middle grip indicated on the left.
    Move the witness line to the wall face indicated by the arrow in the center of the image.

    Note that the witness line automatically snaps to the wall face.

12. With the dimension selected, on the Options bar, change the Prefer to **Wall centerlines**.

13. Select the dimension and activate the grip indicated.
    Drag the witness line to the center of the left wall.

14. 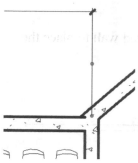 Left click once on the grip on the right side of the dimension to shift the witness line to the wall centerline.

15. *Note how the dimension value updates.*

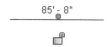

85' - 8"

16.   Select the **Aligned** tool.

Aligned

17. Set **Pick to Individual References** on the options bar.

Wall centerline ▾ Pick: Individual Reference ▾ | Options

18. 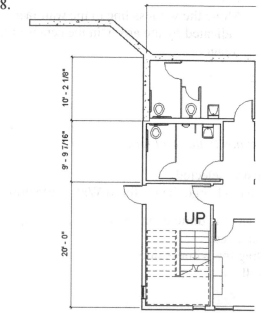 Select the centerlines of the walls indicated.

ESC or Cancel out of the ALIGNED DIMENSION command.

19.

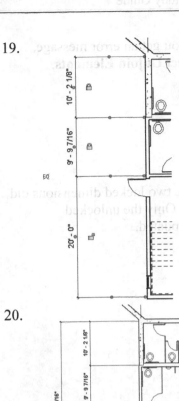

Select the dimension so it highlights.

Select the two locks on the top dimensions to switch the permanent dimensions to *locked* dimensions. This means these distances will not be changed.

You should see two of the padlocks as closed and one as open.

20.

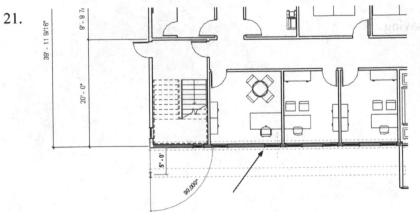

Place an overall dimension using the wall centerlines.

ESC or Cancel out of the ALIGNED DIMENSION command.

21.

Drag the wall indicated down to a new position.

Move down 5'-0".

22.

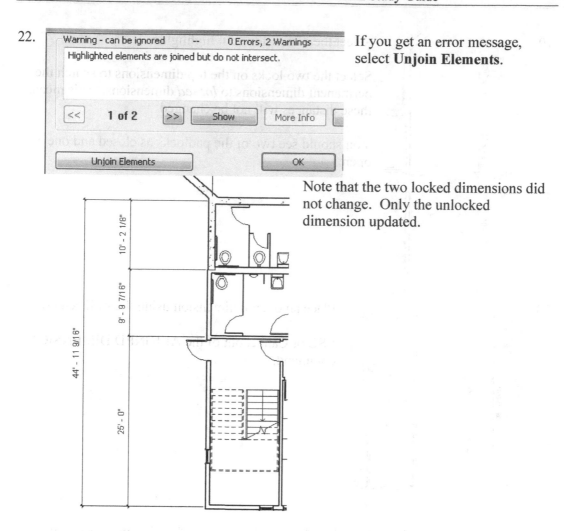

If you get an error message, select **Unjoin Elements**.

Note that the two locked dimensions did not change. Only the unlocked dimension updated.

23. Close without saving.

Exercise 2-20
Modify Dimensions

Drawing Name: **dimtext.rvt**
Estimated Time to Completion: 15 Minutes

Scope
Replace Dimension Text
Restore Dimension Text
Modify Dimension Text

Solution

1.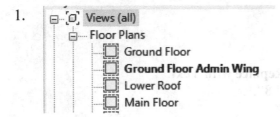
Activate the **Ground Floor Admin Wing** floor plan.

2.

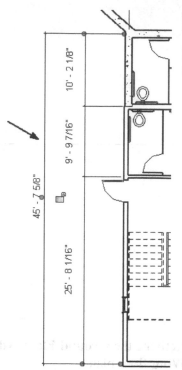

Select the dimension indicated.

Double left click on the dimension text to bring up the dimension text dialog.

3.

Dimension Value

◯ Use Actual Value   45' - 7 5/8"

◉ Replace With Text   45' 8"

Enable **Replace with Text**.
Enter **45′ 8″**.
Click **OK**.

4.

Invalid Dimension Value   ✕

Specify descriptive text for a dimension segment instead of a numeric value.

To change the dimension value for a length or angle of a segment, select the element the dimension refers to, and click the value to edit it.

You will get an error message stating that you cannot change the numeric value without moving the element to correspond with that value.

Click **Close**.

5.

Dimension Value

◯ Use Actual Value   45' - 7 5/8"

◉ Replace With Text   OVERALL

Enable **Replace with Text**.
Enter **OVERALL**.
Click **OK**.

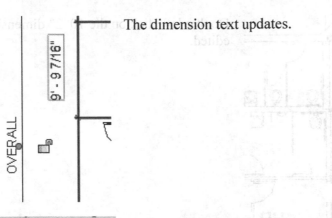

The dimension text updates.

6. Activate the **Manage** ribbon.
Select **Project Units** on the Settings panel.

7. Select the **Length** button under the Format column.

8. Set the Rounding **To the nearest ½"**.
Click **OK** until all dialogs are closed.

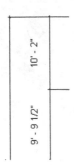

Note that the dimensions update to the new rounding.

9.

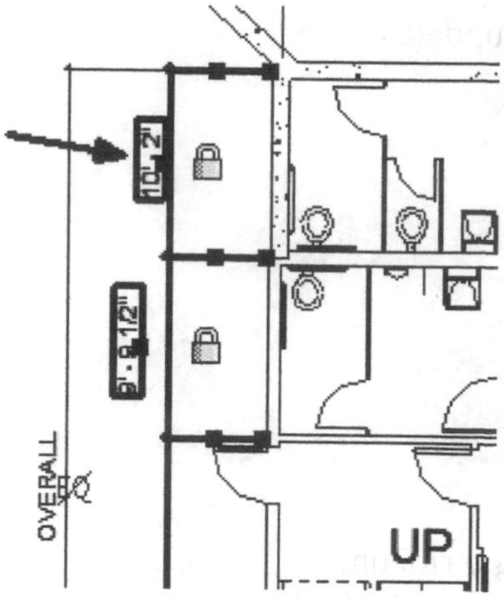

    Double left click on the 10′ 2″ dimension to be edited.

10.

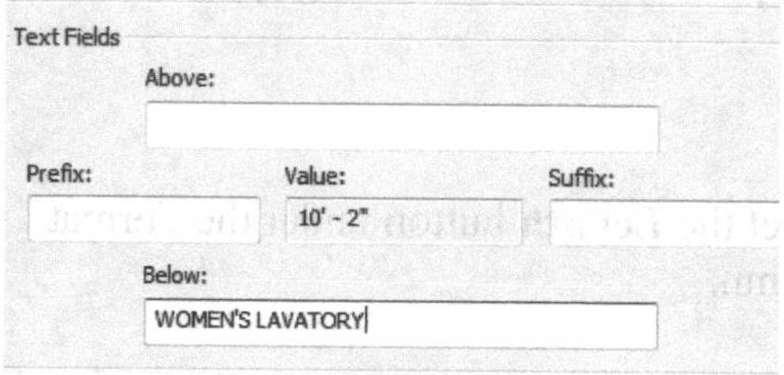

    In the Below field, type **WOMEN'S LAVATORY**. Click **OK**.

11.

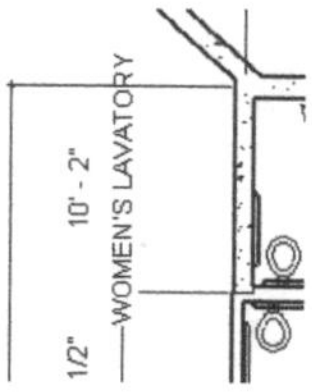

    Note how the dimension updates.

12. 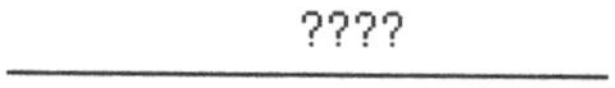 ????

    Double left click on the dimension with **???**.

13.

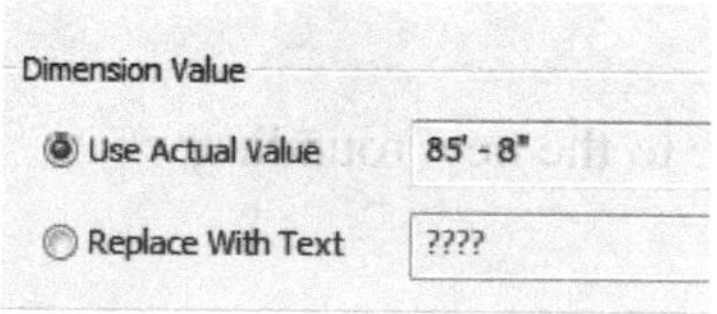

    Enable **Use Actual Value**. Click **OK**.

14. Close without saving.

Exercise 2-21

Convert Temporary Dimensions to Permanent Dimensions

Drawing Name: **i_dimensions.rvt**
Estimated Time to Completion: 10 Minutes

Scope
Place dimensions using different options
Convert temporary dimension to permanent

Solution

1.

Activate the **Ground Floor Admin Wing** floor plan.

2.

Select the wall indicated.

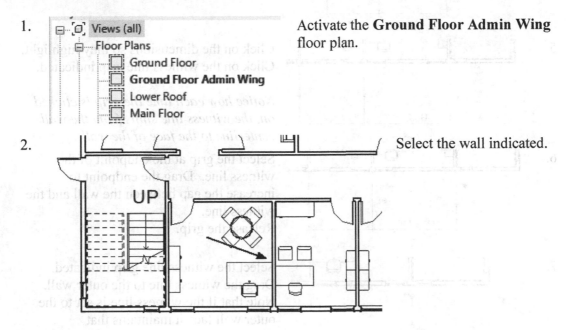

3.

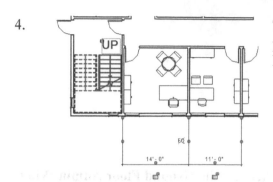

Two dimensions will appear.
There is a small dimension icon visible.
This icon converts a temporary dimension to a permanent dimension.
Left click on this icon.

4.

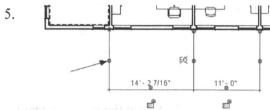

The dimensions are now converted to permanent dimensions.
Drag the dimensions below the view.

5.

Click on the dimensions so they highlight.
Click on the witness line grip indicated.

Notice how each time the grip is clicked on, the witness line shifts from the wall centerline to the face of the wall.

6.

Select the grip at the endpoint of the witness line. Drag the endpoint to increase the gap between the wall and the witness line.
Release the grip.

7.

Select the witness line grip indicated.
Drag the witness line to the outer wall.
Note that if the witness line is set to the outer wall face it maintains that orientation.
Release.

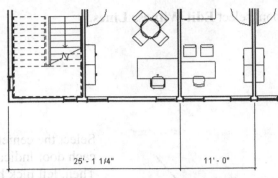

25'- 1 1/4"   11'- 0"

The dimension updates.
Note the new dimension value.

8. Close without saving.

Exercise 2-22
Multi-Segmented Dimensions

Drawing Name: **multisegment.rvt**
Estimated Time to Completion: 10 Minutes

Scope
Add and delete witness lines to a multi-segment dimension

Solution

1.
   Views (all)
   Floor Plans
     Ground Floor
     Lower Roof
     Main Floor
     Main Roof
     Site

   Activate the **Main Floor** floor plan.

2.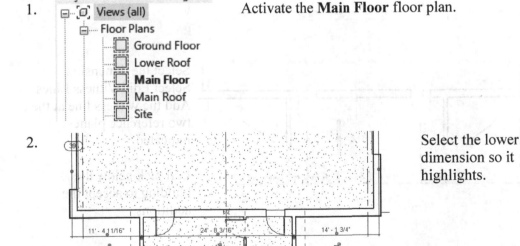

   11'- 4 11/16"   24'- 8 3/16"   14'- 1 3/4"

   Select the lower
   dimension so it
   highlights.

3. 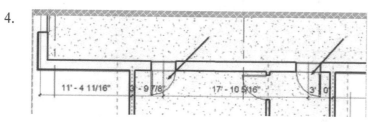 On the ribbon, select **Edit Witness Lines**.

4. 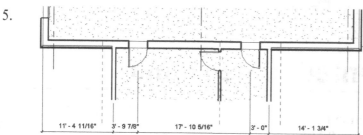 Select the center of each door indicated. Then, left pick below the building to accept the additions.

5. 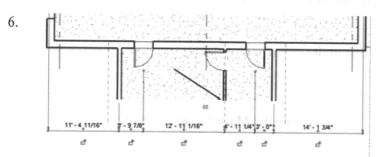 Drag the dimensions down so you can see them easily.

6. 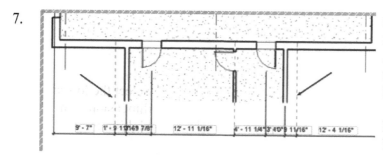 Select the dimension. Select Edit Witness Lines. Add the witness line at the wall indicated.

7. Select the dimension. Select Edit Witness Lines. Add the witness line at the two reference planes indicated.

Left pick below the building to accept the additions.

8.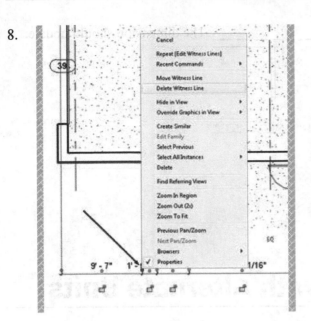

Select the dimension.
Select the grip on the witness line for the left reference plane. Right click and select Delete Witness Line.

9.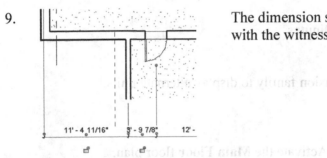

The dimension should update with the witness line removed.

10.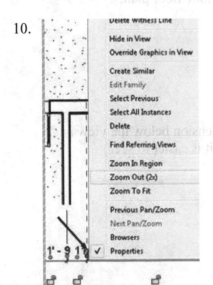

Repeat to delete the witness line on the right reference plane.

11. 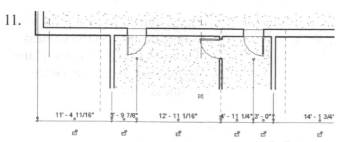   The dimension should update.

12. Close without saving.

Exercise 2-23

Dimension Style with Alternate Units

Drawing Name: **multisegment.rvt**
Estimated Time to Completion: 5 Minutes

Scope
Edit the Type Properties of a Dimension family to display alternate units.

Solution

1.
    ▣ ⌂ Views (all)
       ▣ Floor Plans
           ☐ Ground Floor
           ☐ Lower Roof
           ☐ **Main Floor**
           ☐ Main Roof
           ☐ Site

    Activate the **Main Floor** floor plan.

2.
    Select the dimension below the view and drag down so it is easier to see.

3.  Select the horizontal dimension.

Select **Edit Type** on the Properties palette.

4. 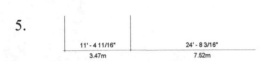 Scroll down to Alternate Units.

Set the Alternate Units to display **Below**.
Set the Alternate Unites Format to **Meters**.
Set the Alternate Units Suffix to **m** with a precision of two decimals.

Click **OK**.

5.
11' - 4 11/16"	24' - 8 3/16"
3.47m	7.52m

Alternate units are now displayed by the dimension.

Close without saving.

Matchline

A match line is a line on a design drawing that projects a location or distance from one portion of the drawing to another portion of the drawing. The matchline indicates where a new view splits off.

You can modify the appearance of a matchline by using Visibility/Graphics overrides.

Exercise 2-24
Using a Matchline

Drawing Name: matchline.rvt
Estimated Time: 30 minutes

This exercise reinforces the following skills:

- ❑ Matchline
- ❑ Sheets
- ❑ Views
- ❑ View Reference Annotation

1.  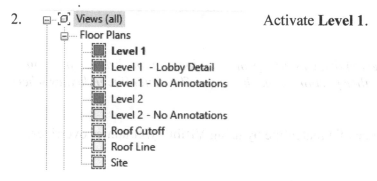 Open *matchline.rvt*.

2. Activate **Level 1**.

3. 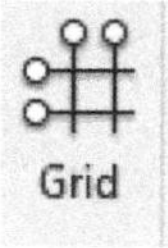 Select the **Grid** tool from the Architecture ribbon.

4. 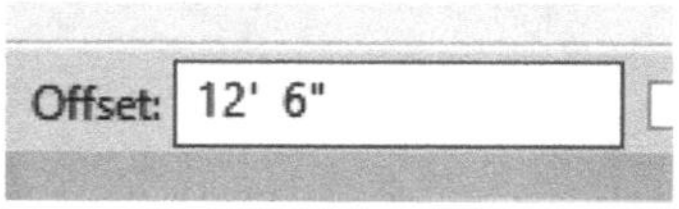 On the Options ribbon, set the Offset to **12' 6"**.

5.

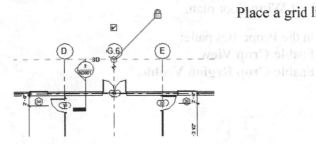

Select **Pick Line** mode on the Draw panel on the ribbon.

6.

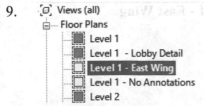

Place a grid line between Grids D & E.

7.

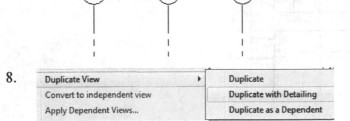

Rename the Grid bubble **D.5**.

8.

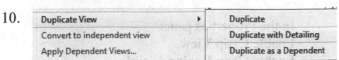

Right click on **Level 1** in the browser.
Select **Duplicate View →
Duplicate with Detailing.**

9.

Views (all)
└─ Floor Plans
   ├─ Level 1
   ├─ Level 1 - Lobby Detail
   ├─ Level 1 - East Wing
   ├─ Level 1 - No Annotations
   └─ Level 2

Rename the view **Level 1-East Wing**.

10.

Duplicate View	▶	Duplicate
Convert to independent view		Duplicate with Detailing
Apply Dependent Views...		Duplicate as a Dependent

Right click on **Level 1** in the browser.
**Select Duplicate View →
Duplicate with Detailing**.

11. Rename the view **Level 1-West Wing**.

12. Activate the view **Level 1-West Wing** floor plan.

In the Properties pane:
Enable **Crop View**.
Enable **Crop Region Visible**.

13.

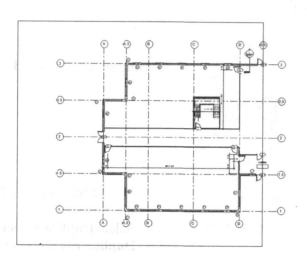

Use the grips on the crop region to only show the west side of the floor plan.

14. Activate **Level 1 - East Wing**.

15. In the Properties pane:
Enable **Crop View**.
Enable **Crop Region Visible**.

16.

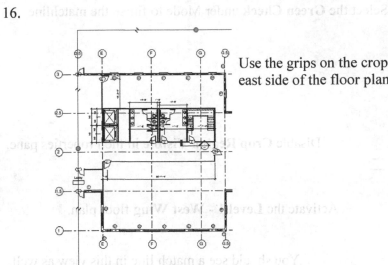

Use the grips on the crop region to only show the east side of the floor plan.

17. | View | Activate the **View** ribbon.

18.

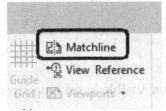

Under the Sheet Composition panel:

Select the **Matchline** tool.

19.

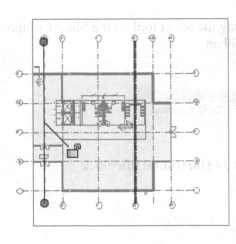

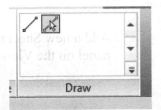

Activate **Pick Line** mode.

Select Grid **D.5**.

20.  Select the **Green Check** under Mode to finish the matchline.

21.  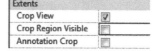 Disable **Crop Region Visible** in the Properties pane.

22. 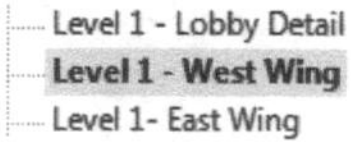 Level 1 - Lobby Detail
    Level 1 - West Wing
    Level 1- East Wing

    Activate the **Level 1 - West Wing** floor plan.

23. 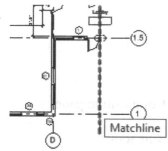 You should see a match line in this view as well.

24. Disable **Crop Region Visible** in the Properties pane.

25.  Add a new Sheet using the Sheet tool on the Sheet Composition panel on the View ribbon.

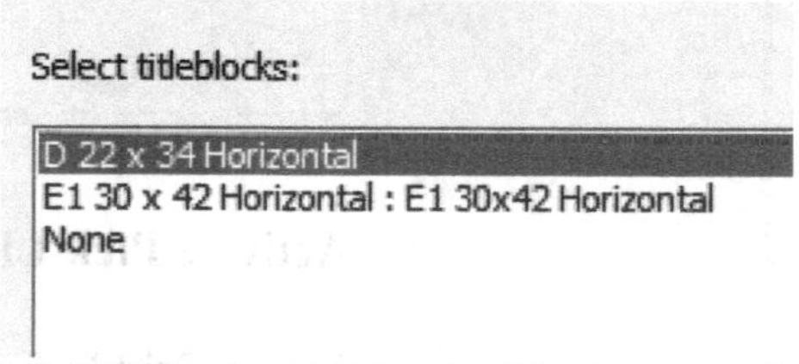

    Select titleblocks:

    D 22 x 34 Horizontal
    E1 30 x 42 Horizontal : E1 30x42 Horizontal
    None

    Highlight the **D 22 x 34 Horizontal** titleblock.
    Click **OK**.

26.

Place the Level 1- West Wing floor plan on the sheet.

Adjust the scale so it fills the sheet.

27.
M. Instructor
J. Student
M. Instructor
J. Student
A106
Level 1 - West Wing

Change the Sheet Number to **A106**.

Name the sheet **Level 1 -West Wing**.

Enter in your name and your instructor's name.

28.

Zoom into the view title bar on the sheet.

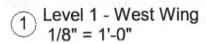

Level 1 - West Wing
1/8" = 1'-0"

29.
Views (all)
Floor Plans
Level
Level 1
Level 1
Level 1
Level 1

Activate the **Level 1** floor plan.

30.

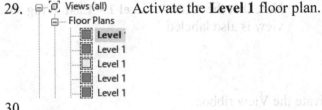

You can see the matchline that was added to the view if you zoom into the D.5 grid line.

31.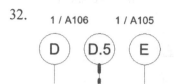

Activate the View ribbon.

Select the **View Reference** tool on the Sheet Composition panel.

This adds an annotation to a view.

32.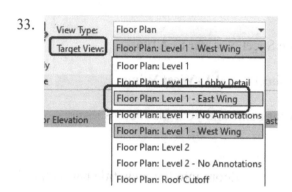

Place the note on the right side of the grid line to correspond with the view associated with the right wing.

Cancel out of the command.

33.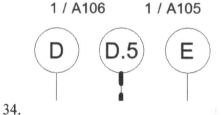

Select the **View Reference** label.

Set the Target View on the ribbon to the **Floor Plan: Level 1 – East Wing.**

1 / A106   1 / A105

The View Reference label updates to the correct sheet number.

34.

Note that the **A106 – Level 1 – East Wing** sheet view is also labeled 1.

35.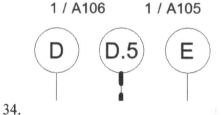

Activate the View ribbon.

Select the **View Reference** tool on the Sheet Composition panel.

36.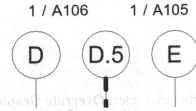

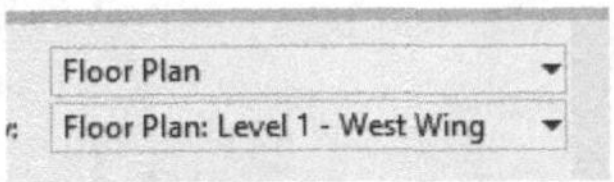

On the ribbon, you can select which view you want to reference.

Select the **Level 1 – West Wing** view.

37.

1 / A106   1 / A105

D   D.5   E

38. Save the file as *ex2-24.rvt*.

Exercise 2-25

Modifying a Matchline Appearance

Drawing Name: matchline_appearance.rvt
Estimated Time: 10 minutes

This exercise reinforces the following skills:

- Matchline
- Visibility/Graphics Overrides

1.    Open *matchline_appearance.rvt*.

2.   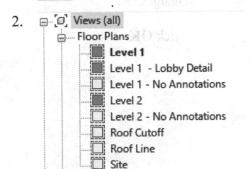   Activate **Level 1.**

3.　On the View ribbon:

Disable **Thin Lines.**

This allows you to see line weights of elements.

4.　Use the mouse to hover over Grid D.5 to locate the Matchline.

Matchline

Select the Matchline.

5.　Right click and select **Override Graphics in View→By Category**.

If you override the appearance by category, the appearance should change in all views where the element is visible. If you override the appearance by element, then the appearance only changes in the active view.

6.　

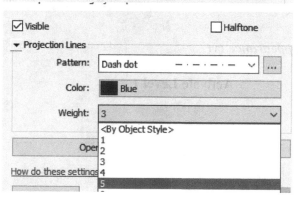

View Specific Category Graphics

Change the Pattern to **Dash dot**.
Change the Color to **Blue**.
Change the Weight to **3**.

Click **Apply** to preview the changes.

7.　Change the Weight to 5.

Click **Apply** to preview the changes.

Click **OK**.

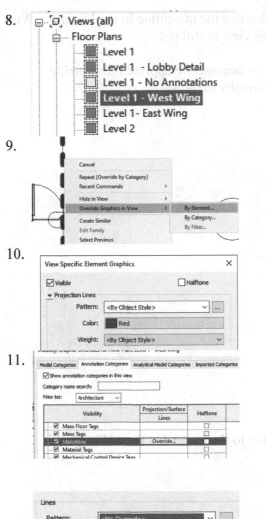

8. Switch to the **Level 1 – West Wing** floor plan view.

9. Select the Matchline.

Right click and select **Override Graphics in View→By Element**.

10. Change the Color to **Red**.

Click **OK**.

11. Type **VV** to launch the Visibility/Graphics dialog.

Select the **Annotation Categories** tab.

Locate the **Matchline** category.
Click **Override**.

Notice that the category shows no override.

12. Change the Pattern to **Dash dot**.
Change the Color to **Blue**.
Change the Weight to **3**.

Click **OK**.

Close the dialog.

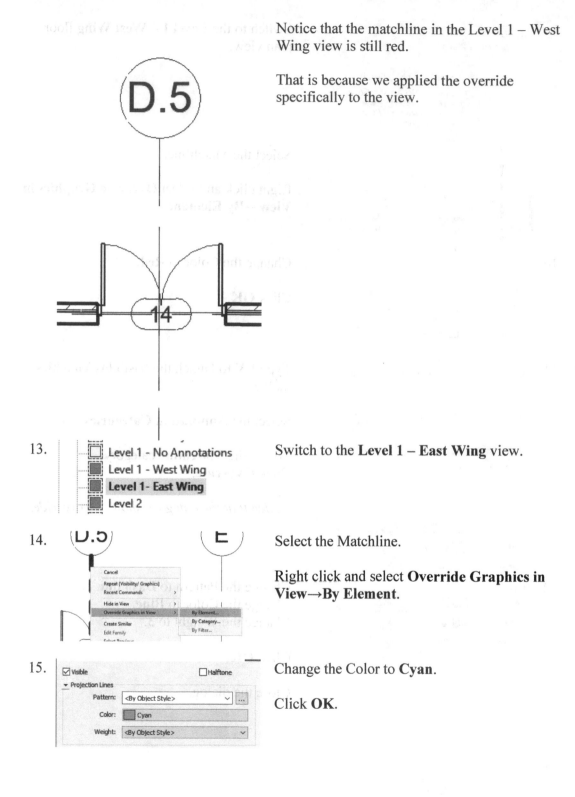

Notice that the matchline in the Level 1 – West Wing view is still red.

That is because we applied the override specifically to the view.

13. Switch to the **Level 1 – East Wing** view.

14. Select the Matchline.

Right click and select **Override Graphics in View→By Element**.

15. Change the Color to **Cyan**.

Click **OK**.

16. 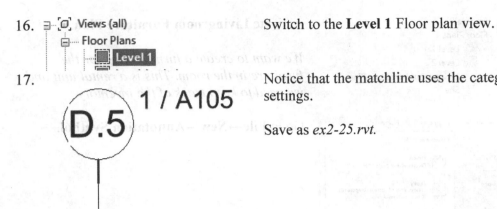 Switch to the **Level 1** Floor plan view.

17. Notice that the matchline uses the category settings.

Save as *ex2-25.rvt*.

Tags

Tags are used to identify elements, such as rooms, doors, and furniture. By tagging an element, you can use that data in schedules. When a tag is created, labels are added to display the value of desired element parameters. These labels display the values for the object's corresponding parameters after the tag is loaded and placed in the project.

Exercise 2-26

Create a Tag

Drawing Name: tags.rvt
Estimated Time: 15 minutes

This exercise reinforces the following skills:

- ❏ Tag
- ❏ Families
- ❏ Categories
- ❏ Labels

1.    Open *tags.rvt*.

2.

```
Views (all)
  Floor Plans
    Level 1
    Level 2
    Livingroom Furniture Plan
    Site
```

Activate **Livingroom Furniture Plan**.

We want to create a furniture tag for the furniture in the room. This is a rental unit and we need to keep track of the inventory.

3.

Go to **File→New→Annotation Symbol**.

4.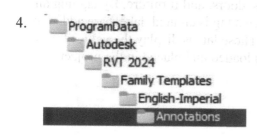

Browse to the *English-Imperial/Annotations* folder under *Family Templates*.

5.

File name:	Generic Tag.rft
Files of type:	Family Template Files (*.rft)

Select the **Generic Tag** template.

6.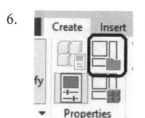

Select **Categories**.

This determines which elements the tag will recognize.

7.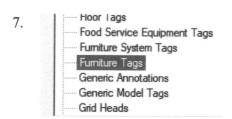

```
Floor Tags
Food Service Equipment Tags
Furniture System Tags
Furniture Tags
Generic Annotations
Generic Model Tags
Grid Heads
```

Highlight **Furniture Tags**.

Click **OK**.

8. Select **Label**.

Labels are associated with parameters.

Click above the intersection of the reference planes to place the label.

9. 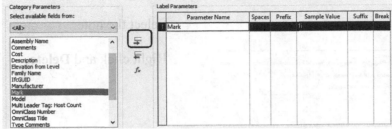 Highlight **Mark.**

Click **Add parameter to Label.**

Click **OK.**

10. Position the label above the tag intersection.

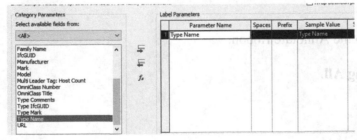

1i

11.  Select **Label**.

Click below the first label that was placed.

12. Highlight **Type Name.**

Click **Add parameter to Label.**

Click **OK.**

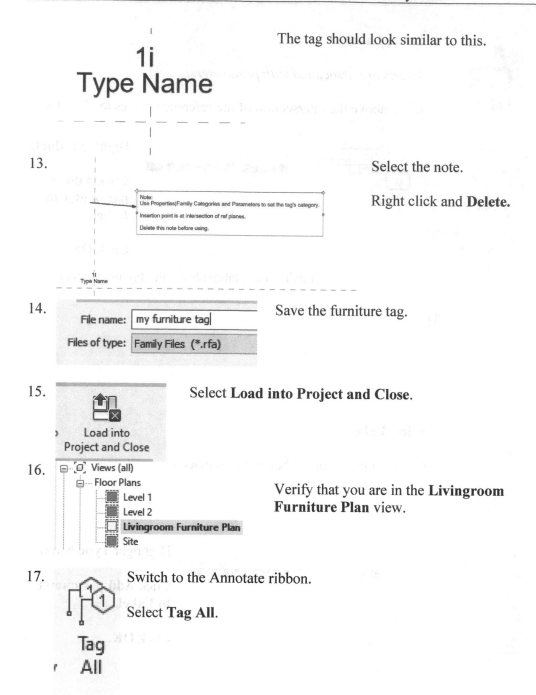

The tag should look similar to this.

1i
Type Name

13. Select the note.

Right click and **Delete**.

Note:
Use Properties|Family Categories and Parameters to set the tag's category.

Insertion point is at intersection of ref planes.

Delete this note before using.

1i
Type Name

14. Save the furniture tag.

File name: my furniture tag

Files of type: Family Files (*.rfa)

15. Select **Load into Project and Close**.

Load into
Project and Close

16. Verify that you are in the **Livingroom Furniture Plan** view.

Views (all)
   Floor Plans
      Level 1
      Level 2
      Livingroom Furniture Plan
      Site

17. Switch to the Annotate ribbon.

Select **Tag All**.

Tag
All

18.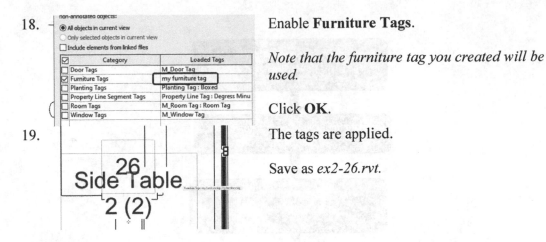

Enable **Furniture Tags**.

Note that the furniture tag you created will be used.

Click **OK**.

19.

The tags are applied.

Save as *ex2-26.rvt*.

Revisions

Revision tracking is the process of recording changes made to a building model after sheets have been issued.

When working on building projects, changes are often required to meet client or regulatory requirements.

Revisions need to be tracked for future reference. For example, you may want to check the revision history to identify when, why, and by whom a change was made. Revit provides tools that enable you to track revisions and include revision information on sheets in a construction document set.

Use the workflow below for entering and managing revisions. On the exam, you may be asked to organize this workflow or how revision clouds work.

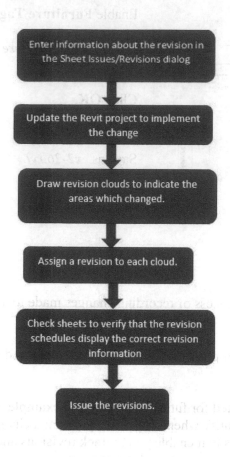

Exercise 2-27
Revision Control

Drawing Name: **i_Revisions.rvt**
Estimated Time to Completion: 15 Minutes

Scope
Add a sheet.
Add a view to a sheet.
Setting up Revision Control in a project.

Solution

1.  Activate the **View** ribbon. Select the **New Sheet** tool on the Sheet Composition panel.

   Sheet

2.  Load... Select the **Load** button.

3. 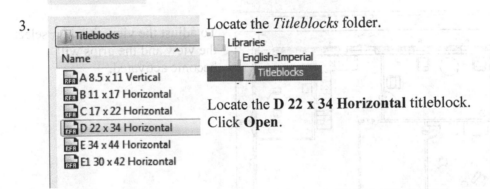 Locate the *Titleblocks* folder.

   Locate the **D 22 x 34 Horizontal** titleblock. Click **Open**.

4. 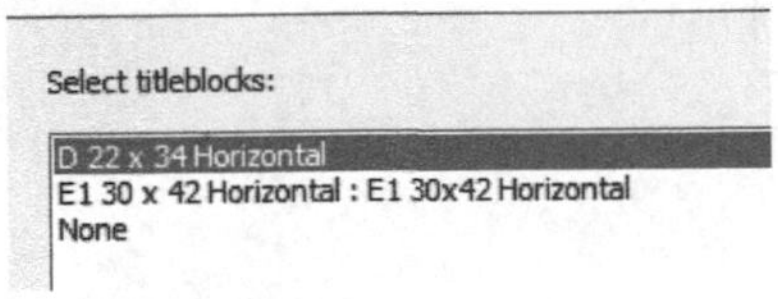 Highlight the **D 22 x 34 Horizontal** titleblock. Click **OK**.

   Select titleblocks:

   D 22 x 34 Horizontal
   E1 30 x 42 Horizontal : E1 30x42 Horizontal
   None

5.

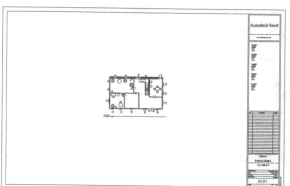

Drag and drop the **Level 1** floor plan onto the sheet.

6.

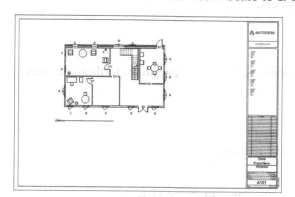

Select the view.
In the Properties pane,
set the View Scale to **1/4″ = 1′-0″**.

7.

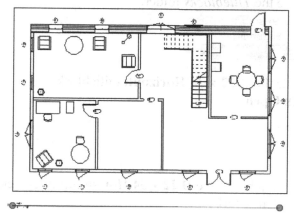

Position the view on the sheet.

8.

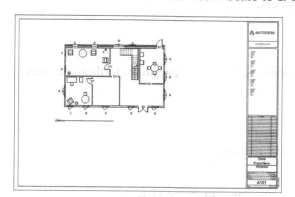

To adjust the view title bar, select the view and the grips will become activated.

9.

Activate the **View** ribbon.

Revisions (

> Enter information about the revision in the Sheet Issues/Revisions dialog

This is the first step in the Revisions workflow.

Select **Revisions** on the **Sheet Composition** panel.

10. This dialog manages revision control settings and history.

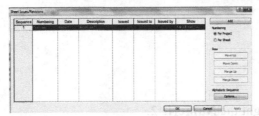

Numbering can be controlled per project or per sheet. The setting used depends on your company's standards.

Numbering
- ⦿ Per Project
- ◯ Per Sheet

Enable **Per Project**.

One Revision is available by default. Additional revisions are added using the **Add** button.

11. The visibility of revisions can be set to **None**, **Tag** or **Cloud and Tag**. Use the **None** setting for older revisions which are no longer applicable so as not to confuse the contractors.

Show
Cloud and Tag

Set the revision to show **Cloud and Tag**.

12. Click **Numbering**.

Customize Numbering

Numbering...

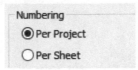

13.

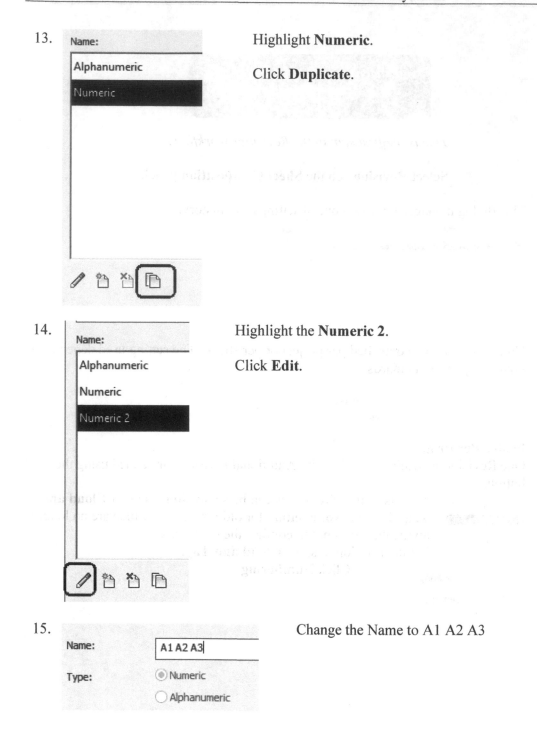

Highlight **Numeric**.

Click **Duplicate**.

14.

Highlight the **Numeric 2**.

Click **Edit**.

15.

Change the Name to A1 A2 A3

16.

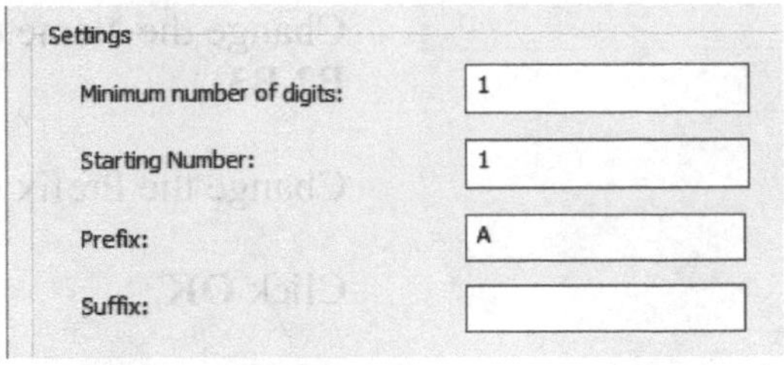

Settings	
Minimum number of digits:	1
Starting Number:	1
Prefix:	A
Suffix:	

Add the Prefix **A**.

Click **OK**.

17.

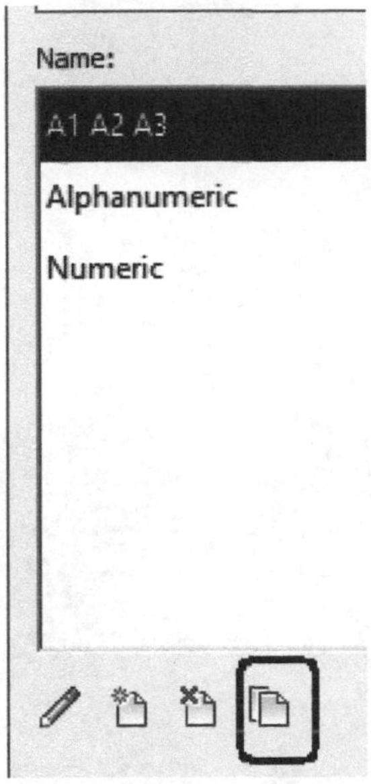

Name:

A1 A2 A3

Alphanumeric

Numeric

Highlight **A1 A2 A3**.

Click **Duplicate**.

18.

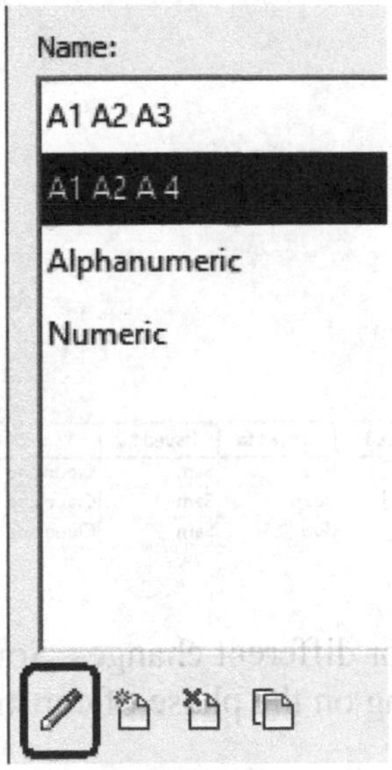

Name:

A1 A2 A3

A1 A2 A 4

Alphanumeric

Numeric

Highlight **A1 A2 A4**.

Click **Edit**.

19.

Name:	B1 B2 B3
Type:	○ Numeric
	○ Alphanumeric

Settings

Minimum number of digits:	1
Starting Number:	1
Prefix:	B
Suffix:	

Change the Name to **B1 B2 B3**.

Change the Prefix to **B**.

Click **OK**.

20.

Name:

A1 A2 A3

Alphanumeric

B1 B2 B3

Numeric

Click **OK**.

21. Enter the three revision changes shown.

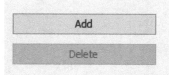

Add

Delete

Use the Add button to add the additional lines.

Sequence	Revision Number	Numbering	Date	Description	Issued	Issued to	Issued by	Show
1	A1	A1 A2 A3	08.02	Wall Finish Change	☐	Joe	Sam	Cloud and Tag
2	B1	B1 B2 B3	08.04	Window Style Change	☐	Joe	Sam	Cloud and Tag
3	A	Alphanumeric ⌄	08.13	Door Hardware Change	☐	Joe	Sam	Cloud and Tag

Notice you can select different Numbering schemes for different changes. Some companies use different numbering schemes depending on the phase of construction.

22. Click **OK** to close the dialog.
    You can delete revisions if you make a mistake. Just highlight the row and click
    Delete.

Sequence	Numbering	Date	Description	Issued	Issued to	Issued by	Show
1	Alphabetic	Date 1	Revision 1	☐			Cloud and Tag
2	Alphabetic	Date 2	Revision 2				Cloud and Tag

Add

Delete

23. Save the project as *ex2-27.rvt*.

Exercise 2-28

Modify a Revision Schedule

Drawing Name: **revision_schedule.rvt**
Estimated Time to Completion: 20 Minutes

Scope
Modify a revision schedule in a title block.

Solution

1. ⊟ 🖼 Sheets (all)
    ⊞ A100 - Cover Sheet
    ⊞ **A101 - Level 1 Floorplan**

    Open the **A101 – Level 1 Floorplan** sheet.

2.

    Zoom in to the **Revision Block** area on the sheet.
    Note that the title block includes a revision schedule
    by default.

3.  Select the title block.
Right click and select **Edit Family**.

4. Select the **Revision Schedule** in the Project Browser.

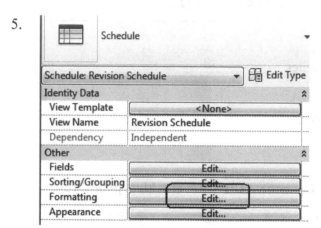

5. Select **Edit** next to Formatting in the Properties pane.

6. Change the First Column Header to **Rev**.

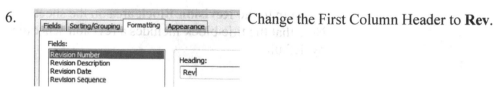

7. Select the Fields tab.
Add the **Issued By** field.

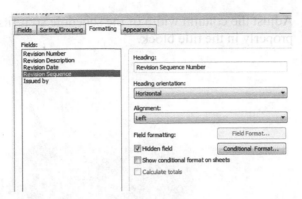

Do **NOT** remove the Revision Sequence field. This is a hidden field.

Note that the Hidden field control is located on the Formatting tab. This is a possible question on the certification exam.

8.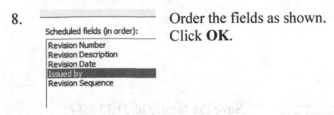

Order the fields as shown.
Click **OK**.

9.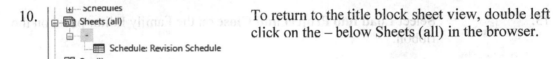

The Revision Schedule updates.

10.

To return to the title block sheet view, double left click on the – below Sheets (all) in the browser.

11.

Adjust the column width of the schedule so it fits properly in the title block.

12.

File name: D 22 x 34 Horizontal - Rev Schedule.rfa

Files of type: Family Files (*.rfa)

Save the family as *D 22 x 34 Horizontal – Rev Schedule.rfa*.

13.

Select **Load into Project and Close** on the Family Editor panel on the ribbon.

14. Click Cancel when you return to the sheet so you don't place a second titleblock.

15.

Properties

D 22 x 34 Horizontal

R

D 22 x 34 Horizontal

D 22 x 34 Horizontal

D 22 x 34 Horizontal - Rev Schedule

D 22 x 34 Horizontal - Rev Schedule

Select the existing titleblock.

Use the Type Selector to switch to the *D 22 x 34 Horizontal – Rev Schedule* title block.

Click in the window to release the selection.

16.

Note the title block updates with the new revision schedule format.

Save as *ex2-28.rvt*.

Legends

A Legend is used to display a list of various model components and annotations used in a project.

Legends fall into three main types: Component, Keynote and Symbol.

Component Legends display symbolic representations of model components. Examples of elements that would appear in a component legend are electrical fixtures, wall types, mechanical equipment and site elements.

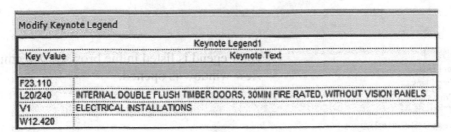

Keynote Legends are schedules which list the keynotes applied in a view on the sheet and what the different key values designate.

Symbol Legends are similar to keynote legends. They list the different annotation symbols used in a view and what they mean.

Legends can be placed on more than one sheet. If you use the same legends across projects, you can use Copy-Paste to copy legends from one project to another. Autodesk also offers a free add-on called Transfer Legend that allows you to copy/transfer legends from a linked Revit file to the host file.

Exercise 2-29

Create a Legend

Drawing Name: **i_Legends.rvt**
Estimated Time to Completion: 30 Minutes

Scope
Create a Legend.
Add to a sheet.

Solution

1.  Activate the **View** ribbon.
Select **Legend** from the Create panel.

2. Set the Name to **Door and Window Legend**.
Set the Scale to **¼″ = 1′-0″**.
Click **OK**.

Name:	Door and Window Legend
Scale:	1/4" = 1'-0"
Scale value 1:	48

3. The Legend is listed in the browser. An empty view window is opened.

- Views (all)
  - Floor Plans
  - Ceiling Plans
  - 3D Views
    - Southeast
  - Elevations (Elevation 1)
  - Legends
    - Door and Window Legend

4.
```
⊞── Detail Items
⊟── Doors
    ⊞── Cased Opening
    ⊞── double glass
    ⊞── Sgl Flush
    ⊞── single flush
```
In the browser, locate all the door families.

5.
```
⊞── Detail Items
⊟── Doors
    ⊟── Cased Opening
        └── 36" x 84"
    ⊟── double glass
        └── 60"x80"
    ⊟── Sgl Flush
        ├── 30" x 80"
        ├── 30" x 84"
        ├── 32" x 84"
        ├── 34" x 80"
        ├── 34" x 84"
        ├── 36" x 80"
        └── 36" x 84"
    ⊟── single flush
        ├── 34" x 80"
        └── 36" x 84"
⊞── Duct Systems
```
Expand each door family to see which types are available.

6.
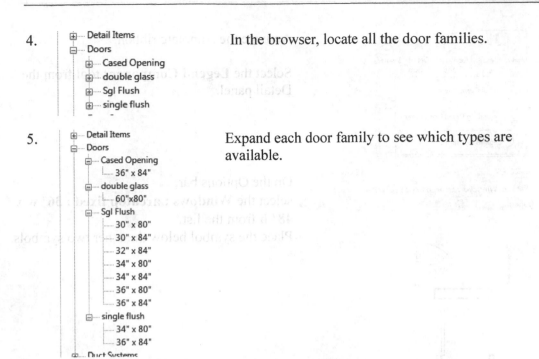
Highlight the **60″ x 80″ Double Glass** door. Drag and drop it into the Legend view.

Right click and select **Cancel** to exit the command.

7.
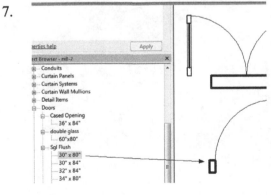
Highlight the **30″ x 80″ Sgl Flush** door.

Drag and drop it into the Legend view.

Right click and select **Cancel** to exit the command.

8.

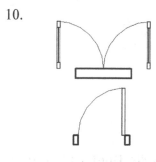

Activate the Annotate ribbon.

Select the **Legend Component** tool from the Detail panel.

9.

Family: Windows : archtop fixed : 36"w x 48"h

Properties

On the Options bar,
select the **Windows : archtop fixed : 36″ w x 48″ h** from the list.

10.

Place the symbol below the other two symbols.

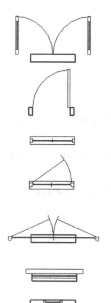

11.

Drag and drop each window symbol into the Legend view. *Symbols can be dragged and dropped from the Project Browser or using the Legend Component tool.*

12. Activate the **Annotate** ribbon.
Select the **Text** tool.

13. DOUBLE GLASS DOOR   Add text next to each symbol.

SINGLE FLUSH DOOR

14. Legend Components : Legend Component - Windows :   Mouse over a symbol to see what family
archtop fixed : 36"w x 48"h (Floor Plan)   type it is, if needed.

15. DOUBLE GLASS DOOR   Add text as shown.

 SINGLE FLUSH DOOR

 ARCHTOP FIXED

CASEMENT

 DOUBLE CASEMENT WITH TRIM

 DOUBLE HUNG WITH TRIM

FIXED

16. Select the **Detail Line** tool from the Detail panel on the Annotate ribbon.

17.

SYMBOL	DESCRIPTION
	DOUBLE GLASS DOOR
	SINGLE FLUSH DOOR
	ARCHTOP FIXED
	CASEMENT
	DOUBLE CASEMENT WITH TRIM
	DOUBLE HUNG WITH TRIM
	FIXED

Add a rectangle.
Add a vertical and horizontal line as shown.
Add the header text.

18. 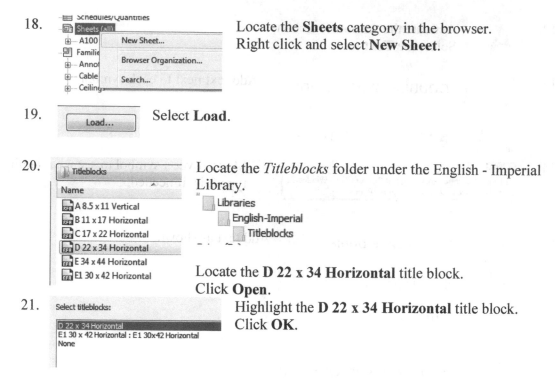 Locate the **Sheets** category in the browser.
Right click and select **New Sheet**.

19. Select **Load**.

20. Locate the *Titleblocks* folder under the English - Imperial Library.

Libraries
English-Imperial
Titleblocks

Locate the **D 22 x 34 Horizontal** title block.
Click **Open**.

21. Highlight the **D 22 x 34 Horizontal** title block.
Click **OK**.

22. Drag and drop the Level 1 floor plan onto the sheet.

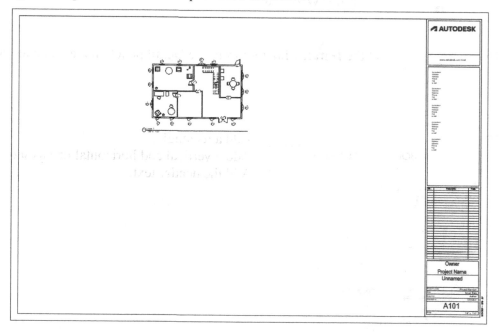

23.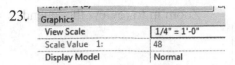

Graphics	
View Scale	1/4" = 1'-0"
Scale Value 1:	48
Display Model	Normal

In the Properties pane, change the View Scale to ¼" = 1'-0".

24. Drag and drop the Door and Window Legend onto the sheet. Place next to the floor plan view.

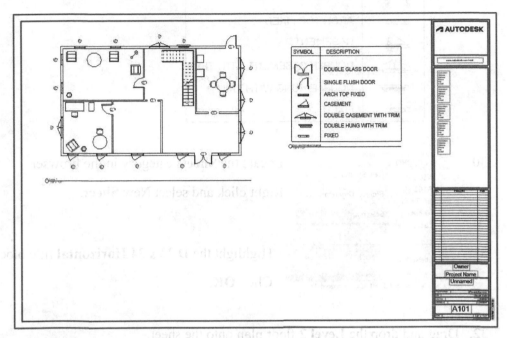

25. Select the legend so it highlights.

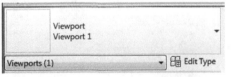

Select **Edit Type** on the Properties pane.

26. [Duplicate...]   Select **Duplicate**.

27. Name: [Viewport with no Title]   Type **Viewport with no Title**. Click **OK**.

28.

Type Parameters

Parameter	
Graphics	
Title	view title
Show Title	No
Show Extension Line	☐
Line Weight	1
Color	■ Black
Line Pattern	Solid

Uncheck **Show Extension Line**.
Set Show Title to **No**.
Click **OK**.

29. Zoom in to review the legend.

SYMBOL
DESCRIPTION
DOUBLE GLASS DOOR
SINGLE FLUSH DOOR
ARCH TOP FIXED
CASEMENT
DOUBLE CASEMENT WITH TRIM
DOUBLE HUNG WITH TRIM
FIXED

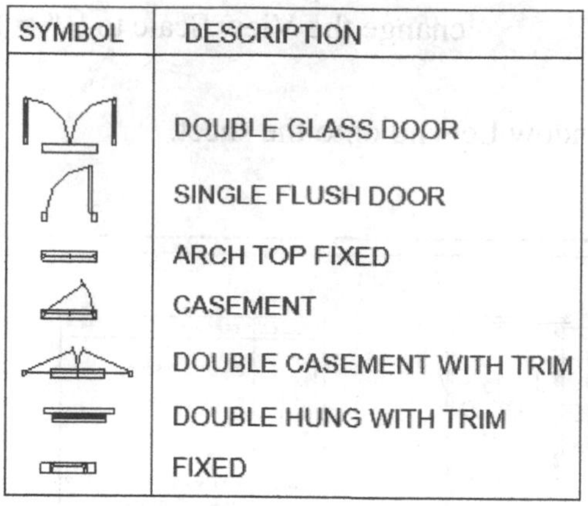

30. Locate the **Sheets** category in the browser.

Right click and select **New Sheet**.

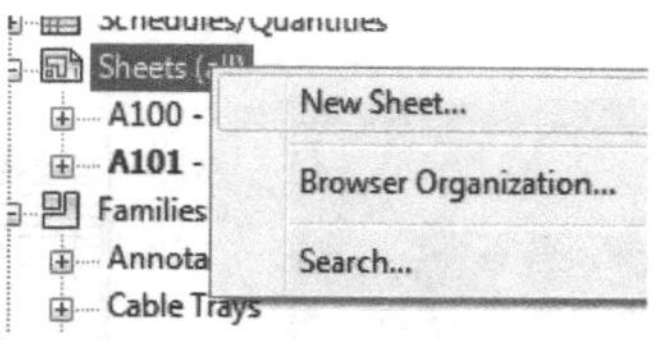

Sheets
A100 -
A101 -
Families
Annota
Cable Trays
New Sheet...
Browser Organization...
Search...

31. Highlight the **D 22 x 34 Horizontal** title block.

Click **OK**.

Select titleblocks:
D 22 x 34 Horizontal
E1 30 x 42 Horizontal : E1 30x42 Horizontal
None

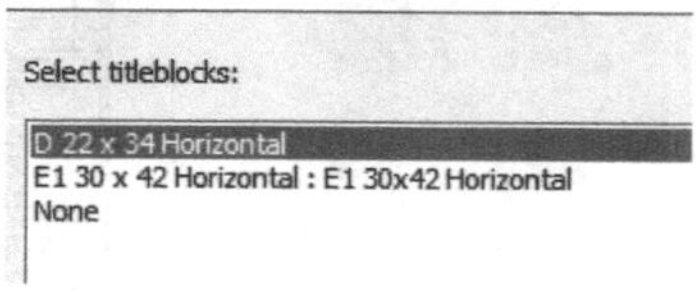

32. Drag and drop the **Level 2** floor plan onto the sheet.

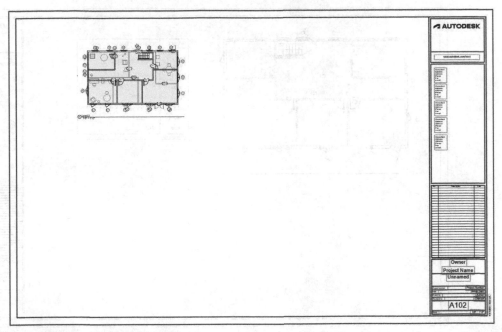

33.

Viewports (1)	
Graphics	
View Scale	1/4" = 1'-0"
Scale Value 1:	48
Display Model	Normal

34. Drag and drop the Door and Window Legend onto the sheet.
Place next to the view.

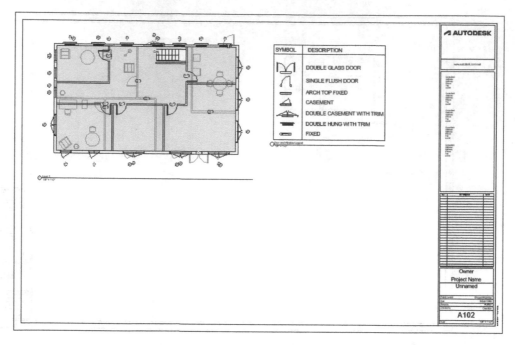

35. Select the legend view and then use the Type Selector in the Properties pane to set whether the view should have a title or not.

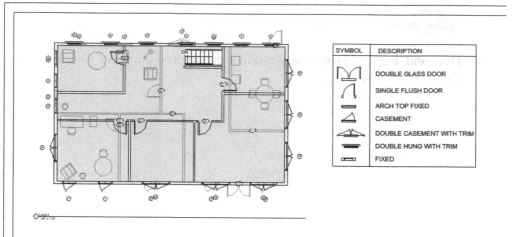

Note that legends can be placed on more than one sheet.

36. Save as *ex2-29.rvt*.

Exercise 2-30

Import a Legend

Drawing Name: **export_legend.rvt, import_legend.rvt**
Estimated Time to Completion: 10 Minutes

Scope

Copy and paste a legend between project files.

Solution

1. Detail Views (Detail)
   Renderings
   Legends
   Schedules/Quantities (all)
   Furniture Schedule

   Open the *import_legend.rvt* file.

   Check in the Project Browser to see if there are any legends.

2. Sheets (all)
   A001 - Title Sheet
   A101 - Site Plan
   A102 - Plans

   Open the **A102 – Plans** sheet.

3. Sheets (all)
   A100 - Cover Sheet
   A101 - Level 1 Floor Plan
   A102 - Level 2 Floor Plan
   Families

   Open the *export_legend.rvt* file.

   Open the **A101 – Level 1 Floor Plan** sheet.

4. Select the legend view.

   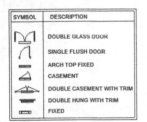

   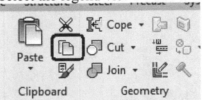

   Select Copy to Clipboard from the ribbon.

5. A001 - Title Sheet   A102 - Plans ✕

   Use the tabs to switch to the *import_legend.rvt* file.

6. Sheets (all)
   A001 - Title Sheet
   A101 - Site Plan
   A102 - Plans

   Open the **A102 – Plans** sheet.

7. 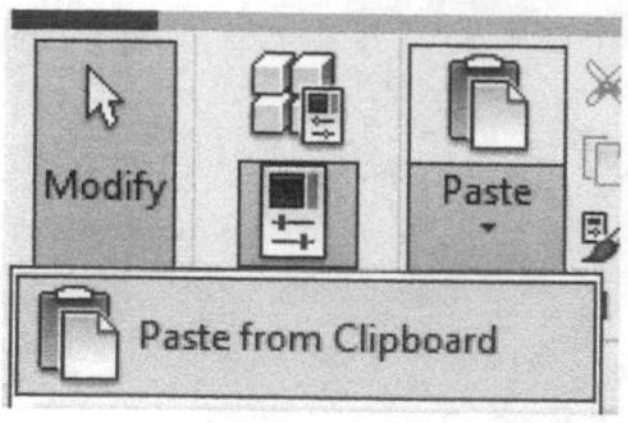 Switch to the Modify ribbon.

    Click **Paste from Clipboard.**

8. 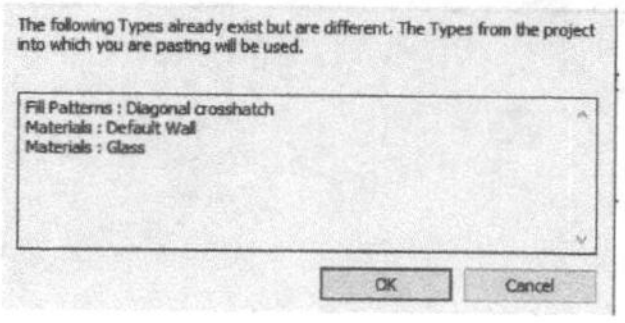 Click **OK**.

9. 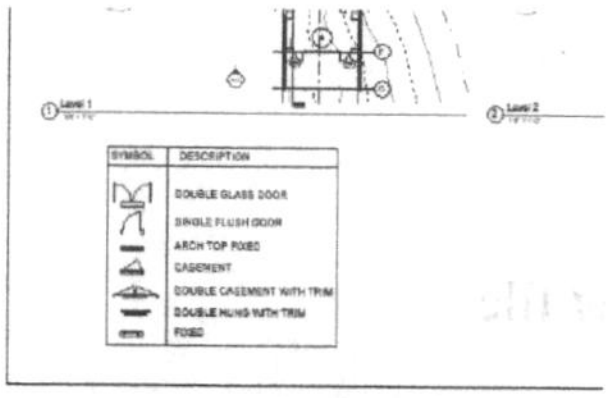 Place the legend view on the sheet.

10. 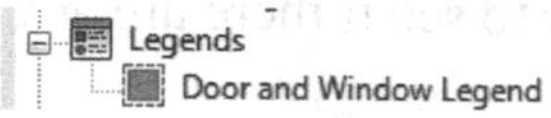 Notice that the legend is now available in the Project Browser as well.

11. Save as *ex2-30.rvt*.

Exercise 2-31

Create a Keynote Legend

Drawing Name: keynote legend.rvt [m_ keynote legend.rvt]
Estimated Time: 30 minutes

This exercise reinforces the following skills:

- ❑ Keynote Legend

1.   Open *keynote legend.rvt [m_ keynote legend.rvt]*.

2.
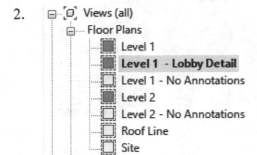
Activate the **Level 1 - Lobby Detail** view.

3.
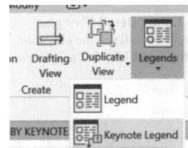
Activate the View ribbon.

Select **Legends→Keynote Legend**.

4.
Name:
FINISH SCHEDULE KEYS

OK   Cancel

Type **FINISH SCHEDULE KEYS**.

Click **OK**.

5.   Click **OK** to accept the default legend created.

6.   Legends
        FINISH SCHEDULE KEYS   The legend appears in the Project Browser.

7.

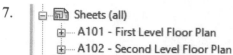

Open the **Lobby Keynotes** sheet.

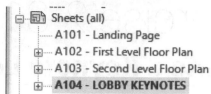

8.   Drag and drop the FINISH SCHEDULE KEYS onto the sheet.

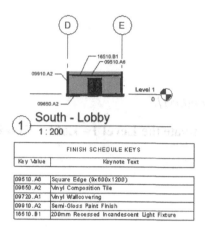

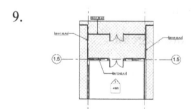

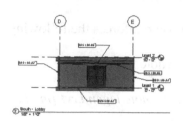

9.

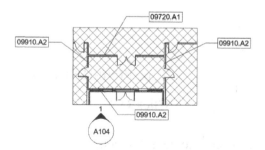

Activate the **Level 1 – Lobby Detail** floor plan. Zoom in to the view and you will see the elevation marker now indicates the Sheet Number and View Number.

10.

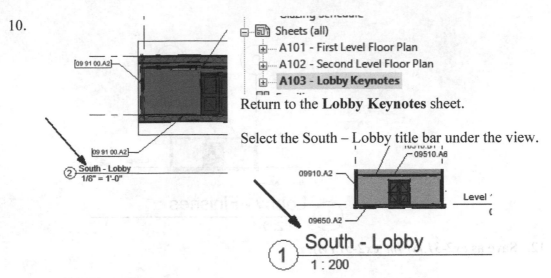

Return to the **Lobby Keynotes** sheet.

Select the South – Lobby title bar under the view.

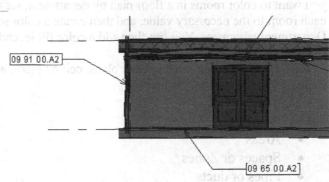

11.

Associated Datum	None
Identity Data	
View Template	<None>
View Name	South - Lobby
Dependency	Independent
Title on Sheet	Lobby - Finishes
Sheet Number	A603
Sheet Name	Lobby Keynotes
Referencing Sheet	A603
	1

On the Properties pane:
Change the Title on the Sheet to **Lobby – Finishes**.

The view title updates.

Notice that the view name remains the same.

2-141

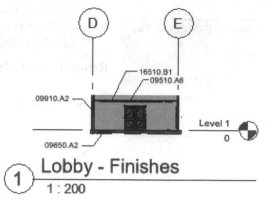

12. Save as *ex2-31.rvt [m_ex2-31.rvt]*.

Color Schemes

Colors schemes are useful for graphically illustrating categories of spaces, rooms or areas.

For example, you can create a color scheme by room name, area, occupancy, or department. If you want to color rooms in a floor plan by department, set the Department parameter value for each room to the necessary value, and then create a color scheme based on the values of the Department parameter. You can then add a color fill legend to identify the department that each color represents.
Color schemse can be placed in floor plans, ceiling plans, and sections.
Color schemes can be applied to:

- Rooms
- Areas
- Spaces or Zones
- Pipes or ducts

The display of the color legend corresponds to a type which can be customized. In a view, only the rooms visible in that view will be displayed. However, the color scheme includes all rooms in the project. If you want the color legend to only display the rooms shown in the view, use the properties option to display using By View.

Exercise 2-32
Color Schemes

Drawing Name: **i_Color_Scheme.rvt**
Estimated Time to Completion: 15 Minutes

Scope
Create an area plan using a color scheme.
Place a color scheme legend.
Modify the appearance of a color scheme legend.

Solution

1. Views (all)
    Floor Plans
    01 - Entry Level
    01 - Entry Level-Rooms
    02 - Floor
    03 - Floor

    Activate the **01- Entry Level** floor plan.

2. | Color Scheme Location | Background |
    | Color Scheme | \<none\> |
    | System Color Schemes | Edit... |
    | Default Analysis Display Style | None |

    In the Properties pane, scroll down to the **Color Scheme** field.
    Click **\<none\>**.

3. Schemes
    Category:
    Rooms
    (none)
    Name
    Department

    Select the **Rooms** category.
    Highlight **Name**.

4. Scheme Definition
    Title: Area Legend
    Color: Area

    In the Title field, enter **Area Legend**.
    Under Color, select the **Area scheme**.

5. Colors are not preserved when changing which parameter is colored. To color by a different parameter, consider making a new color scheme.

    Click **OK**.

6.

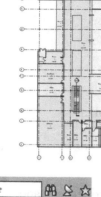

Note that different colors are applied depending on the square footage of the rooms.

Click OK.

7.

The rooms fill in according to the square footage.

8.

Activate the Annotate ribbon.
Select the **Color Fill Legend** tool on the Color Fill panel.

9.   Place the legend in the view.

10.    Select the legend that was placed in the display window. Select **Edit Scheme** from the Scheme panel on the tab on ribbon.

11. Change the Fill Pattern for 47 SF **to Crosshatch-Small**.

	Value	Visible	Color	Fill Pattern	Preview	
1	47 SF	☑	RGB 156-185	Crosshatch-small ▾	▦	
2	58 SF	☑	PANTONE 3	Solid fill		
3	64 SF	☑	PANTONE 6	Solid fill		

12. Change the hatch patterns for the next few rows. Click **OK**.

	Value	Visible	Color	Fill Pattern	Preview	
1	47 SF	☑	RGB 156-185	Crosshatch-small	▦	
2	58 SF	☑	PANTONE 3	Crosshatch		
3	64 SF	☑	PANTONE 6	Diagonal crosshatch		
4	76 SF	☑	RGB 139-166	Diagonal down ▾		
5	90 SF	☑	PANTONE 6	Solid fill		
6	94 SF	☑	RGB 096-175	Solid fill		

13.

Note that the legend updates.

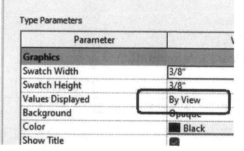

Area Legend

☐ 47 SF

☐ 58 SF

☐ 64 SF

☐ 76 SF

14. Select the Color Fill Legend. On the Properties pane, select **Edit Type**.

15. Note that the Values Displayed are **By View**.

Type Parameters	
Parameter	
Graphics	
Swatch Width	3/8"
Swatch Height	3/8"
Values Displayed	By View
Background	Opaque
Color	■ Black
Show Title	☑

16. Change the font for the Text to **Tahoma**. Enable **Bold**.

Text	
Font	Tahoma
Size	3/16"
Bold	☑
Italic	☐
Underline	☐

17. Change the font for the Title Text to **Tahoma**. Enable **Bold**. Enable **Italic**. Click **OK**.

Title Text	
Font	Tahoma
Size	1/4"
Bold	☑
Italic	☑
Underline	☐

18. Note how the legend updates.

Area Legend
☐ 47 SF
☐ 58 SF
☐ 64 SF

19. Close without saving.

Exercise 2-33

Color Scheme By Department

Drawing Name: **i_Color_Scheme_2.rvt**
Estimated Time to Completion: 10 Minutes

Scope
Create a floor plan using a color scheme.
Place a color scheme legend using the Department parameter.

1. 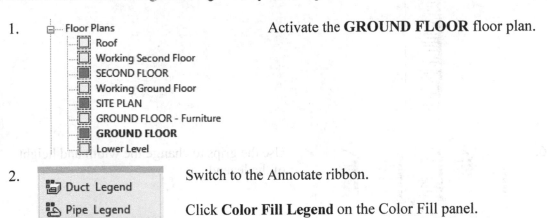 Activate the **GROUND FLOOR** floor plan.

2. Switch to the Annotate ribbon.

   Click **Color Fill Legend** on the Color Fill panel.

3. Click to place the legend below the view.

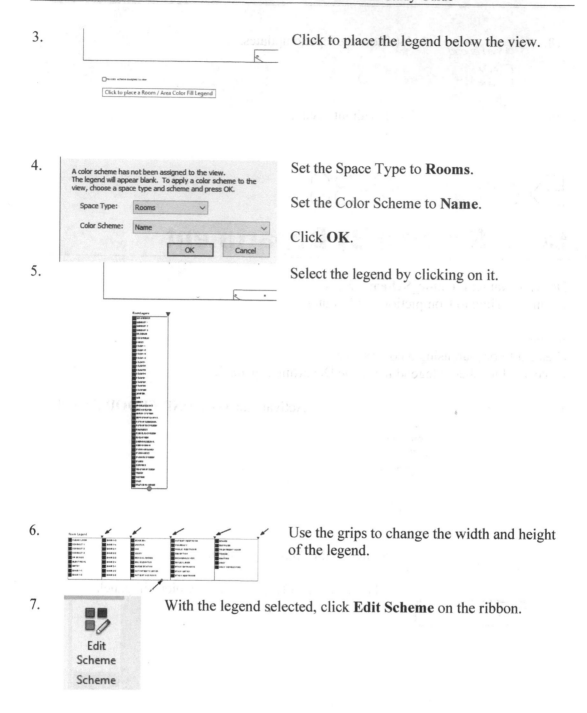

4. Set the Space Type to **Rooms**.

Set the Color Scheme to **Name**.

Click **OK**.

5. Select the legend by clicking on it.

6. Use the grips to change the width and height of the legend.

7. With the legend selected, click **Edit Scheme** on the ribbon.

Edit
Scheme

Scheme

8.

Schemes
Category:
Rooms
(none)
Name
Department

Highlight **Department**.

Click **OK**.

9.

Department Lege

☐ Circulation

☐ Core

☐ Exam

☐ Nursing

The legend updates.

The color fills changes on the floor plan.

10.

☐ Schedules/Quantities (all)
　☐ Area Schedule (Gross Bu
　☐ Area Schedule (Rentable
　■ DOOR SCHEDULE
　■ **ROOM SCHEDULE**

Open the **ROOM SCHEDULE** view.

11.

J	K
n Style	Department
	Circulation
	Circulation
	Core
	Circulation
	Core
	Core
	Core
	Circulation
	Nursing
	Core
	Circulation
	Core
	Nursing

The last column in the schedule is Department.

This is what controls how the color fill schedule is organized and how colors are applied.

Department is a default parameter for Rooms. You do not need to create it, but you do need to fill in the values you want to use in your schedule.

12. Save as *ex2-33.rvt*.

Drafting Views

The following is a sample drafting view created using the 2D detailing tools in Revit Architecture. This is not a 3D view.

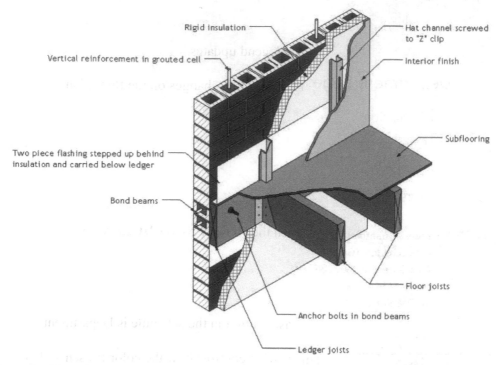

Use a drafting view to create unassociated, view-specific details that are not part of the modeled design.

Rather than create a callout and add details to it, you may want to create detail conditions where the model is not needed (for example, a carpet-transition detail which shows where carpet switches to tile, or roof-drain details not based on a callout on the roof). For this purpose, create a drafting view.

In a drafting view, you create details at differing view scales (coarse, medium, or fine) and use 2D detailing tools: detail lines, detail regions, detail components, insulation, reference planes, dimensions, symbols, and text. These are the exact same tools used in creating a detail view. However, drafting views do not display any model elements.

When you create a drafting view in a project, it is saved with the project.

When using drafting views, consider the following:

- Similar to other views, drafting views are listed in the Project Browser under Drafting Views.
- All of the detailing tools used in detail views are available to you in drafting views.
- Any callouts placed in a drafting view must be reference callouts.
- Although not associated with the model, you can still drag the drafting views from the browser onto a drawing sheet.
- Name drafting views so they are organized neatly in the Project Browser

Many companies will create a library of drafting views that are used across projects. For example, threshold details for doors or sill details for windows. These details provide valuable construction information for the subcontractor.

The following are some examples of drafting view type naming:
- 00-Accessibility-ADA
- 01-General Requirements
- 02-Site
- 03-Concrete
- 04-Masonry
- 05-Metals
- 06-Walls
- 07-Roof
- 08-Door & Windows
- 09-Finishes
- 10-Specialties
- 11-Misc
- 12-Casework
- 13-Stairs Ramps Elevators
- 14-Mechanical
- 15-Plumbing
- 16-Electrical

Filled Regions

The Filled Region tool creates a 2-dimensional, view-specific graphic with a boundary line style and fill pattern within the closed boundary. The tool is useful for defining a filled area in a detail view or for adding a filled region to an annotation family.

The filled region is parallel to the view's sketch plane.
A filled region contains a fill pattern. Fill patterns are of 2 types: Drafting or Model.
Drafting fill patterns are based on the scale of the view. Model fill patterns are based on
the actual dimensions in the building model.

A filled region created for a detail view is part of the Detail Items category. Revit lists the
region in the Project Browser under Families ➤ Detail Items ➤ Filled Region. If you
create a filled region as part of an annotation family, Revit identifies it as a Filled Region,
but does not store it in the Project Browser.

Exercise 2-34

Create a Drafting View

Drawing Name: **drafting_view.rvt**
Estimated Time to Completion: 5 Minutes

Scope
Create a drafting view.
Load detail items.
Add text.

Solution

1.   Activate the **View ribbon.**

    Click **Drafting View.**

2.   Type **06- Interior Wall GWB** in the Name field.
    Set the Scale to **1:25.**

    Click **OK.**

3.   Note that the Drafting View category is added to
    the Project Browser and the view is listed under
    Coordination.

    The view is blank. *Remember a drafting view is a*

2D view that is independent of the model.

4.
Detail Level
Medium
Visibility/Graphics Overrides
Edit...
Discipline
Architectural
Visual Style
Hidden Line

Change the Discipline to **Architectural** in the Properties palette.

Drafting Views (Detail)
Architectural
06- Interior Wall GWB

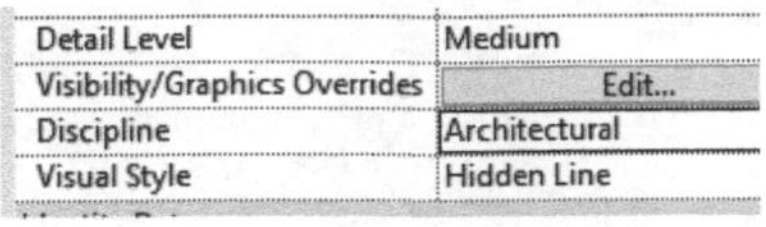

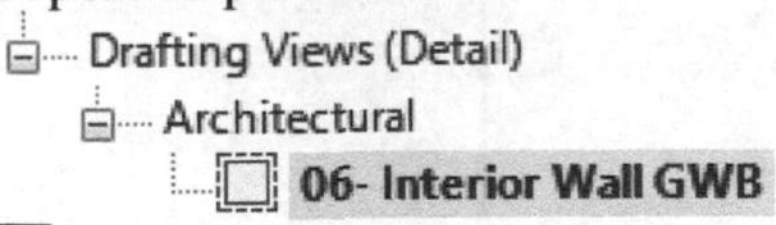

The Project Browser will update.

5.
Annotate
Analyze
Detail
Line
Region
Filled Region

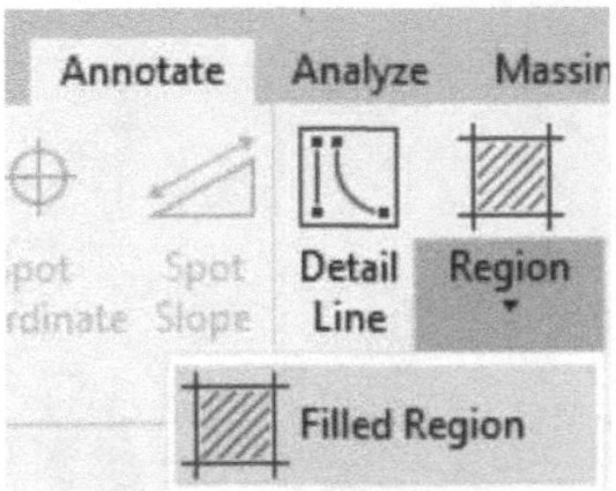

Switch to the **Annotate** ribbon.

Select the **Filled Region** tool.

6.
Filled region
Insulation

Select the **Filled Region: Insulation** type from the Properties palette.

7.
Draw

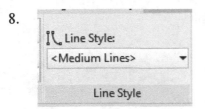

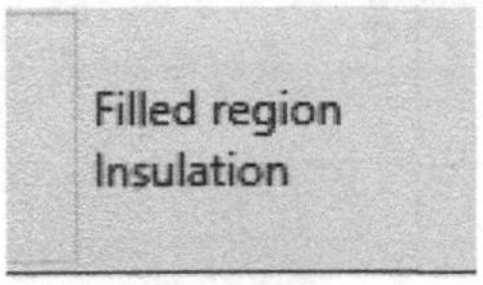

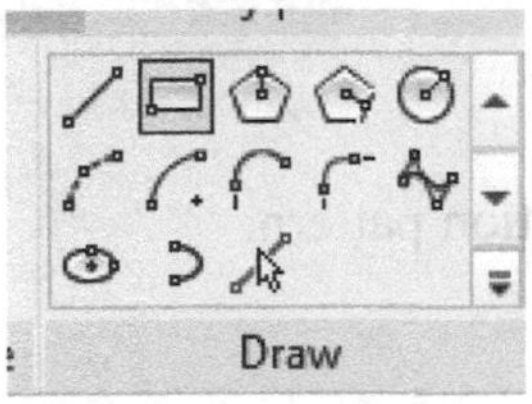

Select the **Rectangle** tool from the Draw panel.

8.
Line Style:
<Medium Lines>
Line Style

Set the Line Style to **Medium Lines.**

9.   Draw a 0.6 x 2100 rectangle.

10.    Click the Green Check to complete the rectangle.

11.   A rectangle is placed with the insulation pattern.

This is a custom insulation hatch pattern. If you would like to use it in future projects, you can use Transfer Project Standards and select Filled Region Types and Fill Patterns.

12.  Select **Detail Component**.

13. Properties

   M_Rough Cut Lumber-Side 25x100mm

   Select **M_Rough Cut Lumber – Side 25 x 100 mm** using the Type Selector.

14. 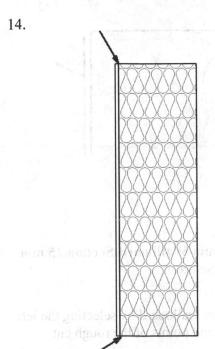 Draw the rough lumber by selecting the left bottom and top end points of the filled region.

15. Select the rough lumber detail item.

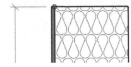

16. 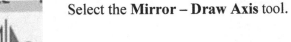 Select the **Mirror – Draw Axis** tool.

17.

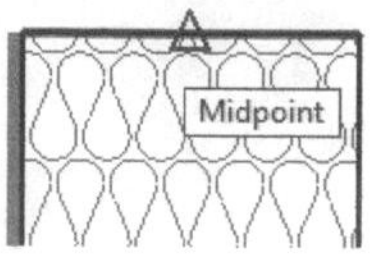

Select the Midpoint of the top of the filled region to start the axis.

18.

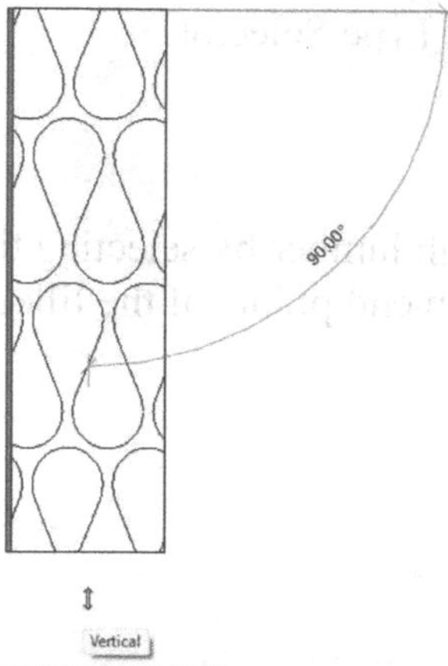

Draw the axis straight down the middle of the filled region.

There are now pieces of rough cut lumber on both sides of the insulation.

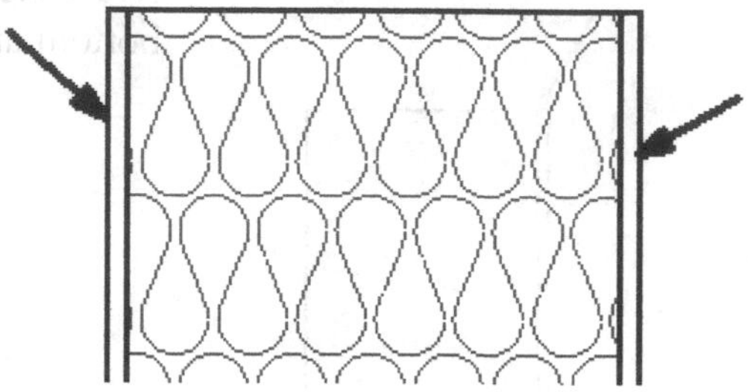

19.

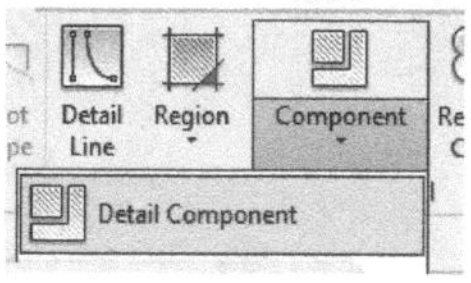

Select **Detail Component**.

20.

M_Gypsum Wallboard-Section 25mm

Select **M_Gypsum Wallboard-Section 25 mm** using the Type Selector.

21.

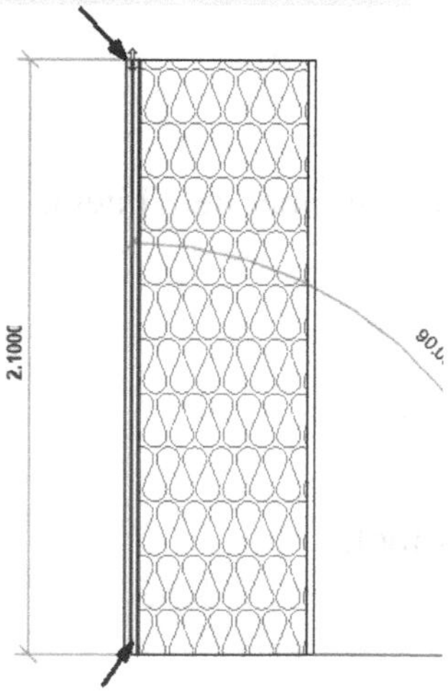

Place the gypsum wallboard by selecting the left bottom and top end points of the rough cut lumber.

22. Select the gypsum wallboard.

23. 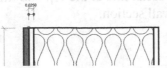 Select the **Mirror – Draw Axis** tool.

24. 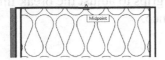 Select the Midpoint of the top of the filled region to start the axis.

25. 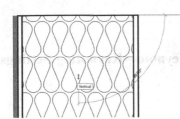 Draw the axis straight down the middle of the filled region.

There are now sections of gypsum wallboard on both sides of the insulation.

26.  Select **Detail Component**.

27. 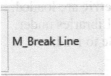 Select **M_Break Line** using the Type Selector.

28.

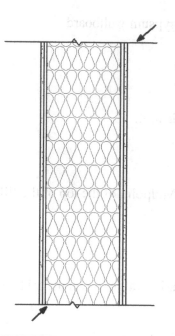

Place a break line at the top and the bottom of the wall section.

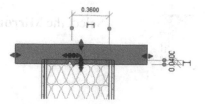

Use the grips to adjust the break line appearance.

29.

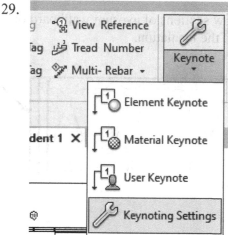

Go to **Keynote→Keynoting Settings**.

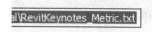

Verify that the **RevitKeynotes_Metric** are loaded. If they are not, then browse to the Libraries under RVT2024 and locate the correct file to load.

30. 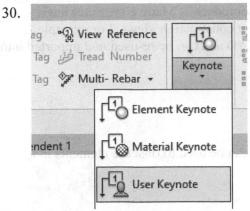 Use the **User Keynote** tool to add keynotes to the view.

31. 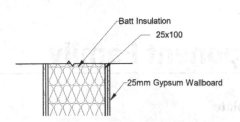 Add the appropriate keynotes to the drafting view.

Save as *ex2-34.rvt*.

Detail Components

Detail components are line-based 2D elements that you can add to detail views or drafting views. They are visible only in those views. They scale with the model, rather than the sheet.

Detail components are not associated with the model elements that are part of the building model. Instead, they provide construction details or other information in a specific view.

For example, in the following drafting view, the studs, insulation, and siding are detail components.

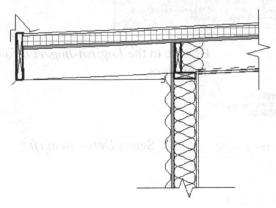

Revit comes with an extensive library of detail components. Many companies have AutoCAD legacy data of detail views for different construction views. For example, threshold and jamb details were drawn in AutoCAD and can be re-used and imported into Revit, so you don't have to recreate them.

Masking Regions

Masking regions are view-specific graphics that can be used to obscure elements in a view.

Exercise 2-35
Create a Detail Component Family

Drawing Name: **new using detail item template**
Estimated Time to Completion: 20 Minutes

Scope
Create a detail component family.

Solution

1.  From the Application menu:

    Select **File→New→Family**.

2. ProgramData
       Autodesk
          RVT 2024
             Family Templates
                English-Imperial

    Browse to the *English-Imperial* folder under Family Templates.

3.  Select *Detail Item.rft*.

    Detail Item line based.rft
    Detail Item.rft
    Division Profile.rft
    Door - Curtain Wall.rft

4. Select the **Reference Plane** tool.

: Reference
Plane

5. Draw two horizontal reference planes and one vertical reference plane.

6. Select the **ALIGNED DIMENSION**.

Aligned A

7. Add dimensions between the reference planes.

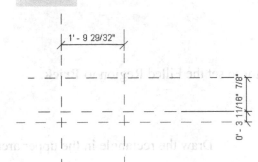

1' - 9 29/32"

0' - 3 11/16" 7/8"

8.

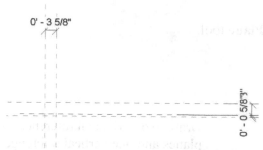

Change the dimensions so that the brick is 3.625" x 3.625" and the space is 5/8" for grout.

9.

Select the **Filled Region** tool on the Create ribbon.

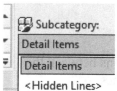

g Filled
ι Region

10.

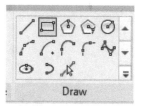

Select the **Rectangle** tool from the Draw panel.

11.

Subcategory:
Detail Items
Detail Items
<Hidden Lines>

Verify that the Subcategory is set to **Detail Items**.

12.

Filled region
Brick

Use the Type Selector to set the Filled Region to **Brick**.

13.

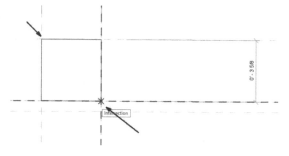

Draw the rectangle in the upper area.

14. 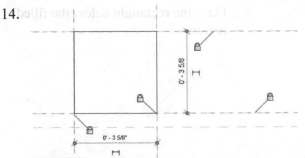 Enable all the locks to constrain the filled region to the reference planes.

15. Green check to complete the filled region.

16. 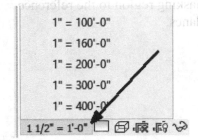 Change the scale to 1 ½" = 1'-0" to see the filled region hatch pattern.

17. 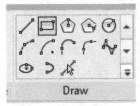 Select the **Masking Region** tool from the Create ribbon.

18. Select the **Rectangle** tool from the Draw panel.

19. 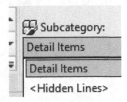 Verify that the Subcategory is set to **Detail Items**.

20.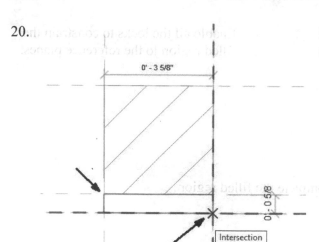

Place the rectangle below the filled region.

21.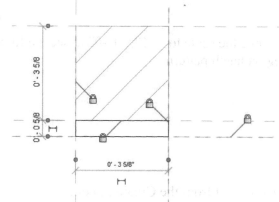

Enable the locks to constrain the masking region to the reference planes.

22.

Green check to complete the masked region.

23. 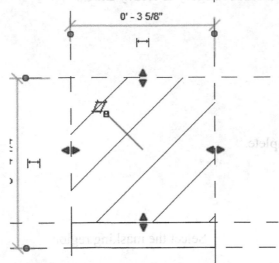 Select the filled region.

24.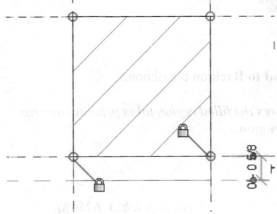

Select **Edit Boundary**.

25. Hover the mouse over one of the outer lines.

Click TAB to select all four lines.

26. Change the Subcategory to Heavy **Lines**.

27. Green check to complete.

28. 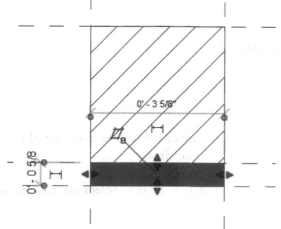 Select the masking region.

29. 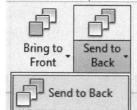 Click **Send to Back** on the ribbon.

This ensures the filled region takes priority over the masking region.

30.  Save as *brick_3_625.rfa*.

Exercise 2-36

Place a Repeating Detail Component Family

Drawing Name: **repeating detail.rvt**
Estimated Time to Completion: 20 Minutes

Scope
Add a repeating detail to a view.

Solution

1. ⊟— Sections (Detail)
    ▢ Callout of Section 1
    ▢ Callout of Section 2
    ▢ **Repeating Detail**

    Open the **Repeating Detail** section view.

2.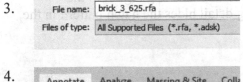

    Detail components need to be loaded BEFORE you select the tools to place them.

    Switch to the Insert ribbon.

    Select **Load Family.**

3. File name: brick_3_625.rfa
    Files of type: All Supported Files (*.rfa, *.adsk)

    Select the *brick_3_625.rfa* file.

    Click **Open**.

4.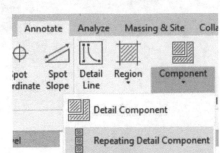

    Switch to the Annotate ribbon.

    Select **Repeating Detail Component**.

5.

The Repeating Detail has one type called **Brick**.

Click **Edit Type.**

6.

Click **Duplicate.**

7.

Type **Repeating Detail – Brick 3 5/8".**

Click **OK**.

8.

Set the Detail to **brick_3_625**.

Set the Layout to **Maximum Spacing**.

Set the Spacing to **4"**.

Click **OK**.

Type Parameters	
Parameter	
Pattern	
Detail	brick_3_625
Layout	Maximum Spacing
Inside	☐
Spacing	0' 4"
Detail Rotation	None

9.

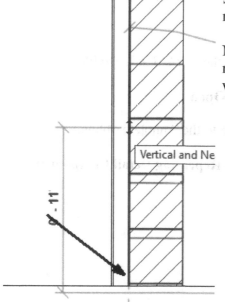

Select the bottom of the wall and move the mouse up the wall to see the brick.

Notice how the masking region in the repeating detail hides the hatch pattern in the wall.

10. 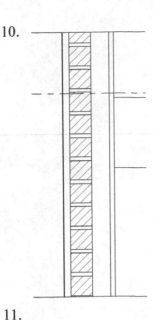 Click above the wall to complete the placement of the repeating detail.

Click **ESC** to exit the command.

11. Save as *ex2-36.rvt*.

Detail Groups

A group cannot contain both model and detail (view-specific) elements. If you select both types of elements and then try to group them, the detail elements are placed into an attached detail group for that model group. For example, if you want to create a furniture group that includes a tag, you need to create a detail group. You create the model group which includes all the model elements, then attach the annotation elements.

You can opt to show/hide the annotation elements in a detail group.

Exercise 2-37
Create a Detail Group

Drawing Name: **detail groups.rvt**
Estimated Time to Completion: 15 Minutes

Scope
Create an attached detail group.
Create a model group.
Modify a model group.
Mirror an attached detail group.

Solution

1.

Activate the **GROUND FLOOR – Exam Rooms** floor plan.

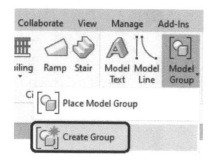

The first exam room has furniture and casework installed.

The furniture and specialty equipment (patient bed) have been tagged to make it easier to manage schedules.

We want to create an attached detail group that includes the furniture, casework, and specialty equipment as well as the tags.

2.

On the Architecture ribbon:

Select **Model Group→Create Group**.

3.

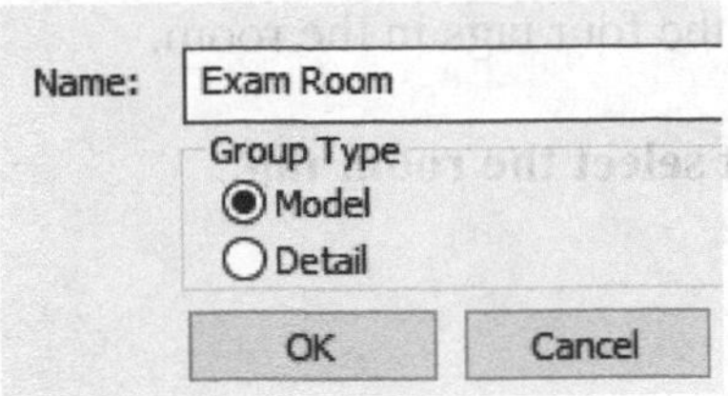

Type **Exam Room**.

Enable **Model**.

Click **OK**.

4.

Click **Add**.

5.

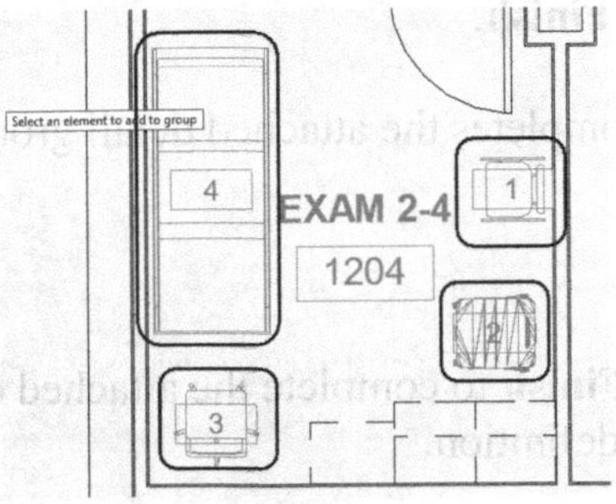

Select the furniture and specialty equipment.

6.

Click **Attach**.

7.

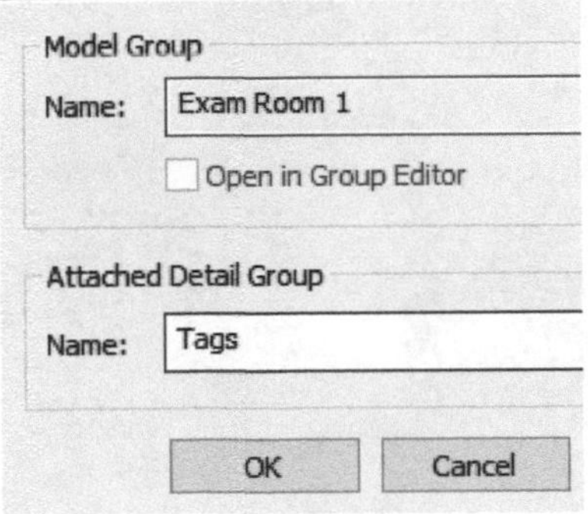

Name the Attached Detail Group **Tags**.

Click **OK**.

8.

Select the four tags in the room.

Do not select the room tag.

9.

Select **Finish**.

This completes the attached detail group.

10.

Click **Finish** to complete the attached detail group definition.

11.

Notice that you can select either the model group or the attached detail group by using the TAB key to select.

12.

Hold down the CTL key to select both the model and attached detail group.

13. Select the **Mirror-Draw Axis** tool.

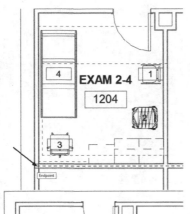

Select the center location line of the wall below the exam room as the mirror axis.

14.

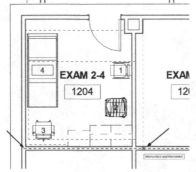

Start the axis at the endpoint and then select a point to the right and horizontal to the first point.

15.

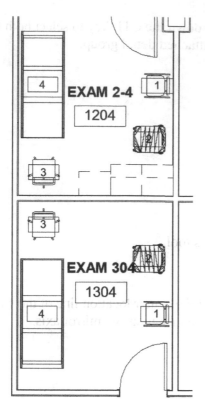

The model group and attached detail group were mirrored.

Are the tags oriented correctly?

Save as *ex2-37.rvt*.

Practice Exam

1. When you add text notes to a view using the Text tool, you can control all of the following except:

    A. The display of leader lines
    B. Text Alignment
    C. Text Font
    D. Whether the text is bold or underlined

2. To use a different text font, you need to:

    A. Use the Text tool.
    B. Create a new Text family and change the Type Properties.
    C. Create a new Text family and change the Instance Properties.
    D. Modify the text.

3. _____ are the measurements that display in the drawing when you select an element. These dimensions disappear when you complete the action or deselect the element.

    A. Temporary
    B. Permanent
    C. Listening
    D. Constraints

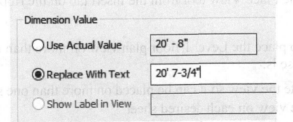

4. You click on a permanent dimension to change the value. You enable Replace With Text and type in the desired value. Then you click OK. What happens next?

    A. An error message appears advising that you cannot modify a dimension value using text
    B. The dimension updates with the new value you entered
    C. The dimension updates and the elements shift location
    D. The dimension updates, but the element remains in the same position

5. To add a new sheet to a project: (select all that apply)

   A. Right click on the Sheets category in the Project Browser and select New Sheet
   B. Select the New Sheet tool on the View tab on the ribbon
   C. Select New Sheet tool on the Annotation tab on the ribbon
   D. Select New Sheet tool on the Manage tab on the ribbon

6. To change the size of a sheet used in a Revit project:

   A. Right click on the sheet in the Project Browser and select Properties
   B. Highlight the sheet in the Project Browser and select Edit Type in the Properties panel
   C. Select the title block on the sheet, select a new title block using the Type Selector
   D. Highlight the sheet in the Project Browser and use the Type Selector to change the size

7. To add a plan view to a sheet: (select all that apply)

   A. Drag and drop a view from the Project Browser on to a sheet
   B. Select the Place View tool from the View tab on the ribbon
   C. Select the Insert View tool from the Insert tab on the ribbon
   D. Select the Place View tool from the Insert tab on the ribbon

8. You want to place the Level 1 floor plan view on more than one sheet. The best method to use is:

   A. Duplicate the view so it can be placed on more than one sheet
   B. Drag the view on each desired sheet
   C. Highlight the view on the sheet, select Edit Type on the Properties Panel, select Duplicate
   D. Highlight the view on the sheet, click CTL+C to copy, switch to the other sheet, click CTL+V to paste

9. Color Schemes can be placed in the following views (select 3):

    A. Floor plan
    B. Ceiling Plan
    C. Section
    D. Elevation
    E. 3D

10. To define the colors and fill patterns used in a color scheme legend, select the legend, and Click Edit Scheme on the:

    A. Properties palette
    B. View Control Bar
    C. Ribbon
    D. Options Bar

11. A legend is a view that can be placed on:

    A. A plan view
    B. A drafting view
    C. Multiple Plan Regions
    D. Multiple Sheets

12. To change the font used in a Color Scheme Legend:

    A. Select the legend, then select Edit Scheme from the ribbon.
    B. Select the text in the legend, double click to edit.
    C. Select the legend, then select Edit Type from the Properties pane.
    D. Select the legend, right click and select Edit Family.

13. Filled regions can be placed in the following views: (select all that apply)

    A. Floor Plan
    B. Elevation
    C. Section
    D. Detail View

14. Filled regions: (select all that apply)

    A. Are two-dimensional elements
    B. View-specific (only visible in the view where they are created and placed)
    C. Use fill patterns
    D. Can be placed in 3D sections

15. Detail Components are:

    A. System Families
    B. Mass Families
    C. Annotation Families
    D. Loadable Families

16. The Repeating Detail tool places an array of _____

    A. Annotations
    B. Filled Regions
    C. Detail Views
    D. Detail Components

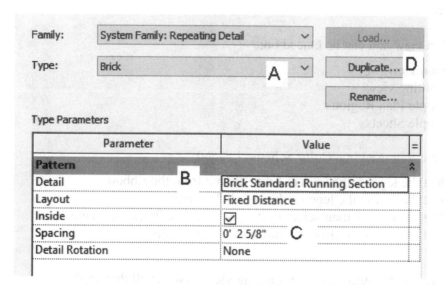

17. Specify the location to select to use a different detail component for the repeating detail.

18. Tagging elements by category:

    A. Allows you to select which elements will be tagged by selecting the element
    B. Allows you to select all the elements in a category (doors/windows/rooms/etc) by simply selecting which tags to be applied in a view
    C. Allows you to select which tag to be placed and then select the element to be tagged one at a time
    D. Applies the category tag to all the selected elements in the entire project

19. What happens when you change the sequence of revisions in the Sheet Issues/Revisions dialog?

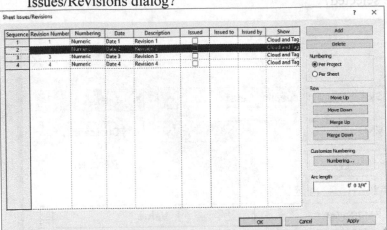

A. The Revision Number does NOT change because the Numbering is Per Project.

B. The Revision Number only changes for revisions marked as Issued.

C. The Revision Number for revisions NOT on sheets changes accordingly.

D. The Revision Number assigned to each revision changes accordingly.

20. What happens when you click the icon below a temporary dimension?

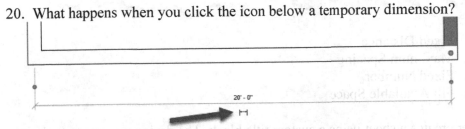

A. Makes the temporary dimension permanent.
B. Deletes the dimension.
C. Locks the dimension.
D. Activates the dimension value to modify it.

21. A designer is creating a repeating detail family. The designer wants the spacing between each CMU block to be the same in the repeating detail. Which Layout method should be selected?

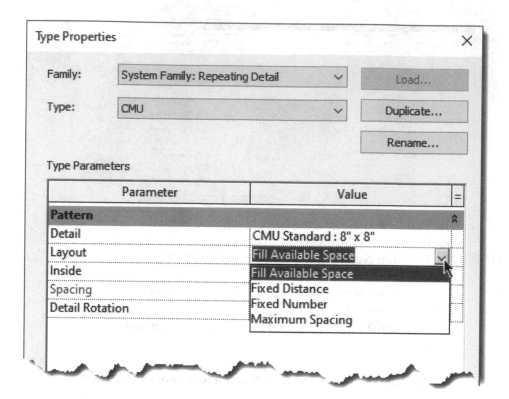

    A. Fixed Distance
    B. Maximum Spacing
    C. Fixed Number
    D. Fill Available Space

22. A designer creates a sheet using a custom title block. The designer wants to include a parameter to indicate the building area. How can this be accomplished?

    A. Add the parameter to the Sheets category as a project parameter.
    B. Add the parameter to the Title Blocks category as a project parameter.
    C. Open the title block in the Family Editor and add the parameter as a label.
    D. Open the sheet and add the parameter as text inside of angle brackets.

23. Which of the following methods can be used to organize a color fill legend?

    A. Name, Department, Space, Room.
    B. Wall face, floor face, or ceiling face.
    C. Spaces only. Color fill plans are used for HVAC zoning.
    D. Rooms, either by Name or Department

24. What is a detail component?
    A. Detail components are Revit families that offer more flexibility than filled regions.
    B. Detail components are the same thing as model components, but drawn from the Annotate tab.
    C. Detail components are Revit families that are only used when creating construction details.
    D. Detail components are 2D Revit families that can be placed in drafting views or detail views to add information to the model.

25. T/F   Revisions in Revit can only be added to sheets that have revision clouds.

26. A designer needs to adjust the line weight for all walls in one floor plan view. Which dialog should the designer use?

    A. View Specific Element Graphics
    B. Graphic Display Options
    C. Object Styles
    D. Visibility/Graphic Overrides

27. What can be accomplished in the Visibility/Graphic Overrides dialog?

    A. Change the line styles of individual elements in the project.
    B. Change the color of individual elements in the current view.
    C. Control the visibility of element categories in the current view.
    D. Control the line weights of element categories in the current project.

28.   A designer is adding windows to exterior walls. The full-height windows are
       visible, but the smaller windows are not showing in the view. What should
       the designer check to fix this issue?

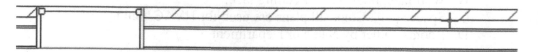

        A. Detail Level
        B. Visual Style
        C. Visibility/Graphic Overrides
        D. View Range

29. T/F Object Styles control the visibility of each view independently.

ANSWERS: 1) D; 2) B; 3) A; 4) A; 5) A & B; 6) C; 7) A & B; 8) A 9) A,B, C, & D; 10) C; 11) D; 12) C; 13) A.B.C.&D; 14) A,B, & C; 15) D; 18) D; 19) D; 20) A; 21) A; 22) C; 23) D; 24) D; 25) T; 26) D; 27) C; 28) D; 29) F;

Collaboration and Coordination

This lesson addresses the following certification exam questions:

- Configure worksharing and initiate worksharing
- Link and manage various file formats
- Perform basic maintenance on project files
- Manage and configure phasing
- Create, configure, and manage design options
- Use the Copy/Monitor process and features

In most building projects, you need to collaborate with outside contractors and with other team members. A mechanical engineer uses an architect's building model to lay out the HVAC (heating and air conditioning) system. Proper coordination and monitoring ensure that the mechanical layout is synchronized with the changes that the architect makes as the building develops. Effective change monitoring reduces errors and keeps a project on schedule.

Worksets are used in a team environment when you have many people working on the same project file. The project file is located on a server (a central file). Each team member downloads a copy of the project to their local machine. The person is assigned a workset consisting of building elements that they can change. If you need to change an element that belongs to another team member, you issue an Editing Request which can be granted or denied. Workers check in and check out the project, updating both the local and server versions of the file upon each check in/out.

Project sharing is the process of linking projects across disciplines. You can share a Revit Structure model with an MEP engineer.

You can link different file formats in a Revit project, including other Revit files (Revit Architecture, Revit Structure, Revit MEP), CAD formats (DWG, DXF, DGN, SAT, SKP), and DWF markup files. Linked files act similarly as external references (XREFs) in AutoCAD. You can also use file linking if you have a project which involves multiple buildings.

It is recommended to use linked Revit models for

- Separate buildings on a site or campus.
- Parts of buildings which are being designed by different design teams or designed for different drawing sets.
- Coordination across different disciplines (for example, an architectural model and a structural model).

Linked models may also be appropriate for the following situations:

- Townhouse design when there is little geometric interactivity between the townhouses.
- Repeating floors of buildings at early stages in the design, where improved Revit model performance (for example, quick change propagation) is more important than full geometric interactivity or complete detailing.

You can select a linked project and bind it within the host project. Binding converts the linked file to a group in the host project. You can also convert a model group into a link which saves the group as an external file.

Worksharing

Worksets are used in a team environment when you have many people working on the same project file. The project file is located on a server (a central file). Each team member downloads a copy of the project to their local machine. The person is assigned a workset consisting of building elements that they can change. If you need to change an element that belongs to another team member, you issue an Editing Request which can be granted or denied. Workers check in and check out the project, updating both the local and server versions of the file upon each check in/out.

In the certification exam, you may be asked a question regarding how to control the display of worksets or how to identify elements in a workset.

Other Hints:

- Name any sheets you create. That way you can distinguish between your sheet and other sheets.
- Create a view that is specific to your changes or workset, so that you have an area where you can keep track of your work.
- Create one view for the existing phase and one view for the new construction phase. That way you can see what has changed. Name each view appropriately.
- Use View Properties to control the phase applied to each view.
- Check Editing Requests often.
- Duplicate any families you need to modify, rename, and redefine. If you modify an existing family, it may cause problems with someone else's workset.

Note: The worksets exercises only work if you have access to the internet.

Exercise 3-1
Worksets

Drawing Name: **worksets.rvt**
Estimated Time to Completion: 15 Minutes

Scope

Use of Worksets
Workset Visibility

Solution

1.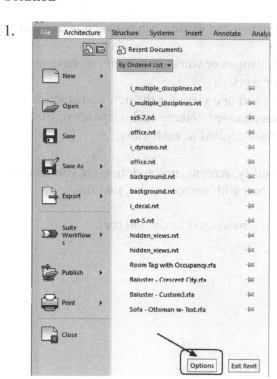

Before you can use Worksets, you need to set Revit to use your name.

Close any open projects.

Go to **Options**.

2.
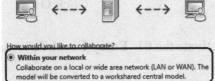

If you are using an Autodesk account, you will see the user name assigned to that account. To use a different name, you need to modify your Autodesk profile to use the desired name. If you sign out, you may lose access to your Revit license.

3. 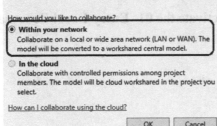 Select the **Collaborate** ribbon.

4. Select the **Collaborate** tool.
There will be a slight pause while Revit checks to see if you are connected to the Internet.

5. Enable **Within your network**.
Click **OK**.

Collaborate in Cloud is only available for those users who purchase a separate subscription for Collaboration for Revit.

6. Select **Worksets** on the Worksets panel.

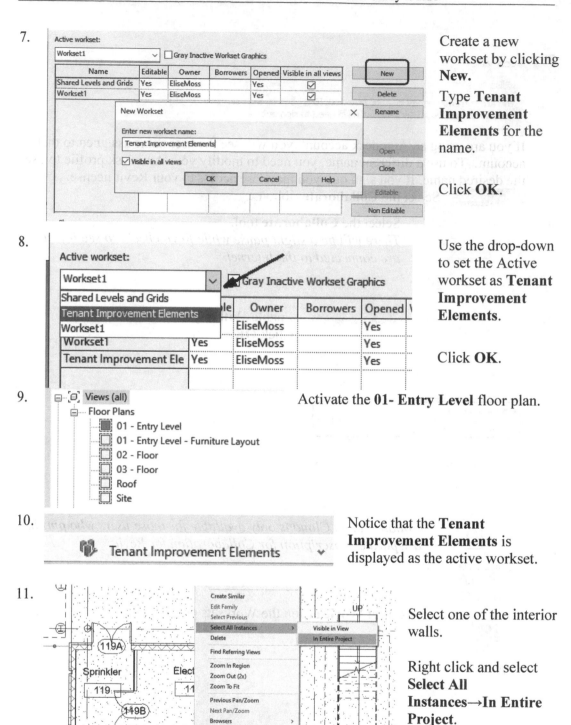

7. Create a new workset by clicking **New.**

    Type **Tenant Improvement Elements** for the name.

    Click **OK.**

8. Use the drop-down to set the Active workset as **Tenant Improvement Elements**.

    Click **OK.**

9. Activate the **01- Entry Level** floor plan.

10. Notice that the **Tenant Improvement Elements** is displayed as the active workset.

11. Select one of the interior walls.

    Right click and select **Select All Instances→In Entire Project.**

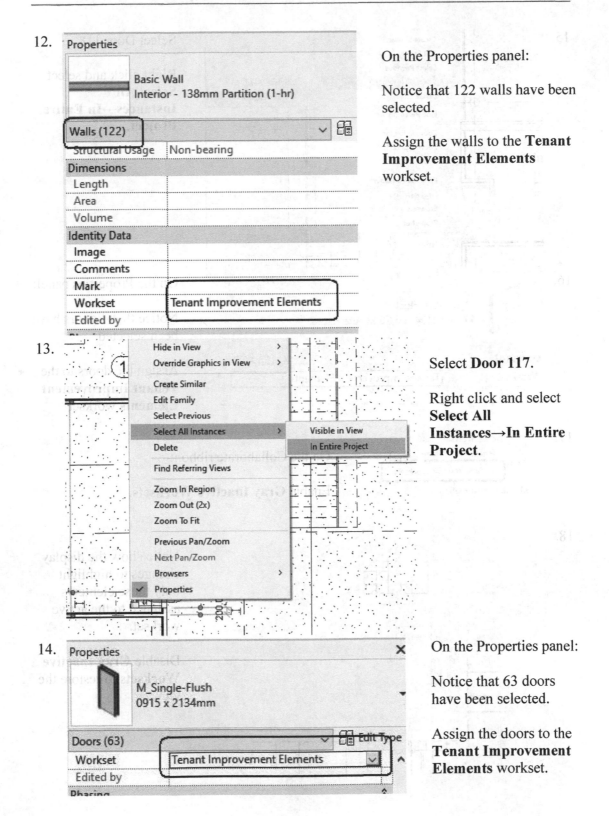

12. On the Properties panel:

Notice that 122 walls have been selected.

Assign the walls to the **Tenant Improvement Elements** workset.

13. Select **Door 117**.

Right click and select **Select All Instances→In Entire Project**.

14. On the Properties panel:

Notice that 63 doors have been selected.

Assign the doors to the **Tenant Improvement Elements** workset.

15.

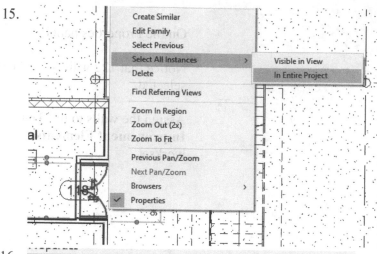

Select Door 118.

Right click and select **Select All Instances→In Entire Project**.

16.

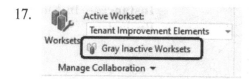

On the Properties panel:

Notice that 4 doors have been selected.

Assign the doors to the **Tenant Improvement Elements** workset.

17.

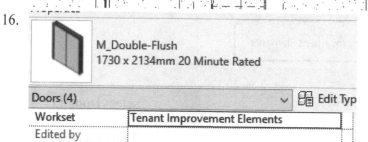

On the Collaborate ribbon:

Enable **Gray Inactive Worksets**.

18.

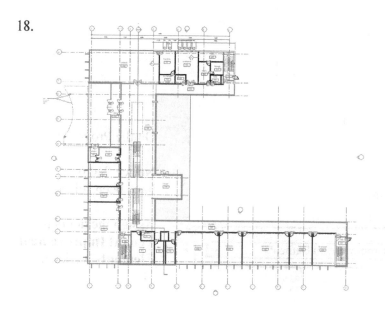

Notice how the display changes to highlight only those elements assigned to the active workset.

Disable **Gray Inactive Worksets** to restore the view.

19.

Select **Worksets** on the Worksets panel.

If a dialog appears asking to save to Central, click **OK**.

20.

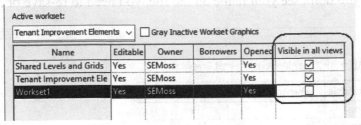

Uncheck Visible in all views for Workset 1.

Click **OK**.

21.

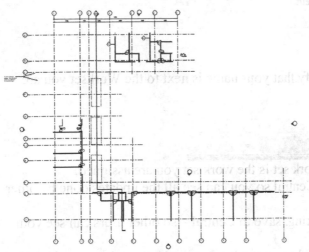

Notice how the view changes.

22. Save the file.

23.

This is the first time that the project has been saved since Worksharing was enabled. This project will therefore become the central model. Do you want to save this project as the central model?

If you want to save the file as the central model with a different name and/or different file location, click No and use the Save As command.

Click **No**.

Save as *ex3-1.rvt*.

Workset Checklist

1. Before you open any files, make sure your name is set in Options.
2. The main file is stored on the server. The file you are working on should be saved locally to your flash drive or your folder. It should be named Urban House [Your Name]. If you don't see your name on the file, you need to re-save or re-load the file.

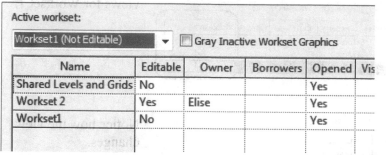

3. Go to Worksets and verify that your name is next to the Workset you are assigned.

4. Verify that the Active work set is the work set you are assigned.
5. Re-load the latest from Central so you can see all the updates done by other users.
6. When you are done working, save to Central – so other users can see your changes.
7. When you close, relinquish your work sets, so others can keep working.

The Worksharing Display Options allow you to color code elements in a view to distinguish between which elements different team members own as well as identifying which worksets different elements have been assigned.

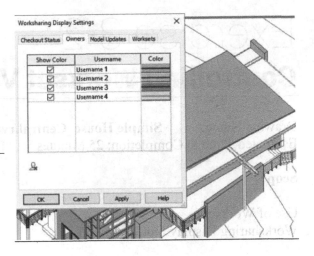

Workset Visibility

In a workshared project, you can control the visibility of worksets in project views.

Note: To improve performance, hide worksets that are not required for current work in the local model.

When you create a workset, use the Visible in all views option of the New Workset dialog to indicate whether the workset displays in all views of the model. This setting is reflected in the Visible in all views column of the Worksets dialog.

This global setting defines the default behavior for each workset in project views. You can override the visibility of each workset for individual views.

All elements that are not in the active workset can display as gray in the drawing area.

Temporary elements, such as temporary dimensions and controls, do not display in gray. This option has no effect on printing, but it helps to prevent adding elements to an undesired workset.

Exercise 3-2
Controlling Workset Visibility

Drawing Name: **Simple House_Central.rvt**
Estimated Time to Completion: 25 Minutes

Scope

Use of Worksets
Worksharing Display Options

Solution

1. Open the **Simple House_Central** file.

2.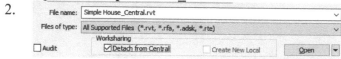

   Enable Detach from Central.

   Click **Open**.

   This turns off worksharing in the file and converts the file to a local file.
   Click **Detach and preserve worksets**.

3.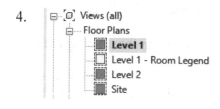

   → Detach and preserve worksets
   You can later save the detached model as a new central model.

   → Detach and discard worksets
   Discards the original workset information, including workset visibility. You can later create new worksets or save the detached model with no worksets.

4. Views (all)
   └ Floor Plans
      ├ **Level 1**
      ├ Level 1 - Room Legend
      ├ Level 2
      └ Site

   Activate **Level 1**.

   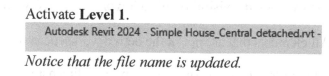
   Autodesk Revit 2024 - Simple House_Central_detached.rvt -

   Notice that the file name is updated.

5. 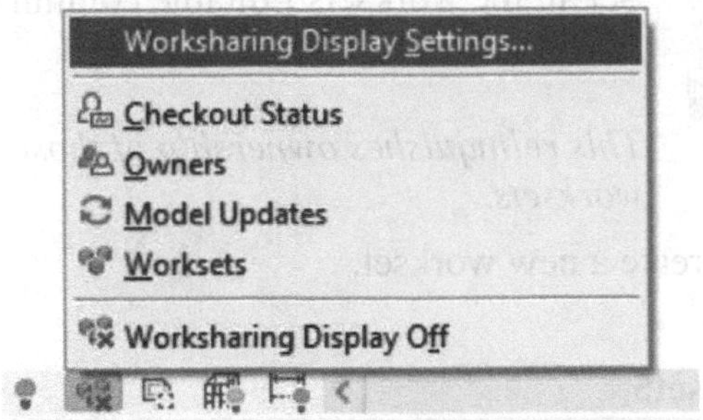 Select **Worksharing Display Settings** on the View Control bar.

6. 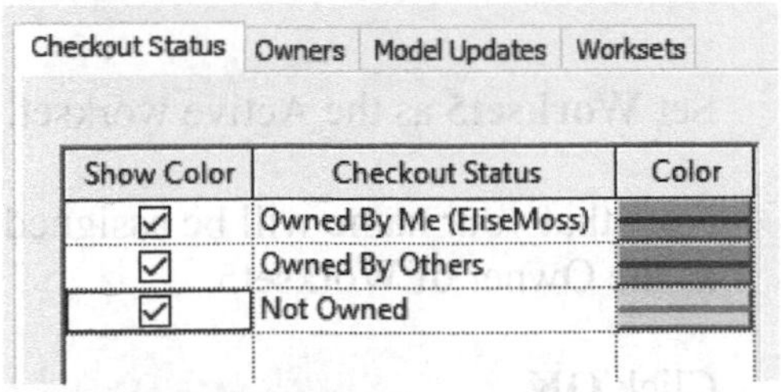 On the Checkout Status tab:

    Enable the **Show Color** display for all the options.

    Click **OK**.

7. 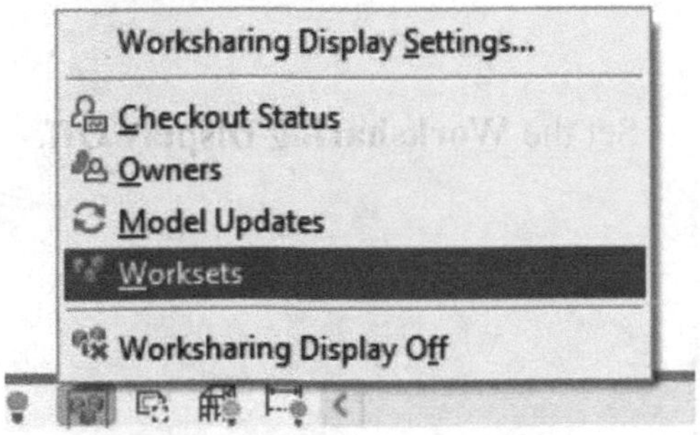 Enable display of **Worksets**.

8. 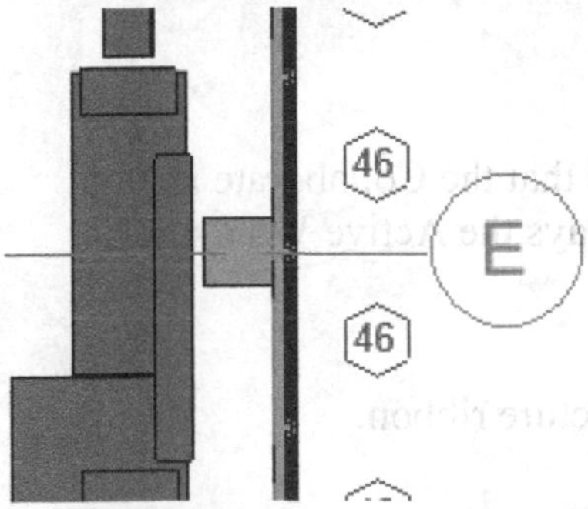 What color is Column Grid E?

    It displays as Green which indicates it is owned by the author.

9. 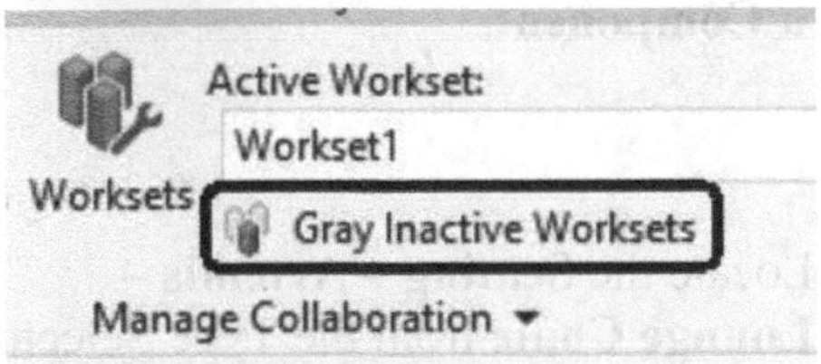 Switch to the Collaborate ribbon.

    Enable **Gray Inactive Worksets**.

10.  Select the **Worksets** tool from the Collaborate ribbon.

11. Set all the worksets Editable column to **No**.

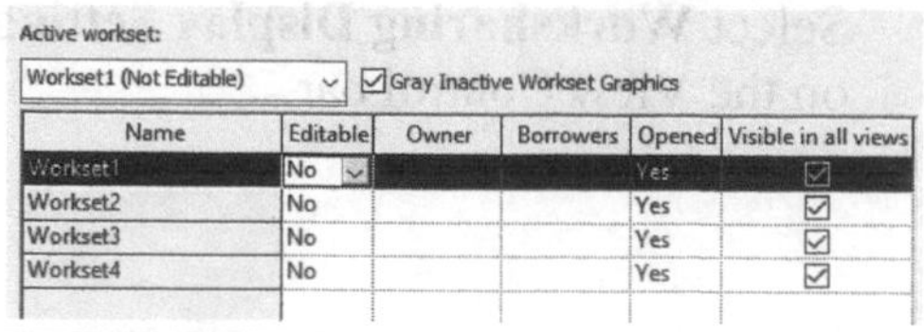

This relinquishes ownership of those worksets.

12. Select **New** to create a new workset.

13. Type **Workset5**.

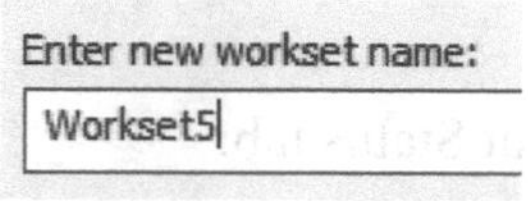

Click **OK**.

14. Set **Workset5** as the Active workset.

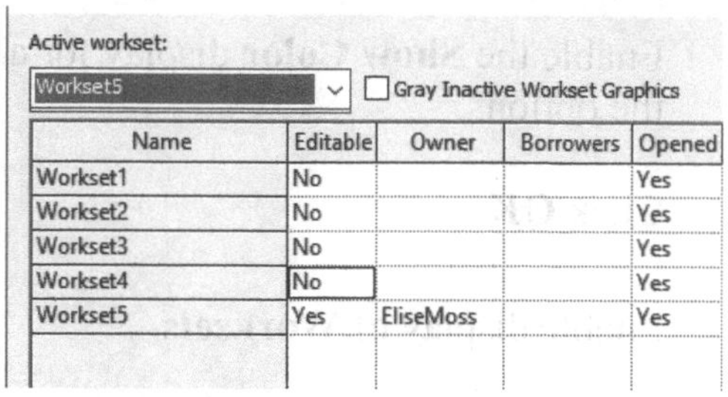

Note that your name will be assigned as the Owner of Workset5.

Click **OK**.

15. Set the **Worksharing Display Off**.

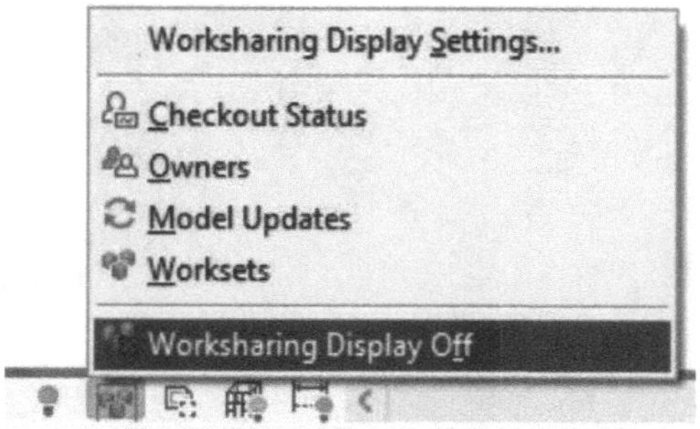

16. Note that the Collaborate ribbon displays the Active Workset.

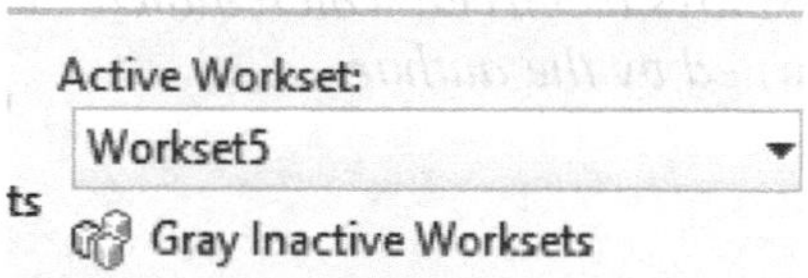

17. Go to the Architecture ribbon.

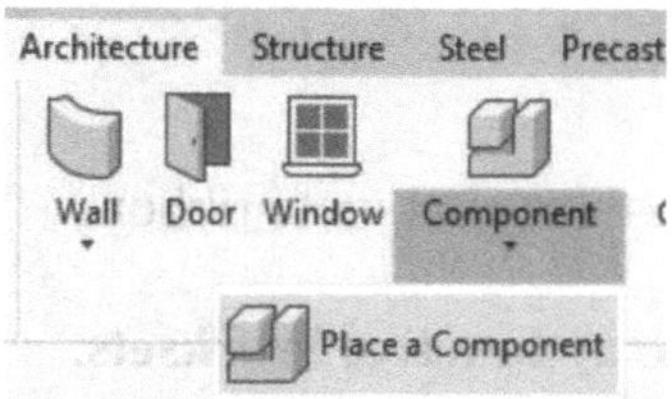

Select **Place a Component**.

18. Locate the **Seating – Artemis – Lounge Chair** from the Type Selector.

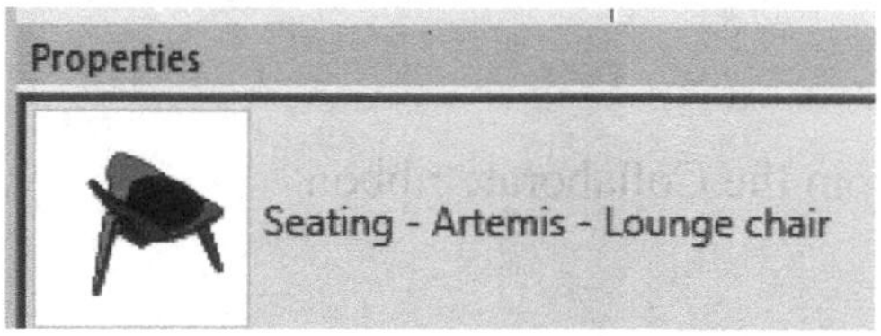

19.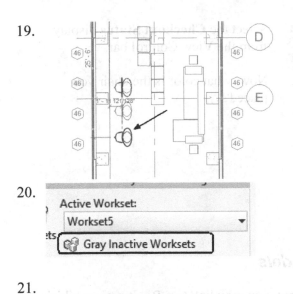

Place a chair below the other two chairs.

Use the SPACE bar to rotate the chair prior to placing.

Note that the chair you placed is not grayed out.

20.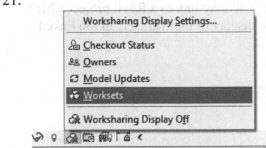

Select the **Gray Inactive Worksets** toggle on the Collaborate ribbon.

Note that the display changes so nothing is grayed out.

21.

Select the **Worksets** display from the View Control bar.

Worksharing Display Settings...

- Checkout Status
- Owners
- Model Updates
- **Worksets**
- Worksharing Display Off

22.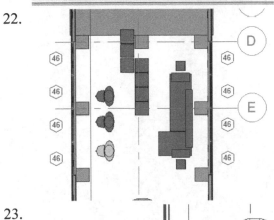

What color is the chair you placed?

23.

Select the **Owners** display from the View Control bar.

Worksharing Display Settings...

- Checkout Status
- **Owners**
- Model Updates
- Worksets
- Worksharing Display Off

Now, what color is the chair you placed?

24.

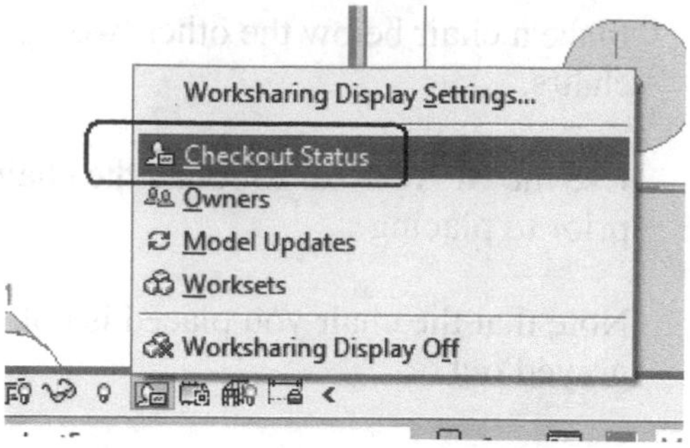

Select the **Checkout Status** display from the View Control bar.

Now, what color is the chair you placed?

25. Close the file without saving.

Manage Worksets in Linked Models

When working in a collaborative environment, you may link to a Revit project which also has worksets defined. You can control the display of the worksets in the linked model.

Exercise 3-3
Worksets in a Linked Model

Drawing Name: **linked_plumbing.rvt**
Estimated Time to Completion: 20 Minutes

Scope

Link to a Revit file
Modifying the display of worksets in a linked file

Solution

1.

Open the Insert ribbon.

Click **Link Revit**.

2.
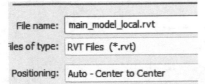
Locate the *main_model_local.rvt* file.
Set the Positioning to **Auto – Center to Center.**

Click **Open**.

3.
Zoom out and you will see the linked file.

4.

Open the Modify ribbon.

Select the **ALIGN** tool.

5.

Select **Grid A** on the host file.
Then, select **Grid A** on the linked file.

6.

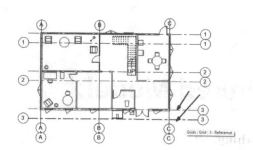

Select **Grid 3** on the host/Reference file. Then, select **Grid 3** on the linked file.

7.

Switch to the Collaborate ribbon.

Click **Worksets**.

8.

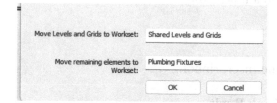

Type **Plumbing Fixtures** for the name of the Workset.

Click **OK**.

9.

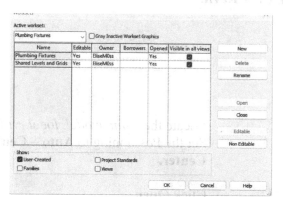

The Worksets dialog appears.

Click **New**.

10.

Type **Main Model**.

Click **OK**.

11.

Click **OK**.

12.

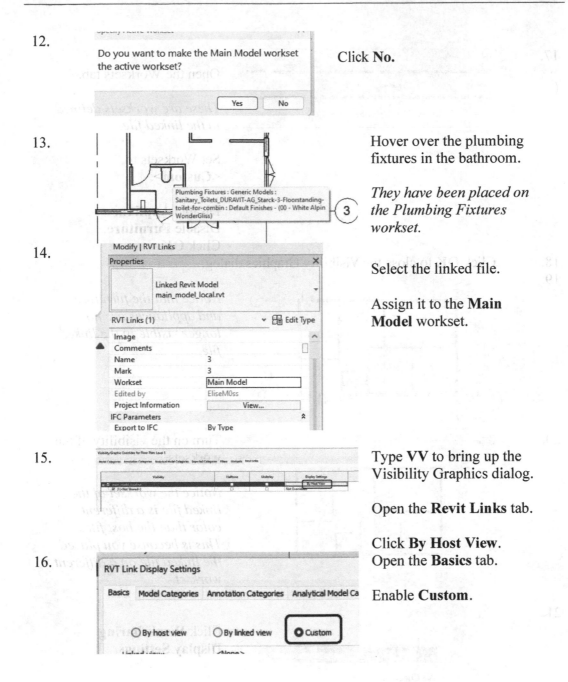

Do you want to make the Main Model workset the active workset?

Yes   No

Click **No.**

13.

Plumbing Fixtures : Generic Models :
Sanitary_Toilets_DURAVIT-AG_Starck-3-Floorstanding-
toilet-for-combin : Default Finishes - (00 - White Alpin
WonderGliss)

3

Hover over the plumbing fixtures in the bathroom.

They have been placed on the Plumbing Fixtures workset.

14.

Modify | RVT Links

Properties   ×

Linked Revit Model
main_model_local.rvt

RVT Links (1)   ∨ ⊞ Edit Type

Image
Comments
Name   3
Mark   3
Workset   Main Model
Edited by   EliseM0ss
Project Information   View...
IFC Parameters   ⊼
Export to IFC   By Type

Select the linked file.

Assign it to the **Main Model** workset.

15.

Visibility/Graphic Overrides for Floor Plan Level 1

Model Categories Annotation Categories Analytical Model Categories Imported Categories Filters Worksets Revit Links

	Visibility	Halftone	Underlay	Display Settings
main_model_local.rvt				By Host View
3 (<Not Shared>)				Not Overridden

Type **VV** to bring up the Visibility Graphics dialog.

Open the **Revit Links** tab.

Click **By Host View**.
Open the **Basics** tab.

16.

RVT Link Display Settings

Basics   Model Categories   Annotation Categories   Analytical Model Ca

○ By host view   ○ By linked view   ● Custom

Enable **Custom**.

3-19

17.

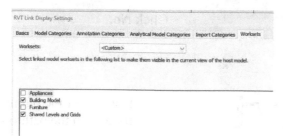

Open the Worksets tab.

These are worksets defined in the linked file.

Set Worksets to **<Custom>.**

Disable **Appliances**.
Disable **Furniture**.
Click **OK**.

18.   Click **OK** to close the Visibility Graphics dialog.

19.

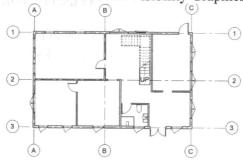

Notice that the furniture and appliances are no longer visible in the linked file.

20.

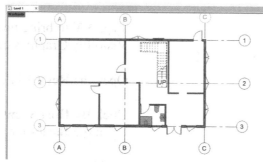

Turn on the visibility of the worksets.

Notice the workset of the linked file is a different color than the host file. This is because you placed the linked file on a different workset.

21.

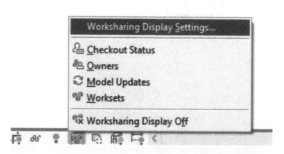

Click **Worksharing Display Settings**.

22.

Show Color	Workset	Color
☑	Main Model	
☑	Plumbing Fixtures	
☑	Shared Levels and Grids	

Checkout Status Owners Model Updates **Worksets**

Change the color of the Main Model workset to **Cyan**.

Click **OK**.

23.

The display updates.

Save as *ex3-3.rvt* and close.

Compacting a Central File

When worksets are active, all the users are synching to a Central File. The process of compacting **rewrites the entire file and removes obsolete parts in order to save space**. Because the Compact process takes more time than a normal save, it is strongly recommended that you only do this when workflow can be interrupted. This option reduces file size of the central model.

Exercise 3-4
Compacting a Central File

Drawing Name: **Simple House_Central.rvt**
Estimated Time to Completion: 35 Minutes

Scope

Use of Worksets
Re-linking Central File
Insert a Component
Synchronizing a Central File

Solution

1. Open the **Simple House_Central** file.

2.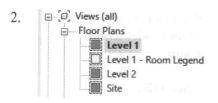

Activate **Level 1**.

- Simple House_Central_EliseM0ss.rvt -

Notice that the file name is updated. A local copy of the file was created with your user name appended to the file name.
This is a possible question on the certification exam.

3.

Switch to the Collaborate ribbon.

Click **Worksets**.

4. Notice the worksets are set not to be editable.

5. Set **Workset 2** as the Active workset.

Enable Workset2 as **Editable**.

Click **OK**.

6.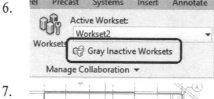

Enable **Gray Inactive Worksets**.

7.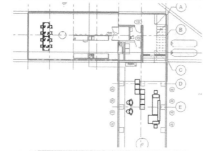

Notice that the furniture is placed on the active workset.

8.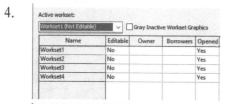

Switch to the Architecture ribbon.

Click **Place a Component**.

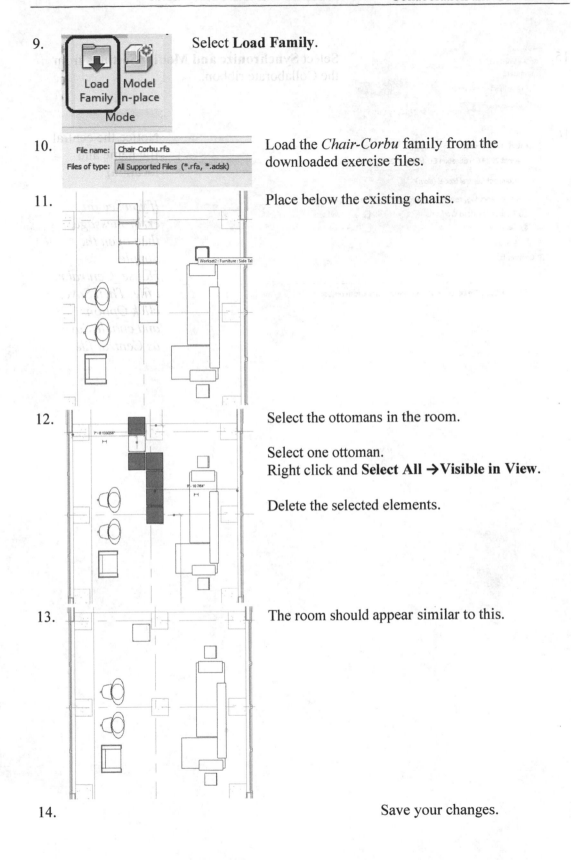

9. Select **Load Family**.

10. Load the *Chair-Corbu* family from the downloaded exercise files.

File name: Chair-Corbu.rfa
Files of type: All Supported Files (*.rfa, *.adsk)

11. Place below the existing chairs.

Workset2 : Furniture : Side Tab

12. Select the ottomans in the room.

Select one ottoman.
Right click and **Select All →Visible in View**.

Delete the selected elements.

13. The room should appear similar to this.

14. Save your changes.

15. 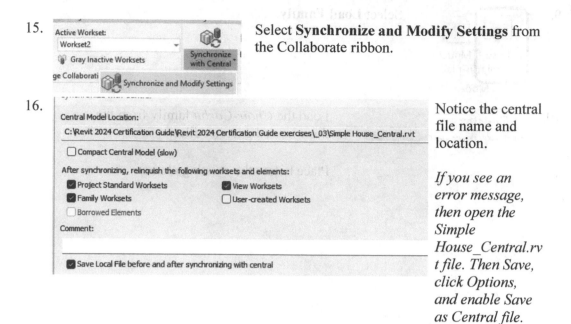 Select **Synchronize and Modify Settings** from the Collaborate ribbon.

16. Notice the central file name and location.

If you see an error message, then open the Simple House_Central.rvt file. Then Save, click Options, and enable Save as Central file.

17.

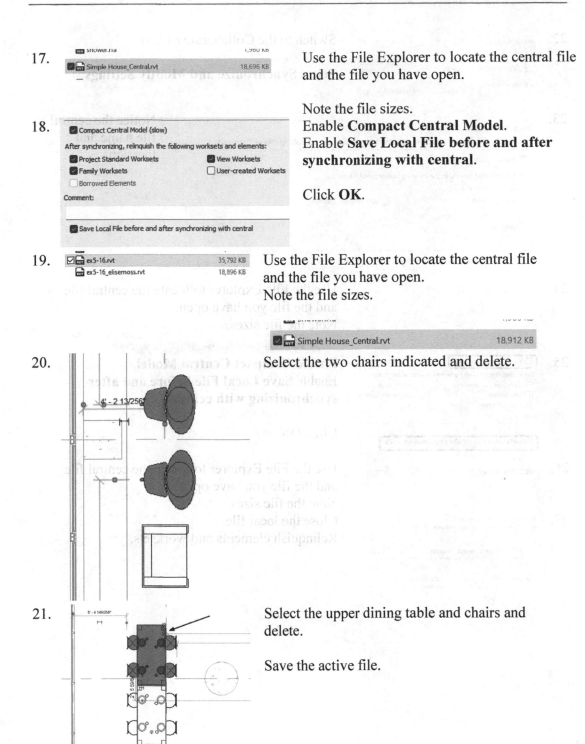

Use the File Explorer to locate the central file and the file you have open.

Note the file sizes.
Enable **Compact Central Model**.
Enable **Save Local File before and after synchronizing with central**.

Click **OK**.

18.

19.

Use the File Explorer to locate the central file and the file you have open.
Note the file sizes.

20.

Select the two chairs indicated and delete.

21.

Select the upper dining table and chairs and delete.

Save the active file.

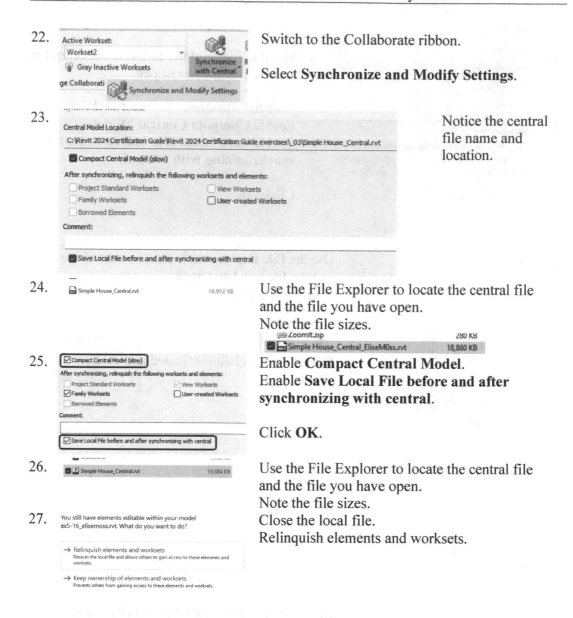

22. Switch to the Collaborate ribbon.

    Select **Synchronize and Modify Settings**.

23. Notice the central file name and location.

24. Use the File Explorer to locate the central file and the file you have open.
    Note the file sizes.

25. Enable **Compact Central Model**.
    Enable **Save Local File before and after synchronizing with central**.

    Click **OK**.

26. Use the File Explorer to locate the central file and the file you have open.
    Note the file sizes.
    Close the local file.
    Relinquish elements and worksets.

27.

Linking Files

Linking models is primarily intended for linking separate buildings, such as those that compose a campus.

You can link architectural models, structural models, and MEP models.

For example, the following site plan shows 4 building models linked to one model.

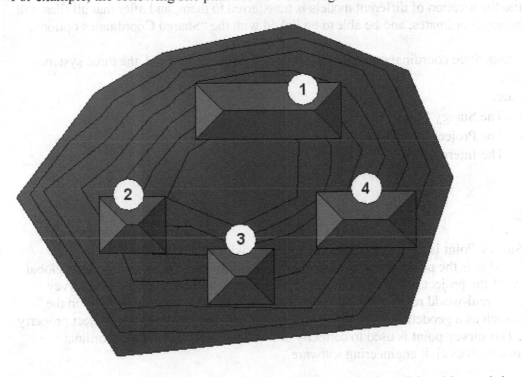

When you link a model into a project, Revit opens the linked model and keeps it in memory. The more links a project contains, the longer it can take to open.

Linked models are listed in the Revit Links branch of the Project Browser.

You can convert linked Revit models to groups, and you can convert groups to linked Revit models.

Shared Coordinates

Coordinates in a file will only be shared, or the files will have shared coordinates after a process of transferring the coordinate system used in a file into another. It does not matter if in two files the location is exactly the same. They will be not sharing coordinates unless the sharing coordinates process has been carried out.
Normally there is one file, and only one, that is the source for sharing coordinates. From this file the location of different models is transferred to them, and after that all files will be sharing coordinates, and be able to be linked with the "Shared Coordinates option".

Revit uses three coordinate systems. When a new project is opened, the three systems overlap.
They are:

- The Survey Point
- The Project Base Point
- The Internal Origin

The Survey Point is represented by a blue triangle with a small plus symbol in the center. This is the point that stores the universal coordinate system, or a defined global system of the project to which all the project structures will be referred. The Survey Point is a real-world relation to the Revit model. It represents a specific point on the Earth, such as a geodetic survey marker or a point of reference based on project property lines. The survey point is used to correctly orient the project with other coordinate systems, such as civil engineering software.

- It is the origin that Revit will use in case of share coordinates between models.
- The origin point used when inserting linked files with the Shared Coordinates option.
- The origin point used when exporting with Shared Coordinates option.
- The origin point to which spot coordinates and spot elevations are referenced, if the Survey Point is the coordinate origin in the type properties.
- The origin point to which level elevation is referred, if the SP is the Elevation Base in the level type properties.
- When the view shows the True North, it is oriented according to the Survey Point settings.

The Project Base Point is represented by a blue circle with an x. The position of this point is unique for each model, and this information is not shared between different models. The Project Base point could be placed in the same location as the Survey Point, but it is not usual to work in that way. This point is used to create a reference for positioning elements in relation to the model itself. By default, the Project Base Point is the origin (0,0,0) of the project. The Project Base Point should be used as a reference point for measurements and references across the site. The location of this point does not affect the shared site coordinates, so it should be located in the model where it makes sense to the model, usually at the intersection of two gridlines or at the corner of a building.

- The origin point to which spot coordinates and spot elevations are referenced, if the Project Base Point is the coordinate origin in the type properties.
- The origin point to which level elevation is referred, if the Project Base Point is the Elevation Base in the level type properties.
- When the view shows the Project North, it is oriented according to the Project Base Point.
- If it is not necessary, it is better not to move this point, so that it always is coincident with the third point: the Internal Origin.
- If we have moved the Project Base Point to a different location, it can be placed back to the initial position by right clicking on it when selected, and use **Move to Startup Location.**

The Internal Origin is now visible starting with the 2022 release of Revit. The Internal Origin is coincident with the Survey Point and the Project Base Point when a new project is started. It is displayed with an XY icon. Prior to this release, users sometimes resorted to placing symbols at the internal origin location to keep track of the origin.

- The project should be modeled in a restricted area around the internal origin point. The model should be inside a 20 miles radius circle around the internal origin, so that Revit can compute accurately.

- This is the origin point that is used when inserting external files using the "Internal Origin to Internal Origin" option.

- This is the origin point that is used when copy/pasting model objects from one file into another using the "aligned" option.

- The origin point to which spot coordinates and spot elevations are referenced, if Relative is the coordinate origin in the type properties.

- The origin point used when exporting with Internal Coordinates option.

- Revit API and Dynamo use this point as the coordinates origin point for internal computational calculations.

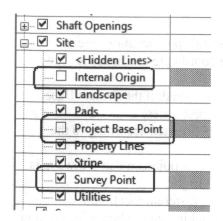

The visibility of these three systems is controlled under the Site category on the Model tab of the Visibility Graphics dialog.

Moving the project base point is like changing the global coordinates of the entire project.

Another way of looking at it is—if you modify the location of the project base point, you are moving the model around the earth; if you modify the location of the survey point, you are moving the earth around the model.

The benefits of correctly adjusting shared coordinates are not limited to aligning models for coordination but can also position the project in the real world if using proper geodetic data from a surveyor landmark or a provided file with the real-world information present.

To fully describe where an object (a building) sits in 3D space (its location on the planet), we need four dimensions:

- East/West position (the X coordinate)
- North/South position (the Y coordinate)
- Elevation (the Z coordinate)
- Rotation angle (East/West/True North)

These four dimensions uniquely position the building on the site and orients the building relative to a known benchmark as well as other landmarks. Revit allows you to assign a unique name to these four coordinates. Revit designates this collection as a **Shared Site**. You can designate as many Shared Sites as you like for a project. This is useful if you are planning a collection of buildings on a campus. By default, every project starts out with a single Shared Site which is named Internal. The name indicates that the Shared Site is using the Internal Coordinate System. Most projects will only require a single Shared Site, but if you are managing a project with more than one building or construction on the same site, it is useful to define a Shared Site for each instance of a linked file.

For example, in a resort, there may be different instances of the same cabin located across the site. You can use the same Revit file with the cabin model and copy the link multiple times. Each cabin location will be assigned a unique Shared Site, allowing each instance of the linked file to have its own unique location in the larger Shared Coordinate system.

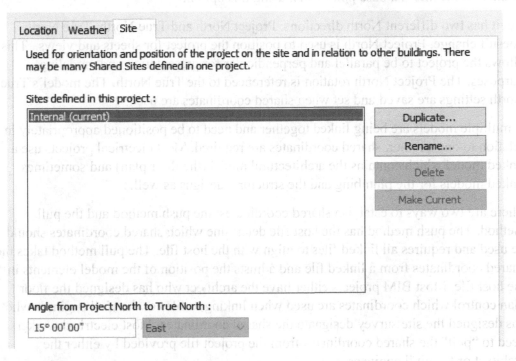

Shared Sites are named using the Location Weather and Site dialog.

The dialog is opened using the Location tool on the Manage ribbon.

Why setting up shared coordinates is important:

- To properly coordinate the project models to work across platforms, including BIM Track.
- To display the project in its real-world location (including BIM and GIS data overlay).
- To link files together that have different base points (arbitrary coordinate systems).

Shared Coordinates allow you to adjust the reference point from project origin to a shared site point, to make it appear in a different position or with a different angle based on the internal base point. It is important to note that nothing **in the Revit model moves when using this system,** even if it looks like it moves and/or rotates on the screen. It is just applying an adjustment to the point of origin, so in fact, all elements keep their relation to the internal base point when using this option.

Revit has two different North directions: Project North and True North. True North doesn't change. Project North is used to position the project for sheets and views. This allows the project to be parallel and perpendicular for presentation and modeling purposes. The Project North rotation is referenced to the True North. The model's True North settings are saved and set when shared coordinates are used.

If multiple models are being linked together and need to be positioned appropriately in relation to one another, shared coordinates are required. Most electrical projects use a linked model which contains the architectural model (the floor plan) and sometimes linked models for the plumbing and the structural designs as well.

There are two ways to establish shared coordinates: the push method and the pull method. The push method has the host file determine which shared coordinates should be used and requires all linked files to align with the host file. The pull method takes the shared coordinates from a linked file and adjusts the position of the model elements in the host file. Most BIM projects either have the architect who has designed the floor plan control which coordinates are used when linking to files or the civil engineer who has designed the site survey designate the shared coordinates. Most electrical workers need to "pull" the shared coordinates from the project file provided by either the architect or the civil engineer.

The steps to pull shared coordinates are as follows:

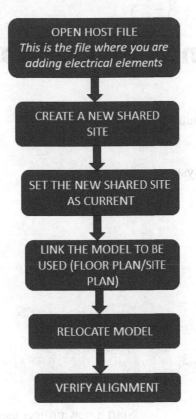

Exercise 3-5
Understanding Shared Coordinates

Drawing Name: new
Estimated Time: 5 minutes

Scope:
- ❑ Shared Coordinate System
- ❑ Site Point
- ❑ Base Point
- ❑ Internal Origin
- ❑ Views
- ❑ Visibility/Graphics

Solution

1.   

Start a new project using the *DefaultMetric.rte* template.

2.      Open the Site floor plan.

3.   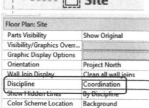   Set the Discipline to **Coordination**.

4. Type **VV** to open the Visibility/Graphics dialog.

On the Model Categories tab:

Enable:
- Internal Origin
- Project Base Point
- Survey Point

These are located under the Site category.

Click **OK**.

The Internal Origin, Project Base Point and Survey Point are all visible.

They are overlaid on top of each other.

5. Save as *ex3-5.rvt*.

Spot Coordinates

Spot coordinates report the North/South and East/West coordinates of points in a project. In drawings, you can add spot coordinates on floors, walls, toposurfaces, and boundary lines. Coordinates are reported with respect to the survey point or the project base point, depending on the value used for the Coordinate Origin type parameter of the spot coordinate family.

Exercise 3-6
Placing a Spot Coordinate

Drawing Name: spot_coordinate.rvt.
Estimated Time: 20 minutes

Scope:
- ❑ Spot Coordinate
- ❑ Specify Coordinates at a Point
- ❑ Report Shared Coordinates

Solution:

We can add a spot coordinate symbol to the project base point to keep track of its location.

1. 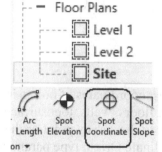 Open the **Site** floor plan.

2. 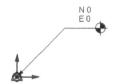 Switch to the Annotate ribbon.

    Select the **Spot Coordinate** tool on the Dimension panel.

3. Select the Project Base Point.

    Left click to place the spot coordinate.
    That is the circular symbol.

    Cancel out of the command.

4.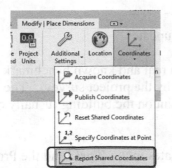

Switch to the Manage ribbon.

Select **Report Shared Coordinates** on the Project Location panel.

Select the Project Base Point.

5.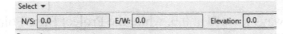

The coordinates are displayed on the Options bar.

Cancel out of the command.

6.  Select the Project Base Point.

Notice that the coordinates that are displayed are blue. This means they are temporary dimensions which can be edited.

7.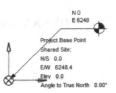

Click on the **E/W** dimension.
Change it to **6248.4.**
Click **ENTER**.

8.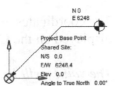

The location of the project base point shifted to be 6248.4 m east of the internal origin/survey point.

Notice that the spot coordinate value updated.

9.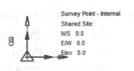

Left click on the Survey Point.
That's the triangle symbol.
Notice that dimensions are black, which means they cannot be edited.

10.

Drag the survey point away from the internal origin (the XY icon).

Notice that the Project Base Point value updates.

Remember the Survey Point represents a real and known benchmark location in the project (usually provided by the project survey). The Project Base Point is simply a known point on the building (usually chosen by the project team).

Think of the Survey Point as the coordinates in the *World* and the Project Base Point as the local building coordinates.

The Survey Point is always located at 0,0, while the Project Base Point is relative to the Survey Point.

It is called a shared coordinate system because it represents the coordinate system of the world around us and is shared by all the buildings on the site.

11.

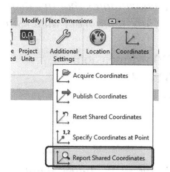

Select **Report Shared Coordinates** on the Project Location panel.

Select the Internal Origin (the XY icon).

12.

The coordinates are displayed on the Options bar.

The values will be different from mine.

Notice that the Internal Origin is located relative to the Survey Point. The Internal Coordinate System represents the building's "local" coordinates.

Cancel out of the command.

13.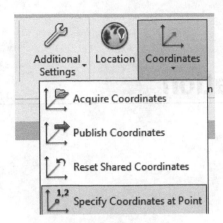

Usually, you will be provided the coordinate information for a project from the civil engineer.

Select **Specify Coordinates at Point**.

Select the Survey Point (the triangle symbol.).

14.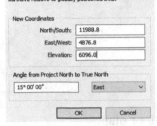

Fill in the Shared Coordinates:

For North/South: **11988.8.**
For East/West: **4876.8.**
For Elevation: **6096.**
For Angle from Project North to True North: **15° East**.

Click **OK**.

15.

Zoom out and you will see that the Survey Point's position has shifted.

Note the coordinates for the Project Base Point updated.

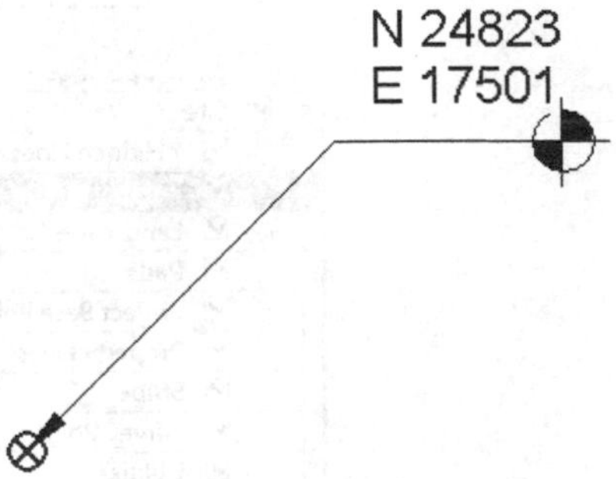

N 24823
E 17501

16. Save as *ex3-6.rvt*.

Exercise 3-7
Understanding Location

Drawing Name: Simple Building.rvt
Estimated Time: 20 minutes

Scope:
- ❑ Linking Files
- ❑ Shared Coordinate System
- ❑ Survey Point
- ❑ Base Point
- ❑ Internal Origin
- ❑ Visibility/Graphics
- ❑ Spot Elevation

Solution:

1. 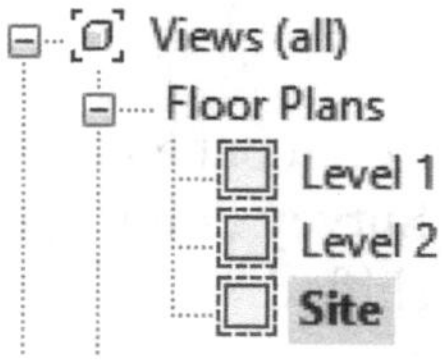
```
☐··[O] Views (all)
   ☐···· Floor Plans
         ·····[☐] Level 1
         ·····[☐] Level 2
         ·····[☐] Site
```
Open the **Site** floor plan.

2.
The Survey Point and the Project Base Point are visible, but not the Internal Origin.

Open the Visibility/Graphics dialog by typing **VV**.

3.
On the Model Categories tab:

Enable: **Internal Origin**

    This is located under the Site category.

Click **OK** and close the dialog.

4.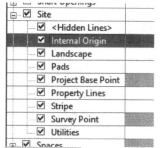
Activate the **Insert** ribbon.

Select **Link Revit**.

5.

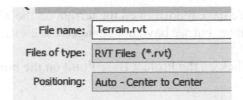

Locate the *Terrain.rvt* file.

Set the Positioning to **Auto – Center to Center**.

Click **Open**.

Many users will be tempted to align the files using Internal Origin to Internal Origin. This is not the best option when we plan to reposition the building on the site plan. Origin to Origin works well when linking MEP files to the Architectural file.

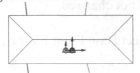

Notice that the linked file's coordinate system is slightly offset from the host file's coordinate system.

Next, we will re-position the linked file into the correct relative position. If we do it this way, we are not moving the building, we are moving the terrain. We will still use the terrain to establish the Shared Coordinate system, which is the "pull" method. The "pull" method pulls the information from the linked file.

6.

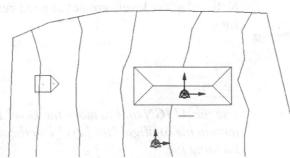

Select the linked file.

Select the **Move** tool on the tab on the ribbon.

Select the bottom midpoint of the building as the start point.

Move the cursor down and type 30' to move the terrain down 30'.

Select the Move tool on the tab on the ribbon. Select the bottom midpoint of the building as the start point.
Move the cursor to the left and type 15' to move the terrain left 15'.

7.

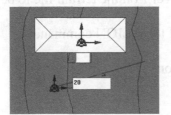

Select the linked file.

Select the Rotate tool on the tab on the ribbon.

Start the angle at the horizontal 0° level. Rotate up and type **20°**.

8.

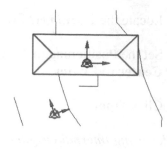

We have repositioned the terrain on the XY plane, but we have not changed the elevation.

Click on the Project Base Point on the building.

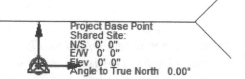

Notice that it has not changed.

9.

Click on the Project Base Point for the terrain file.

It shows as 0,0 still as well.

10.

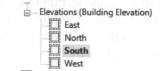

Open the **South** elevation.

11.

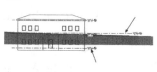

Notice that the levels are not aligned between the files.

12.

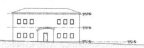

*Use the **ALIGN** tool to move the Level 1 in the terrain file to align with Level 1 in the simple building file.*

Select the **ALIGN** tool on the MODIFY ribbon.

Select the Level 1 that is on the simple building file as the target.
Select the Level 1 on the terrain file to be moved.

Cancel out of the ALIGN command.

13.

Switch to the Annotate ribbon.

Select **Spot Elevation**.

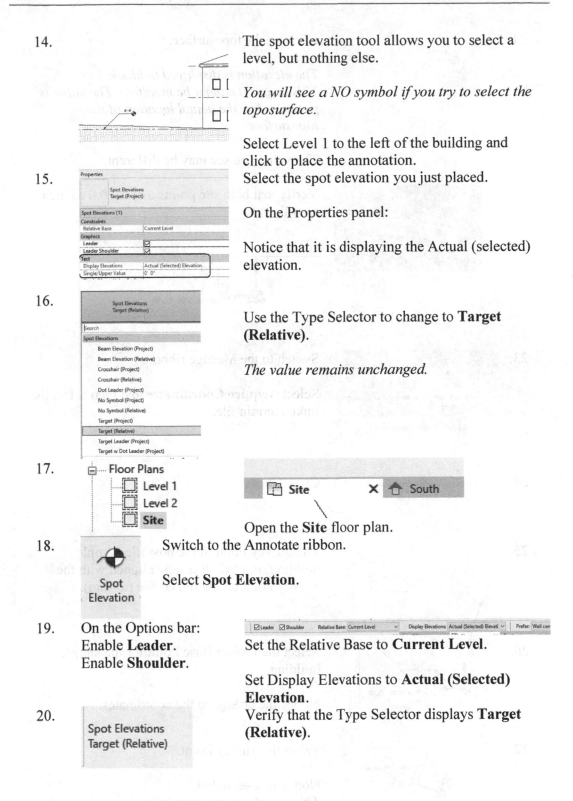

14.

The spot elevation tool allows you to select a level, but nothing else.

You will see a NO symbol if you try to select the toposurface.

Select Level 1 to the left of the building and click to place the annotation.
Select the spot elevation you just placed.

15.

On the Properties panel:

Notice that it is displaying the Actual (selected) elevation.

16.

Use the Type Selector to change to **Target (Relative)**.

The value remains unchanged.

17.

Open the **Site** floor plan.

18.

Switch to the Annotate ribbon.

Select **Spot Elevation**.

19. On the Options bar:
Enable **Leader**.
Enable **Shoulder**.

Set the Relative Base to **Current Level**.

Set Display Elevations to **Actual (Selected) Elevation**.

20.

Verify that the Type Selector displays **Target (Relative)**.

21.

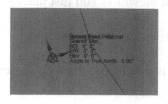

Click on the toposurface.

*The elevation is displayed in black.
This means it cannot be modified. The value is
generated by the actual location of the
toposurface.*

The value you see may be different.

22.

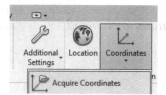

Verify that both site points display 0,0 as their
coordinates.

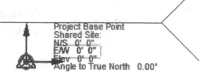

23.

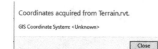

Switch to the Manage ribbon.

Select **Acquire Coordinates** and then select the
linked terrain file.

24.

Click **Close**.

25.

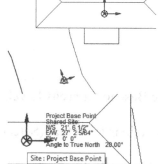

The Survey Point in the host file (simple
building) moved. It is now aligned with the
Survey Point on the linked file (terrain).

26.

Select the Project Base Point for the simple
building.

Notice the change to the coordinates.

27.

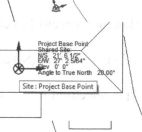

Select the Survey Point.

Notice it is set to 0,0.
*The spot elevation value is unchanged because
the coordinates were acquired from the linked
file and the spot elevation is reporting the*

elevation of the linked file.

28.   Save the file as *e3-7.rvt*.

Exercise 3-8
Linking Files using Shared Coordinates

Drawing Name: terrain.rvt
Estimated Time: 10 minutes

Scope:
- Linking Files
- Shared Coordinate System
- Survey Point
- Base Point
- Internal Origin
- Visibility/Graphics

Solution

1. 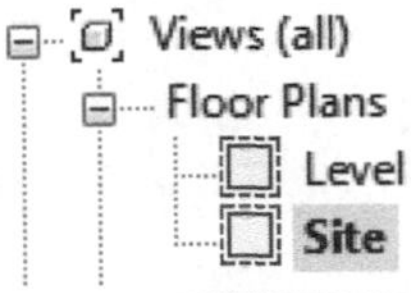   Open the **Site** floor plan.

2. 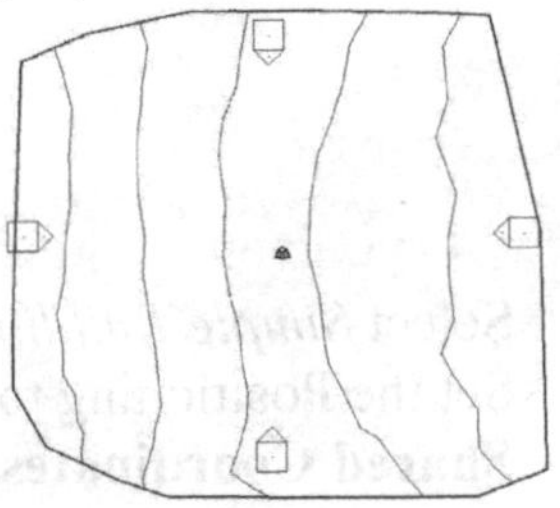   *The Survey Point and the Project Base Point are visible, but not the Internal Origin.*

Open the Visibility/Graphics dialog by typing **VV**.

3. 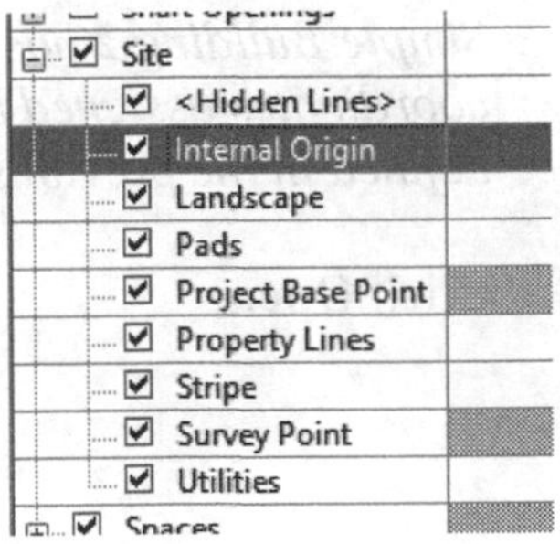   On the Model Categories tab:

Enable: **Internal Origin**

    This is located under the Site category.

Click **OK** and close the dialog.

4.

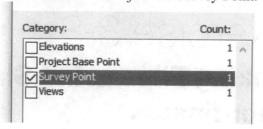

Window around everything.

Use **FILTER** to select just the Survey Point.

Notice that it is still located at 0,0.

5.

Click **ESC** to release the selection.
Select the **Project Base Point**.

It is also located at 0,0.

The file wasn't changed because we acquired the coordinates from the linked file and applied them to the host file.

6.

Switch to the **Insert** ribbon.

Select **Link Revit**.

7.

File name:	Simple Building 2.rvt
Files of type:	RVT Files (*.rvt)
Positioning:	Auto - By Shared Coordinates

Select *Simple Building 2.rvt*.
Set the Positioning to **Auto -By Shared Coordinates.**

Simple Building 2 has the coordinates assigned that we defined in the previous exercise.

Click **Open**.

8.

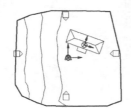

The linked file is positioned in the correct location.

9.

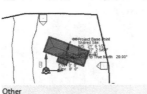

Click on the building.
Notice that Project Base Point is using the Shared Site coordinates.

10.

With the linked file selected, notice in the Properties palette, it is using the Internal Shared Site definition.

11.   Save as *ex3-8.rvt*.

Exercise 3-9
Defining a Shared Site

Drawing Name: terrain 2.rvt
Estimated Time: 30 minutes

Scope
- Linking Files
- Shared Coordinate System
- Survey Point
- Base Point
- Shared Site

Solution

1.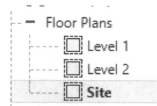

   Open *Building A.rvt.*
   Open the **Site** floor plan.

2.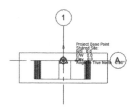

   Verify that the Survey Point and Base Point are located at 0,0.

Project Base Point (1)	
Identity Data	
N/S	0.0
E/W	0.0
Elev	0.0
Angle to True North	0.00°

   If not, select the Project Base Point and change the Properties to 0,0,0 and rotation. 0.

   Close the file.

3. Floor Plans
   - Platform 1
   - Platform 2
   - Platform 3
   - Site (Dimensioned)
   - **Site (Project North)**
   - Site (True North)
   - Site Datum

   Open the *terrain 2. rvt* file.

   Open the **Site (Project North)** floor plan.

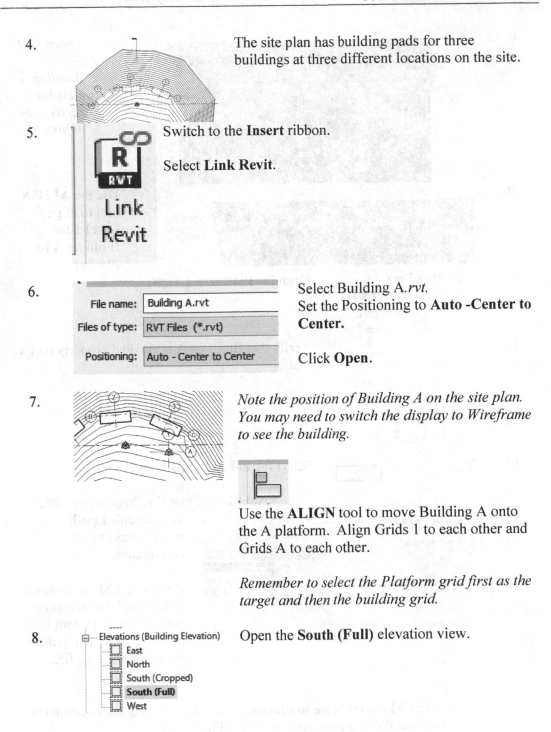

4.   The site plan has building pads for three
     buildings at three different locations on the site.

5.   Switch to the **Insert** ribbon.

     Select **Link Revit**.

6.   Select Building A.*rvt*.
     Set the Positioning to **Auto -Center to
     Center.**

     File name: Building A.rvt

     Files of type: RVT Files (*.rvt)

     Positioning: Auto - Center to Center

     Click **Open**.

7.   *Note the position of Building A on the site plan.
     You may need to switch the display to Wireframe
     to see the building.*

     Use the **ALIGN** tool to move Building A onto
     the A platform. Align Grids 1 to each other and
     Grids A to each other.

     *Remember to select the Platform grid first as the
     target and then the building grid.*

8.   Elevations (Building Elevation)
       East
       North
       South (Cropped)
       South (Full)
       West

     Open the **South (Full)** elevation view.

9.

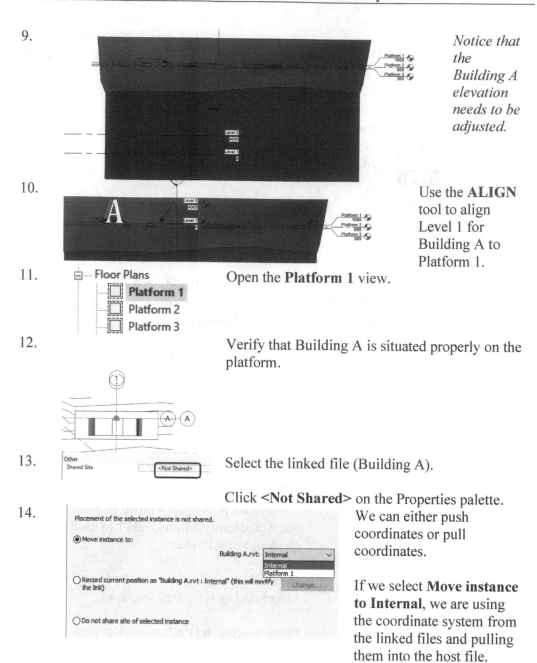

Notice that the Building A elevation needs to be adjusted.

10.

Use the **ALIGN** tool to align Level 1 for Building A to Platform 1.

11.

Open the **Platform 1** view.

12.

Verify that Building A is situated properly on the platform.

13.

Select the linked file (Building A).

14.

Click **<Not Shared>** on the Properties palette. We can either push coordinates or pull coordinates.

If we select **Move instance to Internal**, we are using the coordinate system from the linked files and pulling them into the host file.

If we select **Move instance to Platform 1**…, we are using the coordinates from the host file and pushing them to the linked file.

Regardless of which method we use, after the selection, they will be sharing the same coordinate system.

Since we started with the terrain file, it is standard to let the topo file control which coordinate system is used in a project.

Remember that whichever option you select does not change the linked file or the host file local internal coordinates. We are only defining how we want the files to interact relative to each other.

15.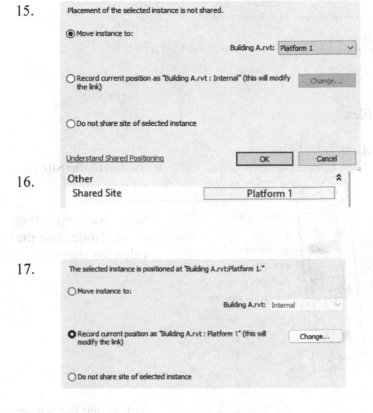

Enable **Move instance to: Platform 1**.

Click **OK**.

16. The Shared Site now displays Platform 1.

Click on the **Platform 1** button.

17. Enable **Record current position as Building A.rvt:Platform 1**.

This saves the new position coordinates back to the Building A file.

Click **OK**.

18.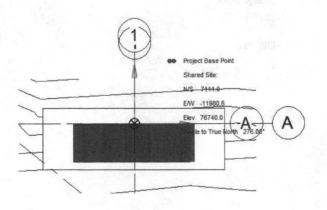

The coordinates on the linked file updates.

Save the file as *ex3-9.rvt*.

19.

You have changed the "current" Position in Building A.rvt. What do you want to do?

→ Save
Saves the new position back to the link.

→ Do not save
Returns to the previously saved position when the link is reloaded or reopened.

→ Disable shared positioning
Retains the current placement of the link and clears the Shared Position parameter.

Click **Save the new position back to the link**.

20. Close all open files.

Open *Building A.rvt*.

21.

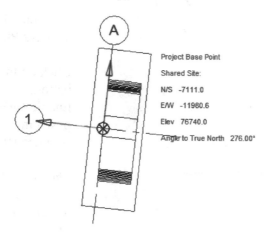

Switch to the **Site** plan view.

Select the Project Base Point. Notice that the value has changed.

22.

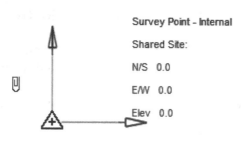

Notice that the Survey Point is still at 0,0.

23.

①

Ⓐ

Open the Level 1 view.

The building is oriented horizontally and vertically on the screen to make it easy to work on.

24.

⊟ [□] Views (all)
   ⊟ Floor Plans
      ▢ Level 1
      ▢ Level 2
      ▢ **Site**

Open the **Site** plan.

25.

R
RVT

Link
Revit

Switch to the **Insert** ribbon.

Select **Link Revit**.

26.

File name:	ex3-9.rvt
Files of type:	RVT Files (*.rvt)
Positioning:	Auto - By Shared Coordinates

Select *ex3-9.rvt*.
Set the Positioning to **Auto - By Shared Coordinates**.

Click **Open**.

27. Notice that the terrain is oriented perfectly to the building.

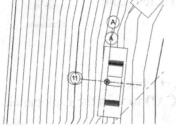

This is because we are using the same shared coordinates between the two files.

Close all files.

Exercise 3-10
Using Project North

Drawing Name: **Import Site.Dwg**
Estimated Time to Completion: 40 Minutes

Scope

Link an AutoCAD file.
Set Shared Coordinates.
Set Project North.

Every project has a project base point ⊗ and a survey point △, although they might not be visible in all views because of visibility settings and view clippings. They cannot be deleted.

The project base point defines the origin (0,0,0) of the project coordinate system. It also can be used to position the building on the site and for locating the design elements of a building during construction. Spot coordinates and spot elevations that reference the project coordinate system are displayed relative to this point.

The survey point represents a known point in the physical world, such as a geodetic survey marker. The survey point is used to correctly orient the building geometry in another coordinate system, such as the coordinate system used in a civil engineering application.

Solution

1. Start a new project file using the **DefaultMetric** template.

    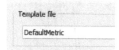

2. Activate the **Site** floor plan.

3. 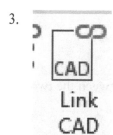   Activate the **Insert** ribbon.
    Select the **Link CAD** tool on the Link panel.

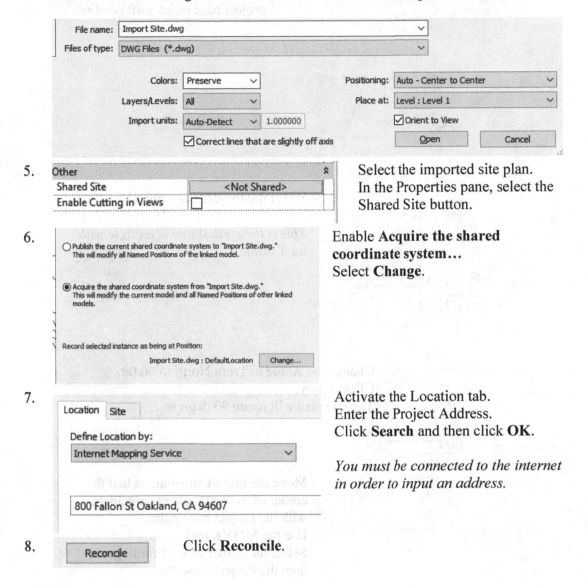

4. Select the *Import Site* drawing. Set Colors to **Preserve**. Set Import Units to **Auto-Detect**. Set Positioning to **Auto - Center to Center**. Click **Open**.

File name: Import Site.dwg

Files of type: DWG Files (*.dwg)

Colors: Preserve

Layers/Levels: All

Import units: Auto-Detect   1.000000

☑ Correct lines that are slightly off axis

Positioning: Auto - Center to Center

Place at: Level : Level 1

☑ Orient to View

Open   Cancel

5. Other ⊼

Shared Site   <Not Shared>

Enable Cutting in Views ☐

Select the imported site plan. In the Properties pane, select the Shared Site button.

6. ○ Publish the current shared coordinate system to "Import Site.dwg." This will modify all Named Positions of the linked model.

◉ Acquire the shared coordinate system from "Import Site.dwg." This will modify the current model and all Named Positions of other linked models.

Record selected instance as being at Position:

Import Site.dwg : DefaultLocation   Change...

Enable **Acquire the shared coordinate system...**
Select **Change**.

7. Location | Site

Define Location by:

Internet Mapping Service

800 Fallon St Oakland, CA 94607

Activate the Location tab.
Enter the Project Address.
Click **Search** and then click **OK**.

You must be connected to the internet in order to input an address.

8. Reconcile   Click **Reconcile**.

9.

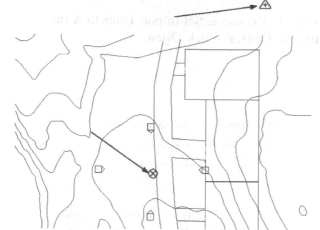

Open the **Site** plan view.

Note that the survey point and project base point shift position.

10.

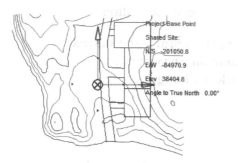

Select the Project Base point.

This is the symbol that is a circle with an X inside.

11.

Project Base Point (1)	
Identity Data	
N/S	-141666.8
E/W	-68891.1
Elev	41404.8
Angle to True North	90.00°

Change the Angle to Truth North to **90.00**.
Click **Apply**.
The site plan will rotate 90 degrees.

12.

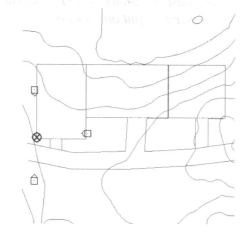

Move the import site plan so that the corner of the rectangle is coincident with the project base point.
Use the **MOVE** tool.
Select the corner of the building and then the Project Base Point.

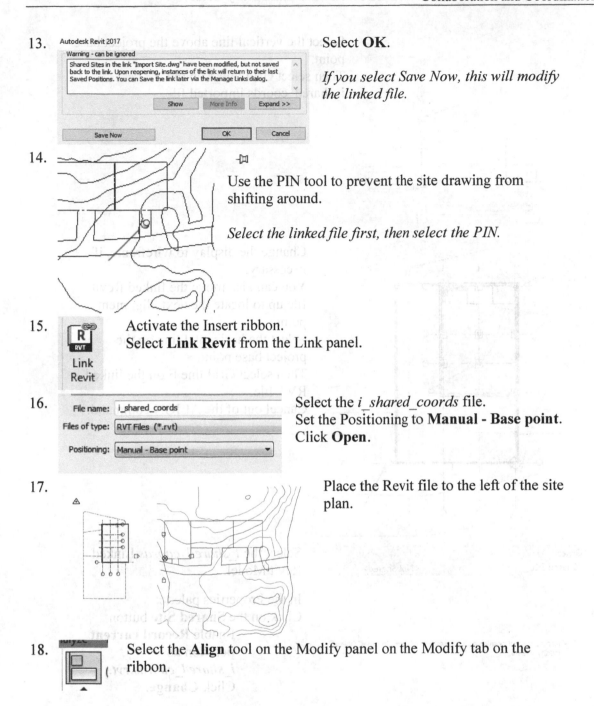

13. Select **OK**.

If you select Save Now, this will modify the linked file.

14. Use the PIN tool to prevent the site drawing from shifting around.

Select the linked file first, then select the PIN.

15. Activate the Insert ribbon.
Select **Link Revit** from the Link panel.

16. Select the *i_shared_coords* file.
Set the Positioning to **Manual - Base point**.
Click **Open**.

17. Place the Revit file to the left of the site plan.

18. Select the **Align** tool on the Modify panel on the Modify tab on the ribbon.

19.

Select the vertical line above the project base point.
Then select Grid line 1 on the i_shared_coords imported file.

RVT Links : Linked Revit Model : i_shared_coords.rvt : 1 : location <Not Shared> : Reference

20.

Change the display to wireframe, if necessary.
You can also move the linked Revit file up to locate the next alignment point.
Select the horizontal line on the project base point.
Then select Grid line E on the linked RVT file.
Cancel out of the ALIGN command.

21.

Other	⌃
Shared Site	<Not Shared>

Select the *i_shared_coords* Linked Revit Model.

In the Properties palette:
Click on the **Shared Site** button.

22.

◉ Record current position as "i_shared_coords.rvt : Internal" (this will modify the link)   [Change...]

Enable **Record current position as** *i_shared_coords.rvt*.
Click **Change**.

23.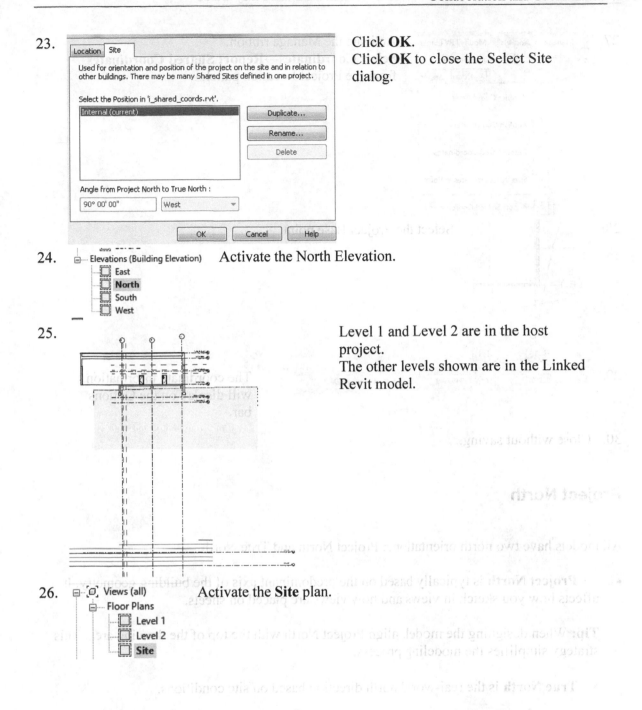

Location | **Site**

Used for orientation and position of the project on the site and in relation to other buildings. There may be many Shared Sites defined in one project.

Select the Position in 'i_shared_coords.rvt'.

Internal (current)

Duplicate...
Rename...
Delete

Angle from Project North to True North :

90° 00' 00"   West

OK   Cancel   Help

Click **OK**.
Click **OK** to close the Select Site dialog.

24. Elevations (Building Elevation)
East
North
South
West

Activate the North Elevation.

25. Level 1 and Level 2 are in the host project.
The other levels shown are in the Linked Revit model.

26. Views (all)
Floor Plans
Level 1
Level 2
Site

Activate the **Site** plan.

27. 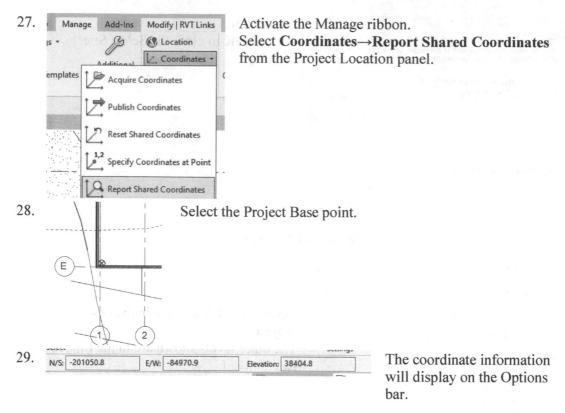 Activate the Manage ribbon.
Select **Coordinates→Report Shared Coordinates** from the Project Location panel.

28. Select the Project Base point.

29. | N/S: -201050.8 | E/W: -84970.9 | Elevation: 38404.8 |

The coordinate information will display on the Options bar.

30. Close without saving.

Project North

All models have two north orientations: Project North and True North.

- **Project North** is typically based on the predominant axis of the building geometry. It affects how you sketch in views and how views are placed on sheets.

Tip: When designing the model, align Project North with the top of the drawing area. This strategy simplifies the modeling process.

- **True North** is the real-world north direction based on site conditions.

Tip: To avoid confusion, define True North only after you begin modeling with Project North aligned to the top of the drawing area and after you receive reliable survey coordinates.

All models start with Project North and True North aligned with the top of the drawing area, as indicated by the survey point △ and the project base point ⊗ in the site plan view.

You may want to rotate True North for the following reasons:

- to represent site conditions
- for solar studies and rendering to ensure that natural light falls on the correct sides of the building model
- for energy analysis
- for heating and cooling loads analysis

Exercise 3-11
Using True North vs Project North

Drawing Name: parcel map.rvt
Estimated Time: 20 minutes

Scope

- ❑ Project North
- ❑ True North

Solution

1.   Views (all)
         Floor Plans
           Level 1
           Level 2
           Site

Open the **Site** floor plan.

2.

Floor Plan: Site	⌄ 🔲 Edit Type
Parts Visibility	Show Original
Visibility/Graphics Overri...	Edit...
Graphic Display Options	Edit...
Orientation	Project North
Wall Join Display	Clean all wall joins

On the Properties panel:

Verify that the Orientation is set to **Project North**.

3.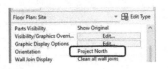

Select **Model Line** from the Architecture ribbon.

4.

Draw a horizontal line from the east quadrant of the cul de sac to the right.

5.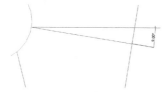

Place an Angular dimension to determine the orientation of the parcel map.

6.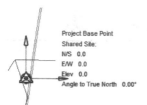

The angle value is 9°.

Delete the model line and the angular dimension.

7.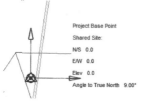

Select the Project Base Point.

8.

Change the Angle to True North to **9°**.

9.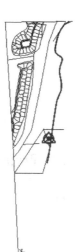

Window around all the elements in the view.

10.  Select **Filter** on the ribbon.

11. Scroll down to the bottom and uncheck:

- Project Base Point
- Survey Point
- Views

Click **OK**.

12. Select **Rotate** on the Modify tab on the ribbon.

13. 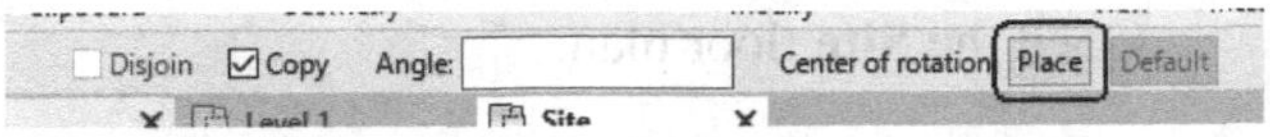 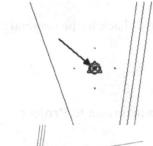 Click **Place** next to Center of Rotation on the Options bar.

Disable **Copy**.

14.  Select the Project Base Point.

15. 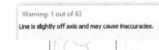 Select a horizontal point to the right of the Project Base Point as the starting point of the angle.

Drag the mouse up to set the end angle at 9°.

16. You can ignore the warning box and close it.

Warning: 1 out of 63

Line is slightly off axis and may cause inaccuracies.

17. 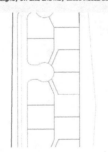 The parcel map now appears to show the parcels oriented orthogonally.

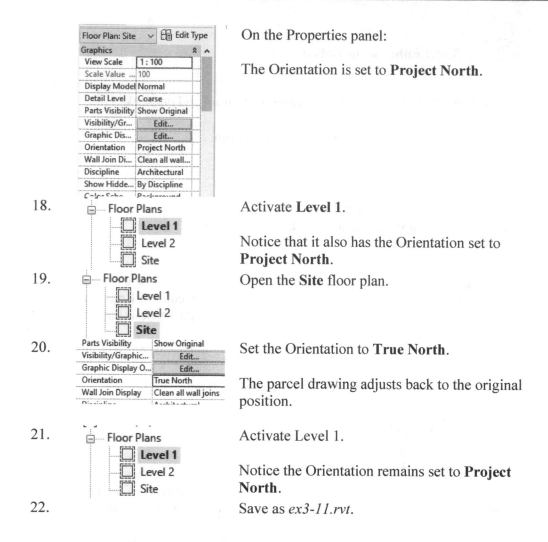

On the Properties panel:

The Orientation is set to **Project North**.

18.   Activate **Level 1**.

Notice that it also has the Orientation set to **Project North**.

19.   Open the **Site** floor plan.

20.   Set the Orientation to **True North**.

The parcel drawing adjusts back to the original position.

21.   Activate Level 1.

Notice the Orientation remains set to **Project North**.

22.   Save as *ex3-11.rvt*.

Control Link Visibility

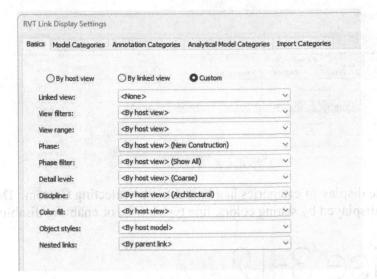

When you link a Revit file to a host project, you have the ability to control how elements and categories are displayed in the the host file without making changes to the linked file.

You can define view filters to change the display of categories to make it easier to identify building components.

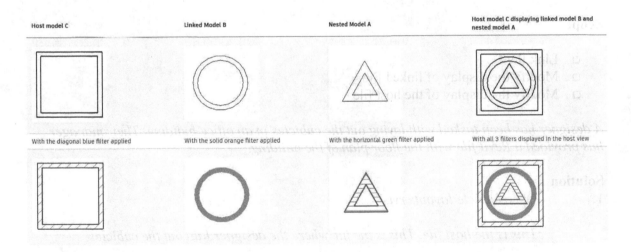

| Host model C | Linked Model B | Nested Model A | Host model C displaying linked model B and nested model A |

| With the diagonal blue filter applied | With the solid orange filter applied | With the horizontal green filter applied | With all 3 filters displayed in the host view |

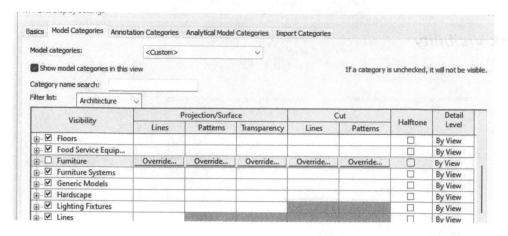

Visibility	Projection/Surface			Cut		Halftone	Detail Level
	Lines	Patterns	Transparency	Lines	Patterns		
⊞ ☑ Floors						☐	By View
⊞ ☑ Food Service Equip...						☐	By View
⊞ ☐ Furniture	Override...	Override...	Override...	Override...	Override...	☐	By View
⊞ ☑ Furniture Systems						☐	By View
⊞ ☑ Generic Models						☐	By View
⊞ ☑ Hardscape						☐	By View
⊞ ☑ Lighting Fixtures						☐	By View
⊞ ☑ Lines						☐	By View

You can control the display of categories in a linked file by selecting Custom. Define how categories will be displayed by setting colors, line types, fills, or enabling/disabling visibility.

Exercise 3-12
Configure Link Display Settings

Drawing Name: cubicle layouts.rvt, office building.rvt
Estimated Time: 15 minutes

Scope

- ❑ Link RVT
- ❑ Modify the display of linked file
- ❑ Modify the display of the host file

A designer has been tasked with laying out the cubicles in an office building. Their manager has provided a Revit file with the floor plan of the building.

Solution

1.   Open *cubicle layouts.rvt.*

      This is the host file. This is the file where the designer laid out the cubicles.

2.

      Manage
      Links

      Switch to the Manage ribbon.

      Click **Manage Links**.

3.

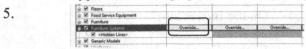

Verify that the *office building.rvt* file is loaded.

Click **OK**.

4. Type **VV** to bring up the Visibility/Graphics dialog.

5.

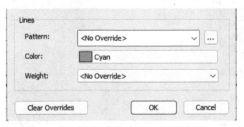

Locate **Furniture Systems** under Model Categories.

Click **Override** in the Lines column. Change the color to **Blue**.

6.

Click **OK.**

7.

Locate **Furniture** under Model Categories.

Click **Override** in the Lines column.

8.

Click the Color button.

Select the **Cyan** color.

Click **OK**.

9.

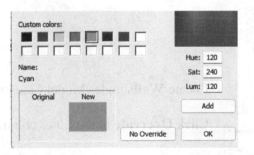

The color should be set to Cyan.

Click **OK**.

10.

Click the Revit Links tab.
Click **By Host View**.

11.

On the Basics tab:

Enable **Custom**.

12.

Open the Model Categories tab.

Set the Model categories to **Custom**.

13.

Type **wall** in the Category name search field.

14.

Locate **Walls** under Model Categories.

Click **Override** in the Lines column.

15.

Click the color button to assign a new color.

16.

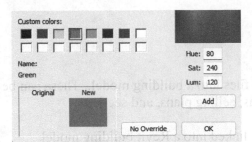

Select the color **Green**.

Click **OK**.

17.

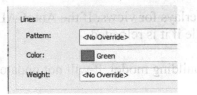

Verify that the color Green is now assigned.

Click **OK**.

18.

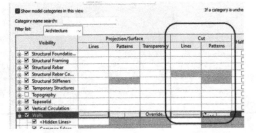

Assign the Color **green** to the Lines column under Cut.

Assign the **Concrete** fill pattern to the column under Cut.

Click **OK**.

Close the dialog box.

19.

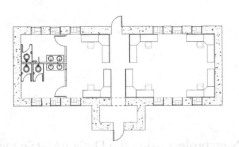

The view updates with the colors and hatch patterns assigned.

Note that the default fill pattern is displayed where the windows are located.

Save as *ex3-12.rvt*.

Extra: *Can you change the colors for the bathroom stalls and sinks? Can you change the color assigned to the doors? How do you determine whether an element belongs to a linked file or a host file?*

Importing Files

Revit allows you to add raster images or PDF files into a building model. These can be inserted into 2D views – elevations, floor plans, ceiling plans, and sections.

AutoCAD DWG files can also be imported or linked into a Revit building model.

Linking AutoCAD files allows them to be used as underlays for views. If the AutoCAD DWG is modified, it will automatically update in the Revit file if it is reloaded.

If you import an AutoCAD file, it is inserted into the building model. It will not update if the external file is changed.

Exercise 3-13

Import DWG

Drawing Name: property survey.dwg
Estimated Time: 10 minutes

Scope

- ❏ Import CAD
- ❏ Property Line
- ❏ Toposolid

Solution

1.   Start a New project using the DefaultMetric template.

   Click **OK**.

2.   Switch to the Insert ribbon.

   Select **Import CAD**.

3.

File name: property survey.dwg
Files of type: DWG Files (*.dwg)

Colors: Preserve
Layers/Levels: All
Import units: feet   1.000000
☑ Correct lines that are slightly off axis

Positioning: Auto - Origin to Internal Origin
Place at: Level : Level 1
☑ Orient to View
Open   Cancel

Select the *property survey.dwg*.

Set Colors to **Preserve**.
Set Positioning to **Auto- Origin to Internal Origin**.
Set Layers/Levels to **All**.
Set Import Units to **feet**.
Click **Open**.

4.

├─ Floor Plans
│   ├─ 🔲 Level 1
│   ├─ 🔲 Level 2
│   └─ 🔲 **Site**

Open the **Site** plan view.

5.

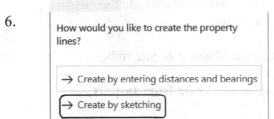

Property Line

Switch to the Massing & Site ribbon.

Select **Property Line**.

6.

How would you like to create the property lines?

→ Create by entering distances and bearings

→ Create by sketching

Select **Create by sketching**.

7.

Draw

Select the **Pick Line** tool.

Select the elements in the property survey drawing.

8.

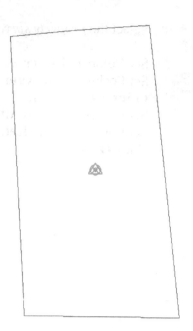

You should see a completed outline.

Select the Green Check on the ribbon to complete.

If you get a warning, zoom in and see if you missed any portions of the outline.

Use **Edit Sketch** to add the missing sections.

9.

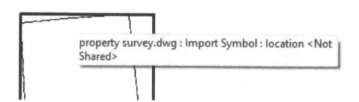

Switch back to the Massing & Site ribbon.

Select **Toposolid →Create from Import**.

10.

Click **Create from CAD** on the ribbon.

Select the imported drawing.

property survey.dwg : Import Symbol : location <Not Shared>

11.

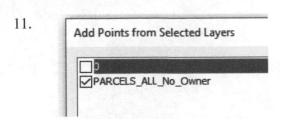

Enable the
PARCELS_All_No_Owner layer.

Click **OK**.

12.

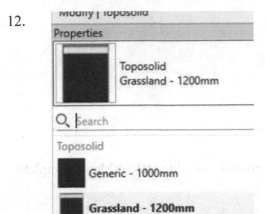

Set the Toposolid to the **Grassland -
1200 mm** type on the Properties panel.

13.

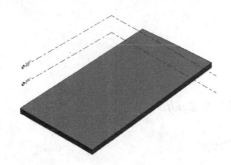

Switch to a 3D view.

Change the display to **Shaded**.

Save as *ex3-13.rvt*.

Exercise 3-14
Import PDF

Drawing Name: bldg._8_2.pdf
Estimated Time: 5 minutes

Scope

❑ Import PDF

Solution

1.

Start a New project using the DefaultMetric template.

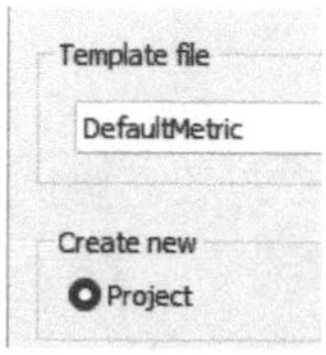

Click **OK**.

2.

Switch to the Insert ribbon.

Select **Import PDF**.

3.

Select *bldg._8_2.pdf*.

Click **Open.**

4.
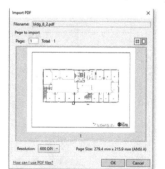
Set the Resolution to **600 DPI**.

Click **OK.**

Left click to place the PDF in the view.

Left click anywhere in the window to release the selection.

5.
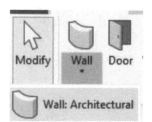
Select the **Wall:Architectural** tool from the Architecture ribbon.

6. Use the Type Selector on the Properties panel to select the **Interior – 79 mm Partition (1-hr)** wall.

7. Set the Location Line to: **Finish Face: Exterior** on the Options bar.

8.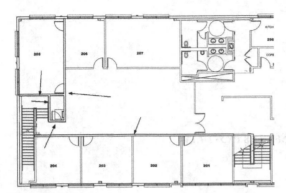

Try tracing over a few of the interior walls.

Notice that you cannot snap to any of the PDF elements.

Exit out of the command.

9. 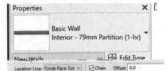 Select the imported pdf.

Toggle **Enable Snaps** on the ribbon so it is ON.

10. Select the **Wall:Architectural** tool from the Architecture ribbon.

11. Use the Type Selector on the Properties panel to select the **Interior – 79 mm Partition (1-hr)** wall.

12. Set the Finish Face to **Exterior** on the Options bar.

13.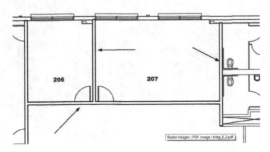

Try tracing over a few of the interior walls.

Notice that now you can snap to the PDF elements to place walls.

14.

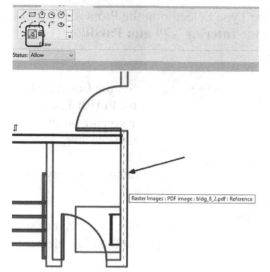

You can also use the **Pick Lines** tool to place walls.

Raster Images : PDF image : bldg_8_2.pdf : Reference

15.

Save as *ex3-14.rvt*.

Exercise 3-15

Import Image

Drawing Name: wrexham.jpg
Estimated Time: 5 minutes

Scope

   ❑ Import Image

Solution

1.   Start a New project using the DefaultMetric template.

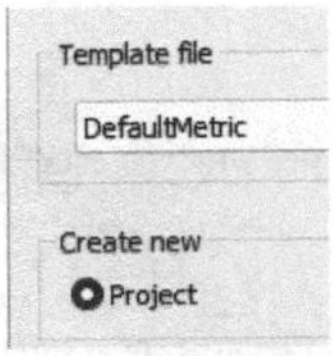

   Click **OK**.

2. 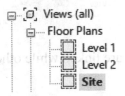   Open the **Site** plan view.

3.

   Switch to the Insert ribbon.

   Select **Import Image**.

4. 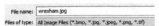   Select *wrexham.jpg*.

   Click **Open**.

   Left click to place in the view.

5.    Use the grips at the corners to enlarge the image.

6.

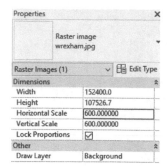

On the Properties panel, change the Horizontal Scale to **600.**

Notice that if Lock Proportions is enabled, the Vertical Scale will be the same as the Horizontal Scale.

7.

Save as *ex3-15.rvt.*

Link Vs. Import

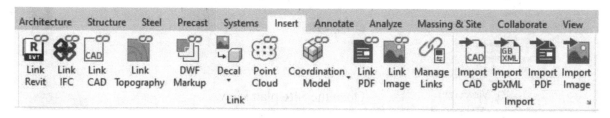

If you study the Insert ribbon, you will notice that some file types are linked while others are imported.

You can link the following file types:

- Revit
- DWG
- DXF
- MicroStation® DGN
- ACIS (SAT)
- OBJ
- Rhinoceros® 3DM
- Trimble® SKP
- STL
- FormIt (AXM)

You can import data from other CAD programs into a Revit model.

Supported CAD formats include:
- DWG
- DXF
- MicroStation® DGN
- ACIS (SAT)
- OBJ
- Rhinoceros® 3DM
- Trimble® SKP
- STL
- FormIt (AXM)

Additionally, you can import PDF, gbXML, and image files.

A link in Revit acts similarly to an external reference to AutoCAD. The linked file exists externally from the host file. If the linked file is updated, the user can reload the linked file and see the updates in the host file. An imported file is static. If changes are made to the imported file, you would have to delete the imported file instance and re-insert it to display the changes.

- If you **import** the file, you can explode the nested xrefs to Revit elements. However, if the xref file is updated after the import, Revit will not automatically reflect changes to the xref file.

- If you **link** the file, Revit automatically updates the geometry to reflect changes to the xref files. However, you cannot explode the nested xrefs to Revit elements.

Exercise 3-16
Linked vs Import

Drawing Name:	cubicle layouts_2.rvt, office building.dwg, office building – revised.dwg
Estimated Time:	20 minutes

Scope

- ❑ Import CAD
- ❑ Linked CAD

A designer has been asked to manage the space planning for a new office building. They must decide which method is better - to import the provided file or link the provided file.

Solution

1.   Open *cubicle layouts_2.rvt.*

2.   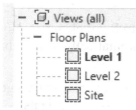 Open the **Level 1** plan view.

3.   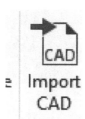 Open the Insert ribbon.

   Click **Import CAD**.

4.   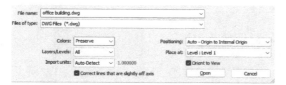 Locate the *office building.dwg* file.

   Set Colors to **Preserve**.
   Set Import Units to **Auto-Detect**.
   Set Positioning to **Auto – Origin to Internal Origin**.

Click **Open**.

5.

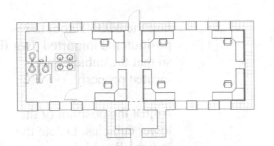

Use the **MOVE** tool to re-position the imported dwg file so that the cubicles are located properly.

You receive an email saying that the floor plan has been modified. You need to replace the imported drawing with an updated file.

6.

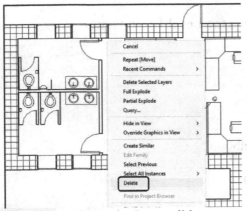

Select the imported file.

Unpin it, if necessary.

Right click and select **Delete**.

7.

Open the Insert ribbon.

Click **Import CAD**.

8.

Locate the *office building - revised.dwg* file.

Set Colors to **Preserve**.
Set Import Units to **Auto-Detect**.
Set Positioning to **Auto – Origin to Internal Origin**.

Click **Open**.

9.

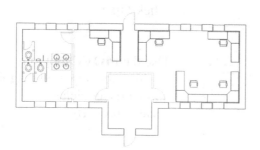

Unpin the imported floor plan.

Use the **MOVE** tool to re-position the imported dwg file so that the cubicles are located properly.

Adjust the position of the lower cubicles. Delete the lower left cubicle.

10.   Save as *ex3-16.rvt*.

11.   Open *cubicle layouts_2.rvt*.

12.

— [□] Views (all)

- — Floor Plans
    - [□] **Level 1**
    - [□] Level 2
    - [□] Site

Open the **Level 1** plan view.

13.

CAD

Link
CAD

Open the Insert ribbon.

Click Link CAD.

14.

File name:	office building.dwg
Files of type:	DWG Files (*.dwg)

Colors: Preserve   Positioning: Auto - Origin to Internal Origin
Layers/Levels: All   Place at: Level : Level 1
Import units: Auto-Detect   1.000000   ☑ Orient to View
☑ Correct lines that are slightly off axis   Open   Cancel

Locate the *office building.dwg* file.

Set Colors to **Preserve**.
Set Import Units to **Auto-Detect**.
Set Positioning to **Auto – Origin to Internal Origin**.

Click **Open**.

15.

Select the linked file.
Unpin it.

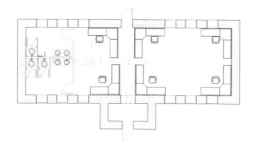

Use the **MOVE** tool to re-position the linked dwg file so that the cubicles are located properly.

You receive an email saying that the floor plan has been modified. You need to replace the imported drawing with an updated file.

16.   Select **Manage Links** from the Insert ribbon.

17.   Open the CAD Formats tab.

Highlight *office building.dwg*.

Click **Reload From**...

Notice that if you enable Preserve graphic overrides, you can control any overrides that have been defined.

Link Name	Status	Positions Not Saved	Size
office building.dwg	Loaded		52.3 KB

Save Positions | Reload From... | Reload | Unload

☑ Preserve graphic overrides

18. | File name: | office building-revised.dwg |
     |-----------|-----------------------------|
     | Files of type: | DWG Files (*.dwg) |

Select the *office building-revised.dwg*.

Click **Open**.

19.   *The linked file is replaced.*

Click **OK**.

Link Name	Status
office building-revised.dwg	Loaded

20.

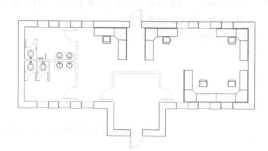

Adjust the position of the linked file.

Adjust the position of the cubicles so they fit properly in the building.

Delete the lower left cubicle.

21.   Save as *ex3-16_linked.rvt*.

Extra: *What are the advantages of importing vs. linking?*

Export to DWG

Many companies continue to use AutoCAD. AutoCAD uses a dwg file extension. Your company may find itself working with an outside vendor that would prefer the drawing sheets or building model in AutoCAD format to make it easier for them to modify and use. If you find yourself in this situation, the hardest thing to control is any revisions to the project. Be sure to have strong communication in place so that you can be informed of any changes in the project regardless of what software is being used. AutoCAD can use 3D models, but all the metadata (parameters and properties will be lost).

American Institute of Architects Standard (AIA)
American Institute of Architects Standard (AIA)
ISO Standard 13567 (ISO 13567)
Singapore Standard 83 (CP83)
British Standard 1192 (BS1192)
Load settings from file...

You can assign each Revit category to a corresponding AutoCAD layer using different standards or create your own standard and load from a *.txt file.

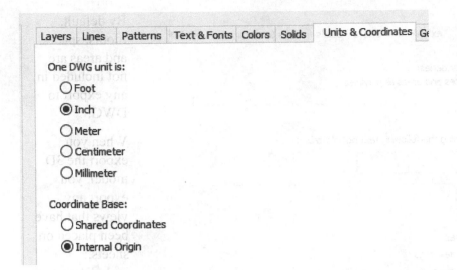

Before you export, make sure you set the correct units for the DWG file and where you want the internal origin to be specified.

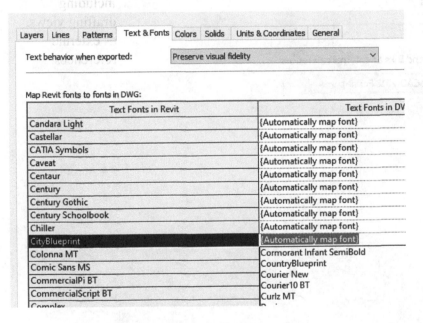

Revit uses Microsoft ttf fonts. AutoCAD uses shx fonts and ttf fonts.

To ensure text and dimensions look proper, you can map the font you are using to a known AutoCAD font.

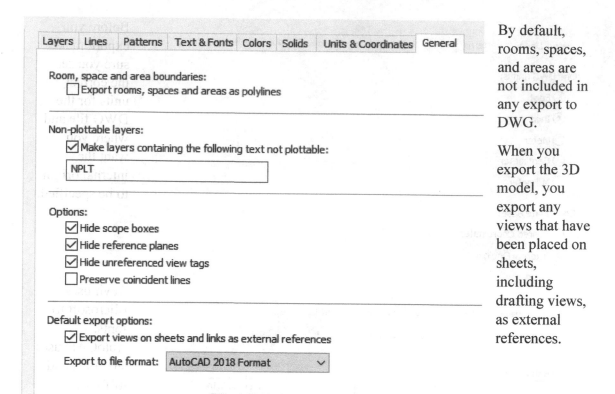

By default, rooms, spaces, and areas are not included in any export to DWG.

When you export the 3D model, you export any views that have been placed on sheets, including drafting views, as external references.

Exercise 3-17

Export to DWG

Drawing Name: **export dwg.rvt**
Estimated Time to Completion: 5 Minutes

Scope
Export sheets to DWG format.

Solution

1.  Go **File→Export→CAD Formats→DWG**.

2. 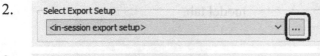 Select **<in-session export setup>**.
   Click the **...** button.
   Each category is assigned an AutoCAD layer.

3. Load layers from standards: **American Institute of Architects Standard (AIA)**.

   Click **OK**.

4. Set Export to **Sheets Only**.

5.

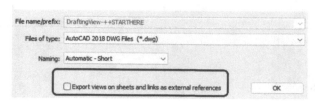

Enable to export only the first sheet A-201.

Click **Next**.

6.

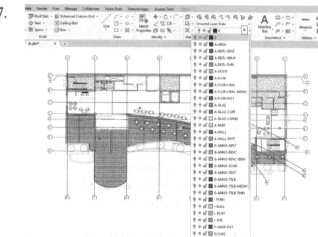

Set the Naming to **Automatic-Short**.

Disable **Export views on sheets and links as external references**.

This will embed any views/links into the drawing file.

Browse to where you want to save the file.

Click **OK**.

7.

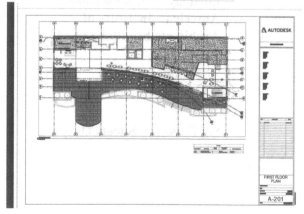

If you have access to AutoCAD, you can open the exported drawing to see how the layers look and how the views are placed.

The floor plan is on the model tab.

8.

The titleblock and schedule are available on the layout tab.

The floorplan is on the layout tab as a view.

9.

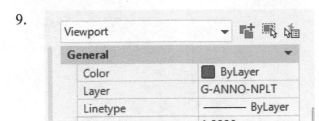

The viewport was automatically placed on the G-ANNO-NPLT layer.

This means the viewport outline will not be plotted.

Errors and Warnings

At any time when working on a project, you can review a list of warning messages to find issues that might require review and resolution.

Warnings display in a dialog in the lower-right corner of the interface. When the warning displays, the element or elements that have an error are highlighted in a user-definable color.

Unlike error messages, warning messages do not prohibit the current action. They just inform you of a situation that may not be your design intent. You can choose to correct the situation or ignore it.

The software maintains a list of warning messages that are displayed and ignored while you are working. The Warnings tool lets you view the list at your convenience to determine if the conditions described in the warnings still exist.

If you receive a file from an outside source, it may be worth your time to review any warnings to determine if there are any issues that may require some resolution.

Exercise 3-18
Review Warnings

Drawing Name: review warnings.rvt
Estimated Time: 60 minutes

Scope

- ❑ Review Warnings
- ❑ Export Warnings
- ❑ Areas
- ❑ Stairs
- ❑ Rooms
- ❑ Floors
- ❑ Element ID

Solution

1.

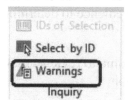

Switch to the Manage ribbon.

Select **Warnings** from the Inquiry panel.

2.

A list of warnings is displayed.

Expand **Warning 1**.

3.

Warnings

⊟ Highlighted elements are joined but do not intersect.
⊟ Warning 1
  ☑ Walls : Basic Wall : Exterior - Insulation on Masonry   id 154427
  ☐ Walls : Basic Wall : Parapet Wall : id 168058
⊞ Warning 2
⊞ Warning 3
⊟ Area separation line is slightly off axis and may cause inaccuracies.
⊞ Warning 4
⊟ Highlighted floors overlap.
⊞ Warning 5
⊟ Actual Number of Risers is different from Desired Number of Risers. Add/remove Risers or change Desired Number of Risers in Stairs Properties.
⊞ Warning 6
⊞ Warning 7

[Show]   [More Info]   [Delete Checked...]

To highlight an element in the graphics window, select it in this tree.

Most standard view commands work without exiting this dialog.

[Export...]   [Close]

Place a check next to the first wall under Warning 1.

Click **Show**.

4.

Error Handling - Viewing Elements   ×

When you view elements, consider the following information:

- You can use standard View commands to see the highlighted elements even while the message dialog is visible.

- The Show command finds views where the relevant elements are visible. Pressing the Show button repeatedly opens each of these views.

- Changing the view to Wireframe might reveal elements unseen in other views.

☑ Do not show me this message again   [Close]

Click **Close**.

Click **Show** again.

5.

All open views that show highlighted elements are already shown. Searching through the closed views to find a good view could take a long time. Continue?

[OK]   [Cancel]

Click **OK**.

6.

A view will open with the wall highlighted.

Click **Close** to close the dialog.

7.

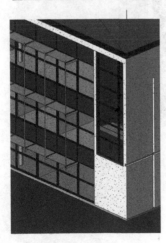

Switch to a 3D view.

8.
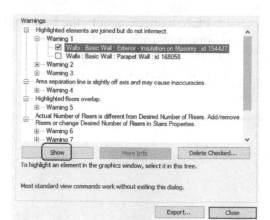

Warnings

Inquiry

Open the Warnings dialog again.

Place a check next to the first wall under Warning 1.

Click **Show**.

9.

The wall highlights.

10.
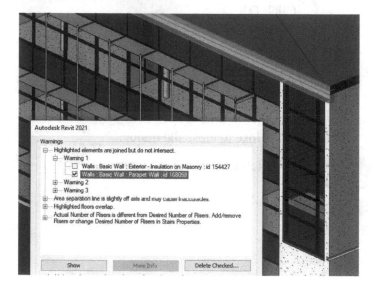

Place a check on the second wall for Warning 1.

The parapet wall highlights.

You can ignore this warning.

11.

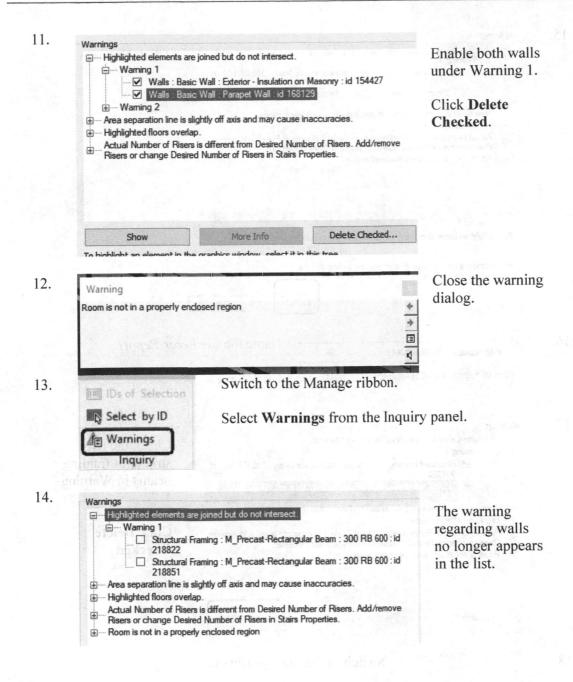

Warnings

- Highlighted elements are joined but do not intersect.
  - Warning 1
    - ☑ Walls : Basic Wall : Exterior - Insulation on Masonry : id 154427
    - ☑ Walls : Basic Wall : Parapet Wall : id 168129
  - Warning 2
- Area separation line is slightly off axis and may cause inaccuracies.
- Highlighted floors overlap.
- Actual Number of Risers is different from Desired Number of Risers. Add/remove Risers or change Desired Number of Risers in Stairs Properties.

| Show | More Info | Delete Checked... |

To highlight an element in the graphics window, select it in this tree.

Enable both walls under Warning 1.

Click **Delete Checked**.

12.

Warning

Room is not in a properly enclosed region

Close the warning dialog.

13.

IDs of Selection

Select by ID

Warnings

Inquiry

Switch to the Manage ribbon.

Select **Warnings** from the Inquiry panel.

14.

Warnings

- Highlighted elements are joined but do not intersect.
  - Warning 1
    - ☐ Structural Framing : M_Precast-Rectangular Beam : 300 RB 600 : id 218822
    - ☐ Structural Framing : M_Precast-Rectangular Beam : 300 RB 600 : id 218851
- Area separation line is slightly off axis and may cause inaccuracies.
- Highlighted floors overlap.
- Actual Number of Risers is different from Desired Number of Risers. Add/remove Risers or change Desired Number of Risers in Stairs Properties.
- Room is not in a properly enclosed region

The warning regarding walls no longer appears in the list.

15.

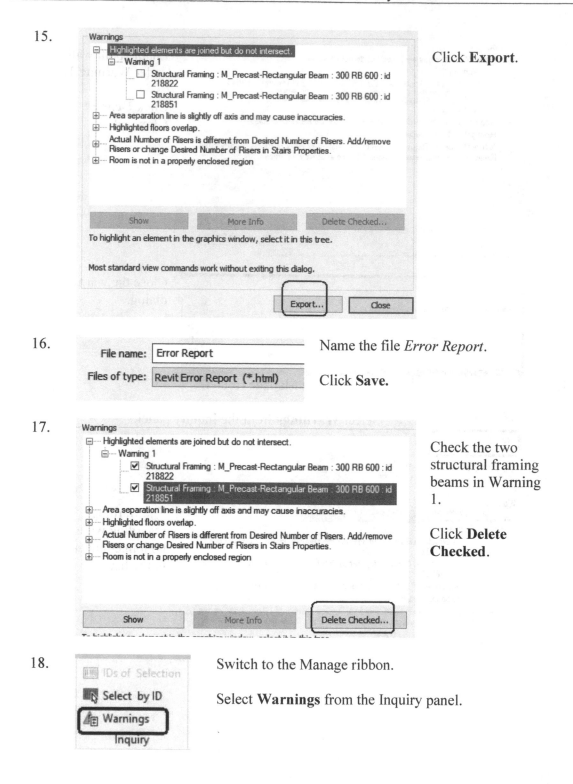

Click **Export**.

16.

Name the file *Error Report*.

Click **Save.**

17.

Check the two structural framing beams in Warning 1.

Click **Delete Checked**.

18.

Switch to the Manage ribbon.

Select **Warnings** from the Inquiry panel.

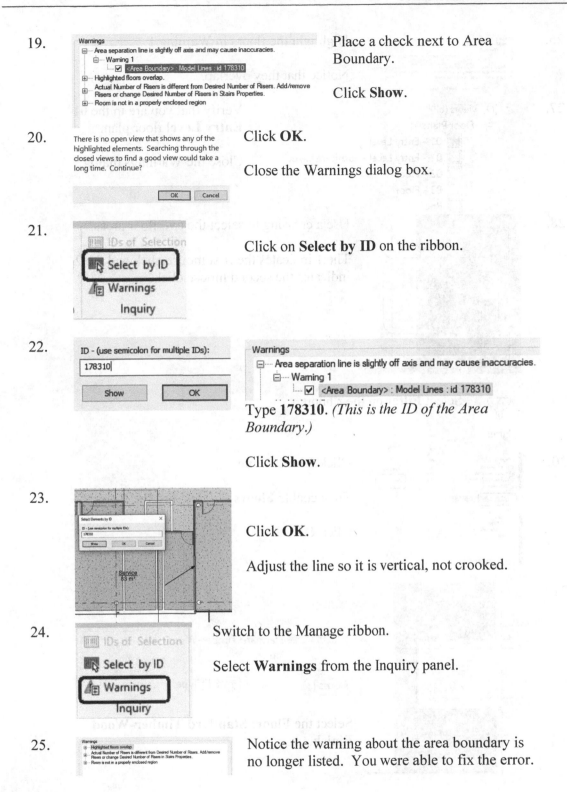

19.

Place a check next to Area Boundary.

Click **Show**.

20.

Click **OK**.

Close the Warnings dialog box.

21.

Click on **Select by ID** on the ribbon.

22.

Type **178310**. *(This is the ID of the Area Boundary.)*

Click **Show**.

23.

Click **OK**.

Adjust the line so it is vertical, not crooked.

24.

Switch to the Manage ribbon.

Select **Warnings** from the Inquiry panel.

25.

Notice the warning about the area boundary is no longer listed. You were able to fix the error.

26. 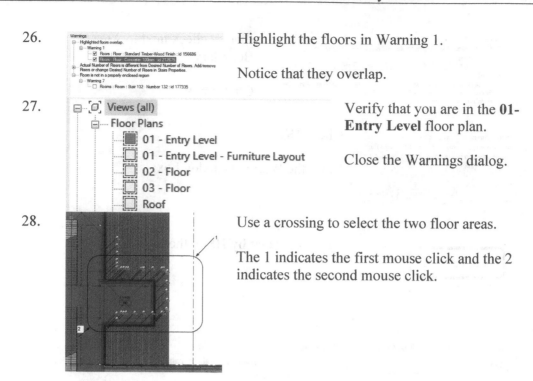 Highlight the floors in Warning 1.

    Notice that they overlap.

27. Verify that you are in the **01-Entry Level** floor plan.

    Close the Warnings dialog.

28. 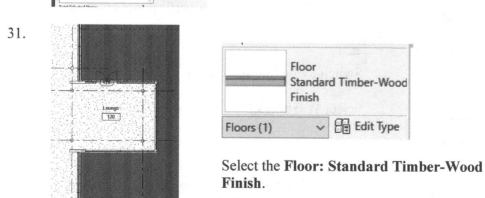 Use a crossing to select the two floor areas.

    The 1 indicates the first mouse click and the 2 indicates the second mouse click.

29. Select **Filter** on the ribbon.

30. Click **Check None**.

    Then enable **Floors** to select just the floors.

    Click **OK**.

31. Floor
    Standard Timber-Wood Finish

    Floors (1)   ⌄   Edit Type

    Select the **Floor: Standard Timber-Wood Finish**.

32. On the ribbon, you will see that there is a warning related to the selected floor.

33. The warning is the warning you already knew about, but it is nice to verify it.

Click **Close** if you opened the Warning dialog.

34. With the **Floor: Standard Timber-Wood Finish** selected, click **Edit Boundary** on the ribbon.

35.

Adjust the boundary so the lines are aligned with the outside edges of the concrete floor.

Green Check to exit the editing mode.

36. Reselect the floor.

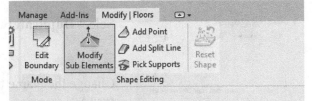

Check to see if the Warning icon appears on the ribbon.

If it does, edit the boundary again and adjust so it no longer overlaps the concrete floor.

37.  Switch to the Manage ribbon.

Select **Warnings** from the Inquiry panel.

38. Expand Warnings 1 through 5.

Can you tell if these warnings apply to only one element or how many elements need to be modified?

39. Place a check next to the Stairs under Warning 1.

Click **Show**.

Close the dialog.

40. Select the stairs.

Scroll down in the Properties panel.

Dimensions	
Width	1625.0
Desired Number of Risers	26
Actual Number of Risers	25
Actual Riser Height	146.2
Actual Tread Depth	300.0

Note that the Desired Number of Risers is 26, but the Actual Number of Risers is 25.

41. Desired Number of Risers is too small. Computed Actual Riser Height is greater than Maximum Riser Height allowed by type.

Change the Desired Number of Risers to 25.

This error will appear.

Click **Close**.

42. Stair 150mm max riser 300mm tread

Click **Edit Type** on the Properties panel.

43. Family: System Family: Stair
Type: 150mm max riser 300mm tread

Select **Duplicate**.

44. Name: 200mm max riser 300mm tread

Change the Name to **200mm max riser 300 mm tread**.

Click **OK**.

45.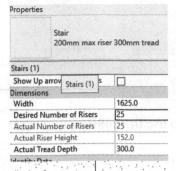

Change the Maximum Riser Height to **200mm**.

Click **OK**.

46.

The Stair now shows the Desired Number of Risers equal to the Actual Number of Risers.

The Actual Riser Height is 152 mm which is about 6 inches high, well within code.

47.

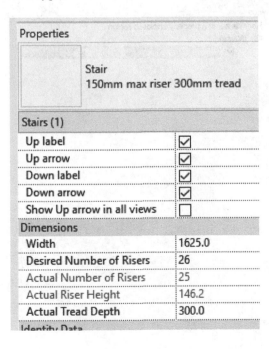

Select the stairs above the stairs you just modified.

This stair has the same error – the desired and actual number of risers do not match.

Use the Type Selector to change the Stair to the new type.

48.

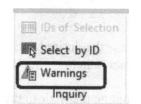

Locate the stair in Room 114.

Right click and select **Select All Instances→In Entire Project**.

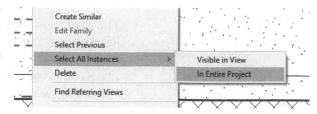

Change the stairs to use the **200mm max riser 300 mm tread**.

Release the selection.

49.

Switch to the Manage ribbon.

Select **Warnings** from the Inquiry panel.

50.

Expand Warning 1.

Click **Show**.

Close the dialog.

51.

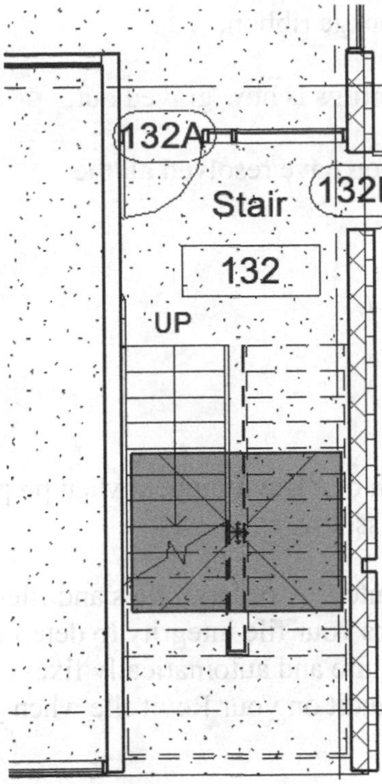

The room is located in Stair 132, but there is no wall on the south side, so the room is not fully enclosed.

52.

Room Separator

Switch to the Architecture ribbon.

Select **Room Separator** on the Room & Area panel.

53.

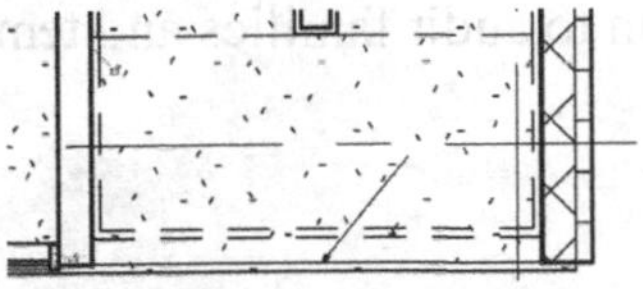

Draw a line between the two vertical lines to enclose the room at the bottom wall.

54.

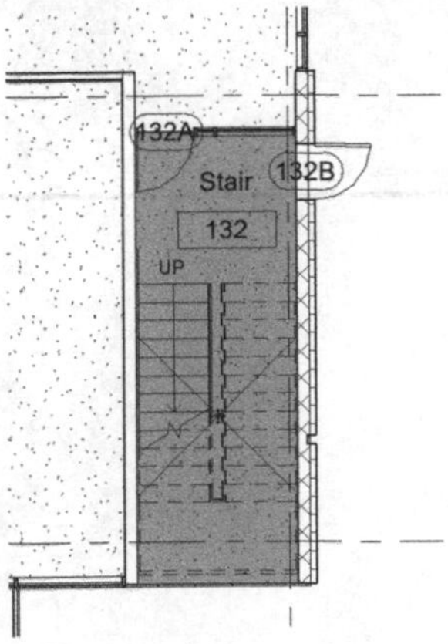

Use a crossing and a filter to select the room again.

The room size has adjusted using the walls and the room separator.

55.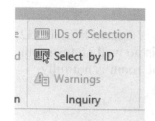

Switch to the Manage ribbon.

Notice that **Warnings** is now grayed out.

This is because you have resolved all the warnings.

56.

Save as *ex3-18.rvt*.

Audit

Use the Audit function periodically to maintain the health of a Revit model, when preparing to upgrade the software, or as needed to locate and correct issues.

Auditing an Autodesk Revit file helps you maintain the integrity of large files and files that more than one user works on. The auditing process checks your file integrity to determine that no corruption has occurred within the file's element structure and automatically fixes any minor file errors that it encounters. You can perform an audit on your Revit file when you open it in the application.

The Audit function scans, detects, and fixes corrupt elements in the model. It does not provide feedback on which elements are fixed.

Auditing a model can be time-consuming, so be prepared to wait while the process completes.

As a best practice, audit the model weekly. If the model is changing rapidly, audit it more often.

In addition to auditing models, you can also use this function to audit families and templates.

Exercise 3-19
Audit a File

Drawing Name: **audit.rvt**
Estimated Time to Completion: 5 Minutes

Scope
Audit a Revit building project

Solution

1.

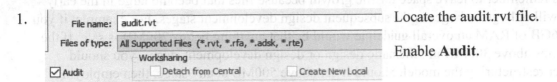

Locate the audit.rvt file.

Enable **Audit**.

2.

This operation can take a long time. Recommended use includes periodic maintenance of large files and preparation for upgrading to a new release.

Click **OK**.

Click **Open**.

3. Save the file and close.

Purge

Purging a Revit model assists in removing unused families, views, and objects from a project. It is recommended that the model should be purged after every submittal and milestone to remove any remaining elements that have accumulated in the project. Revit's automated purging method only removes certain elements within your project; therefore, you still need to go through your models and manually remove any unwanted area schemes, views, groups, and design options. You should also replace in-place families with regular component families to further reduce the file size.

When you consider using *Purge Unused* it is important to understand how this works, mainly because in most cases should you require to remove any unwanted families, objects and views, you would need to employ the 3 step Purge process. The reason for this is when first using *Purge Unused* you can only see the items that are not being currently used in the project. This means that any families that are listed may be using materials that are in the project. So when you purge out a family, its materials now might be unused in the project, and when you next launch the Purge Unused dialog you may see additional materials that can be purged from the project. As a recommendation it is also necessary to select the *Check None* button as your first step when using *Purge Unused*, then look for and select what you would like to purge. This is important because by default everything that can be purged is pre-selected. This means if you have an element in your project template, but you haven't used it yet in the project, it will be removed and you'll have to re-create it or re-load it.

I try to discourage my students from using the Purge tool because they invariably find themselves needing families they have purged and then they have to spend a great deal of time trying to locate them and reload them.

Always remember to leave space for file growth because files that become large in the early stages will only grow larger in the subsequent design development stages. For example, if you have 16GB of RAM an overall guideline would be that anything below 200MB is safe. If the file climbs above 300MB in schematic design or design development stages, you should consider restructuring the model. Should it get above 500MB at any stage, then employing the *Purge Unused* command is strongly suggested.

One way to reduce file size is to use linked files. You can then unload the linked files and reload them when you need to see how elements interact.

Exercise 3-20
Purge a File

Drawing Name: **purge.rvt**
Estimated Time to Completion: 5 Minutes

Scope
Purge a Revit building project

Solution

1.

   Switch to the Manage ribbon.

   Select **Purge Unused** from the Settings panel.

2.

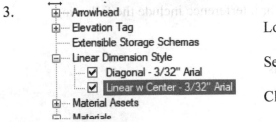

Expand the list.

Notice there is an unused Legend.

This means the legend has not been placed on a sheet, but you may still want to use it.

Browse through the list.

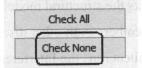

Click **Check None**.

That way you don't accidentally delete an element you may need later.

3.

Locate the **Linear Dimension Style**.

Select the two dimension styles listed.

Click **OK**.

4.   Save as *ex3-20.rvt*.

Interference Checking

The Interference Check tool can find intersections among a set of selected elements or all elements in the model.

Typical Workflow for Interference Checking

This tool can be used during the design process to coordinate major building elements and systems. Use it to prevent conflicts and reduce the risk of construction changes and cost overruns.

The following is a common workflow:

- An architect meets with a client and creates a basic model.

- The building model is sent to a team that includes members from other disciplines, such as structural engineers. They work on their own version of the model. Then the architect links the structural model into his project file and checks for interferences.

- Team members from other disciplines return the model to the architect.

- The architect runs the Interference Check tool on the existing model.

- A report is generated from the interference check, and undesired intersections are noted.

- The design team discusses the interferences and creates a strategy to address them.

- One or more team members are assigned to fix any conflicts.

Elements Requiring Interference Checking

Some examples of elements that could be checked for interference include the following:

- Structural girders and purlins

- Structural columns and architectural columns

- Structural braces and walls

- Structural braces, doors, and windows

- Roofs and floors

- Specialty equipment and floors

- A linked Revit model and elements in the current model

Exercise 3-21
Interference Checking

Drawing Name: **interference_checking.rvt**
Estimated Time to Completion: 15 Minutes

Scope
Describe Interference Checks.
Check and fix interference conditions in a building model.
Generate an interference report.

Solution

1.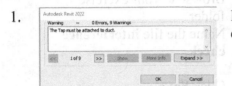
   If this dialog comes up when you open the file,
   click **OK**.

2.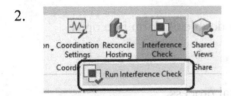
   Activate the **Collaborate** ribbon.
   Select the **Interference Check→ Run
   Interference Check** tool on the Coordinate panel.

3.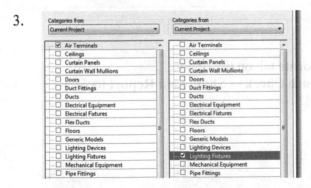
   Enable **Air Terminals** in the left
   panel.
   Enable **Lighting Fixtures** in the right
   panel.

*If you select all in both panels, the check can take a substantial amount of time
depending on the project and the results you get may not be very meaningful.*

4. Click **OK**.

5. Highlight the Lighting Fixture in the first error.

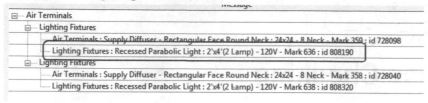

6. 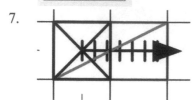 Click the **Show** button.

7. 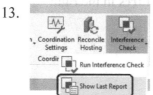 The display will update to show the interference between the two elements.

8.  Select **Export**.

9. 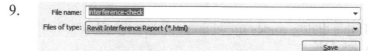 Browse to your exercise folder.
   Name the file **interference check.**

   Click **Save**.

10. Click **Close** to close the dialog box.

11. 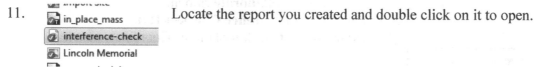 Locate the report you created and double click on it to open.

12. This report can be emailed or used as part of the submittal process.

	A	B
1	Air Terminals : Supply Diffuser - Rectangular Face Round Neck : 24x24 - 8 Neck - Mark 358 : id 728040	Lighting Fixtures : Recessed Parabolic Light : 2'x4'(2 Lamp) - 120V - Mark 638 : id 808320
2	Air Terminals : Supply Diffuser - Rectangular Face Round Neck : 24x24 - 8 Neck - Mark 359 : id 728098	Lighting Fixtures : Recessed Parabolic Light : 2'x4'(2 Lamp) - 120V - Mark 636 : id 808190

End of Interference Report

13. Activate the **Collaborate** tab on the ribbon.
    Select the **Interference Check→ Show Last Report** tool on the Coordinate panel.

14. Highlight the first lighting fixture. Verify that you see which fixture is indicated in the graphics window. Close the dialog.

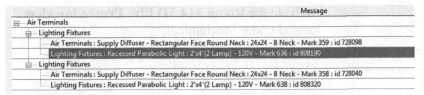

	Message
⊟— Air Terminals	
⊟— Lighting Fixtures	
Air Terminals : Supply Diffuser - Rectangular Face Round Neck : 24x24 - 8 Neck - Mark 359 : id 728098	
Lighting Fixtures : Recessed Parabolic Light : 2'x4'(2 Lamp) - 120V - Mark 636 : id 808190	
⊟— Lighting Fixtures	
Air Terminals : Supply Diffuser - Rectangular Face Round Neck : 24x24 - 8 Neck - Mark 358 : id 728040	
Lighting Fixtures : Recessed Parabolic Light : 2'x4'(2 Lamp) - 120V - Mark 638 : id 808320	

15.

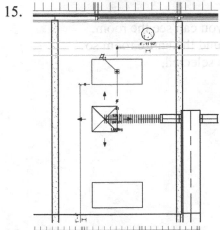

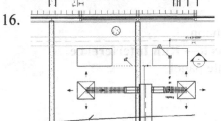

Move the lighting fixture to a position above the air terminal.
Arrange the lighting fixtures and the air diffuser so there should be no interference.

16.

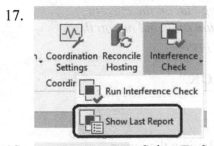

Move the second lighting fixture on the right so it is no longer on top of the air terminal.
Rearrange the fixtures in the room to eliminate any interference.

17.

Activate the **Collaborate** ribbon.
Select the **Interference Check→ Show Last Report** tool on the Coordinate panel.

18. Refresh   Select **Refresh**.

19. The message list is now empty. Close the dialog box.

20.

Q Room 214
— ⊙ Views (Discipline)
└— Mechanical
  ├— FP
  │ └— 3D Views
  │   └— [ ] Room 214 3D Fire Protection
  └— HVAC
    └— Sections (Building Section)
      └— [ ] Room 214 Section

Go to the Project Browser.

Type **Room 214** in the search field.

21. 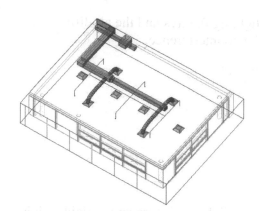 Activate the **Room 214 3D Fire Protection view** located under Mechanical.

22.

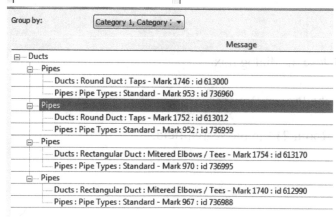

Zoom in so you can see the room. Window around the entire room so everything is selected.

23. Activate the **Collaborate** ribbon.
Select the **Interference Check→ Run Interference Check** tool on the Coordinate panel.

24. *Note that the interference check is being run on the current selection only and not on the entire project.*

Click **OK**.

Categories from		Categories from
Current Selection		Current Selection
☑ Air Terminals		☑ Air Terminals
☑ Duct Fittings		☑ Duct Fittings
☑ Ducts		☑ Ducts
☑ Flex Ducts		☑ Flex Ducts
☑ Mechanical Equipment		☑ Mechanical Equipment
☑ Pipe Fittings		☑ Pipe Fittings
☑ Pipes		☑ Pipes
☑ Sprinklers		☑ Sprinklers

25.

Group by: Category 1, Category : ▼

Message

Review the report.
Click **Close**.

Close all files without saving.

⊟ Ducts
  ⊟ Pipes
    — Ducts : Round Duct : Taps - Mark 1746 : id 613000
    — Pipes : Pipe Types : Standard - Mark 953 : id 736960
  ⊟ Pipes
    — Ducts : Round Duct : Taps - Mark 1752 : id 613012
    — Pipes : Pipe Types : Standard - Mark 952 : id 736959
  ⊟ Pipes
    — Ducts : Rectangular Duct : Mitered Elbows / Tees - Mark 1754 : id 613170
    — Pipes : Pipe Types : Standard - Mark 970 : id 736995
  ⊟ Pipes
    — Ducts : Rectangular Duct : Mitered Elbows / Tees - Mark 1740 : id 612990
    — Pipes : Pipe Types : Standard - Mark 967 : id 736988

Exploding Imported Geometry

When you import a drawing into Revit, you are importing all the elements, such as blocks and external references (xrefs) from the drawing. They are all contained inside an element called an import symbol.

You can explode (disassemble) the import symbol into its next highest level elements: nested import symbols. This is a partial explode. A partial explode of an import symbol yields more import symbols, which, in turn, can be exploded into either elements or other import symbols. This is analogous to exploding in AutoCAD with nested xrefs and blocks. For example, you explode an xref into other xrefs and blocks. Those xrefs and blocks can, in turn, be exploded into more blocks and xrefs.

You can also explode the import symbol immediately into Revit text, curves, lines, and filled regions. This is a full explode.

Note: You cannot explode linked files or an import symbol that would yield more than 10,000 elements.

A good work process if you want to be able to edit and use the elements in a non-Revit CAD file is to import the file into a Revit file, explode it, make the desired changes, save the project. Then, link the Revit file to your host/master file.

Exercise 3-22
Explode a CAD file

Drawing Name: north arrow.dwg, add_symbol.rft
Estimated Time to Completion: 15 Minutes

Scope
Link a DWG file
Import a DWG file.
Use Full explode.
Query
Annotation Symbol

Many architects have accumulated hundreds of symbols that they use in their drawings. This exercise shows you how to take your existing AutoCAD symbols and use them in Revit.

Solution

1.   Locate the *north arrow.dwg* in the Class Files downloaded from the publisher's website.

2.   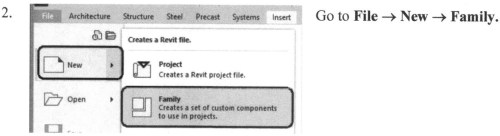   Go to **File → New → Family.**

3.   Browse to the *Annotations* folder.

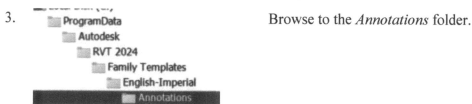

4.   Highlight the *Generic Annotation.rft* file under Annotations.

File name: Generic Annotation
Files of type: Family Template Files (*.rft)

Click **Open**.

5.  Activate the Insert ribbon.

Select **Import CAD**.

6. Locate the *north arrow.dwg* file in the Class Files download.

7.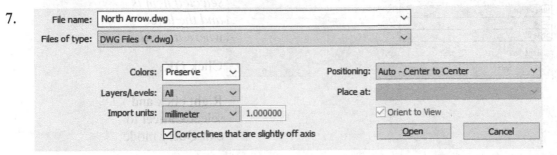

Under Colors:	Enable **Preserve**.
Under Layers:	Select **All**.
Under Import Units:	Select **millimeter**.
Under Positioning:	Select **Auto – Center to Center**.

This allows you to automatically scale your imported data.
Note that we can opt to manually place our imported data or automatically place using the center or origin.

Click **Open**.

8. Import detected no valid elements in the file's Paper space. Do you want to import from the Model space?

The Import tool looks in Paper space and Model space. You will be prompted which data to import.
Click **Yes**.

9. **Zoom In Region** to see the symbol you imported.

10. Pick the imported symbol.

Select **Query** from the ribbon.

11.

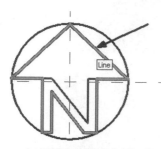

Pick on the line indicated.

12.

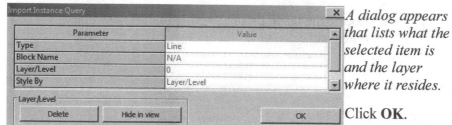

A dialog appears that lists what the selected item is and the layer where it resides.

Click **OK**.

Right click and select **Cancel** to exit Query mode.

13.

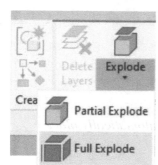

Pick the imported symbol.

Select **Full Explode**.

We can now edit our imported data.

14.

Activate the Create ribbon.

Select the **Filled Region** tool on the Detail panel from the Architecture ribbon.

15.

Enable **Chain** on the status bar.

16.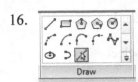
Use the **Pick** tool on the Draw panel to select the existing lines to create a filled arrowhead.

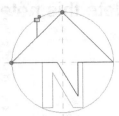

17.
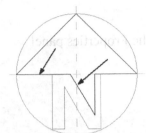
Use the TRIM tool to connect the two lines indicated and close the region outline.

18.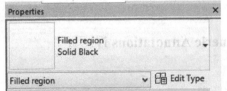

Properties ×

Filled region
Solid Black

Filled region   ⊞ Edit Type

On the Properties palette:
Verify that the Region is set to **Solid Black**.

19. ✕ ✓ Mode

Select the **Green Check** under the Mode panel to **Finish Region**.

20.
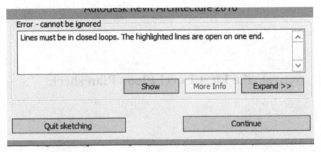

Autodesk Revit Architecture 2010

Error - cannot be ignored

Lines must be in closed loops. The highlighted lines are open on one end.

Show   More Info   Expand >>

Quit sketching   Continue

If you get an error message, check that you have a single closed loop.

21.
Using grips, shorten the horizontal and vertical reference planes if needed.

You will need to unpin the reference planes before you can adjust the link.

Be sure to pin the reference planes after you have finished any adjustments. Pinning the reference planes ensures they do not move.

22. **Delete this note before using**

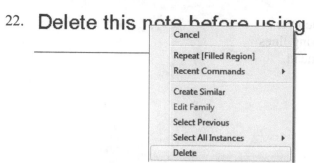

Select the note that is in the symbol.
Right click and select **Delete**.

23.

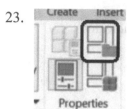

Select **Category and Parameters** under the Properties panel on the Modify ribbon.

24.

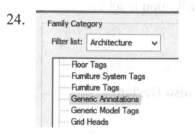

Note that **Generic Annotations** is highlighted.
Click **OK**.

25. File name: north arrow / Files of type: Family Files (*.rfa)

Save the file as a new family called **north arrow**.

26. Open *add_symbol.rvt*.

27.
Sheets (all)
⊞ **A101 - First Level Floor Plan**
⊞ A102 - Second Level Floor Plan
⊞ A103 - Lobby Keynotes
⊞ A104 - Level 1 - East Wing
⊞ A105 - Level 1 - West Wing

Activate the **First Level Floor Plan** sheet.

28. Annotate   Activate the **Annotate** ribbon.

29. Symbol   Select the **Symbol** tool.

30. Select **Load Family**.

31. Locate the *north arrow.rfa* file that was saved to your student folder.

    Click **Open**.

32. Place the North Arrow symbol in your drawing.

    ① Level 1
       1/8" = 1'-0"

33. Save as *ex3-22.rvt*.

Phases

Phases are distinct, separate time periods within the life of a building. Phases can represent either the time periods or how the building appeared during that time period. By default, every Revit project has two phases or time periods already pre-defined. They are named Existing and New Construction.

The most common use of phases is to keep track of the "Before" and "After" scenarios. If you are a recovering AutoCAD user, you probably created a copy of your project and did a "Save As" to re-work the existing building for the proposed remodel. This has a number of disadvantages – not the least of which is the use of external references, the possibility of missing something, and the duplication of data.

If you are working on a completely new building project on a "clean" site, you still might want to use Phases as a way to control when to schedule special equipment on the site as well as crew. You might have a phase for foundation work, a phase for framing, a phase for electrical and so on. You can create schedules based on phases, so you will know exactly what inventory you might need on hand based on the phase.

You determine what elements are displayed in a view by assigning a phase to the view. You can even control colors and linetypes of different phases so you get a visual cue on which elements were created or placed in which phase.

EXISTING PLAN

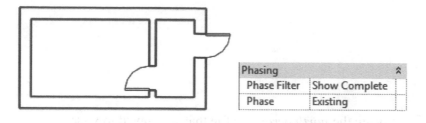

DEMOLITION PLAN

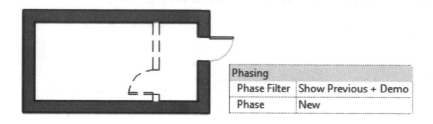

NEW FLOOR PLAN

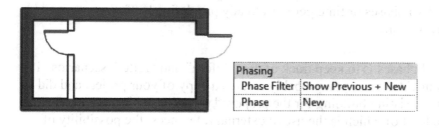

A common error is to create a phase for demolition. This is unnecessary. Instead you can duplicate the view and set one view as the Demolition Plan and one view at the new floor plan.

Exercise 3-23
Phases

Drawing Name: phasing.rvt
Estimated Time to Completion: 75 Minutes

Scope
Properties
Filter
Phases
Rename View
Copy View
Graphic Settings for Phases

Solution

1. 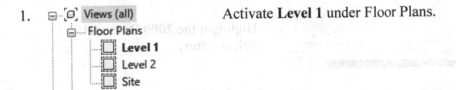   Activate **Level 1** under Floor Plans.

2. 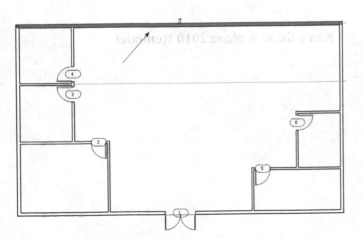   Select the wall indicated.
    It should highlight.

3. 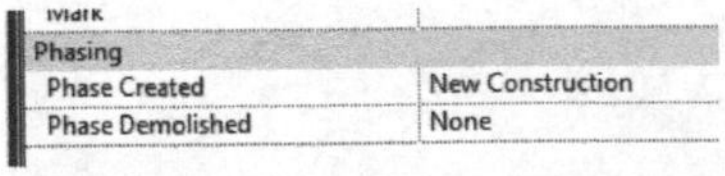 Scroll down to the Phasing category in the
    Properties panel on the upper left.

    This wall was created in the New Construction Phase.
    Note that it is not set to be demolished.

4. Right click and Click **Cancel** to deselect the wall.

5.

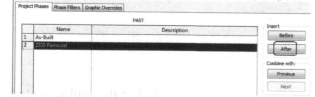

Go to the **Manage** ribbon.
Select **Phases**.

6.

	Name
1	As-Built
2	New Construction

Rename Existing to **As-Built**.

7.

	Name
1	As-Built
2	2000 Remodel

Rename New Construction to **2000 Remodel**.

8.

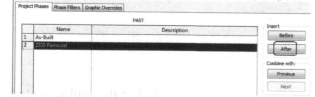

Highlight the **2000 Remodel**.
Select **After**.

9.

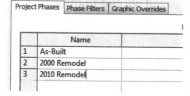

	Name
1	As-Built
2	2000 Remodel
3	2010 Remodel

Name the new phase **2010 Remodel**.

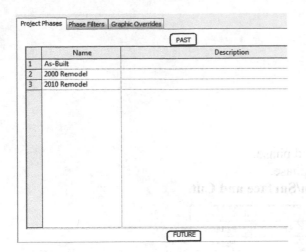

Note that the top indicates the past and the bottom indicates the future to help orient the phases.

10. Select the **Graphic Overrides** tab.

Phase Status	Projection/Surface		Cut		Halftone	Material
	Lines	Patterns	Lines	Patterns		
Existing	———————		———————	Hidden	☐	Phase-Exist
Demolished	----------		----------	Hidden	☐	Phase-Demo
New	———————		———————		☐	Phase-New
Temporary	··············		··············	/////	☐	Phase-Temp

11. Note that in the Lines column for the Existing Phase, the line color is set to gray.

12.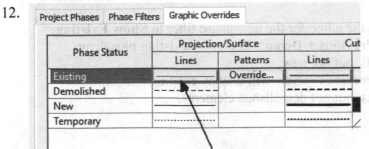

Highlight **Existing**. Click in the **Lines** column and the Line Graphics dialog will display.

Projection/Surface is what is displayed in the floor plan views.
Cut is the display for elevation or section views.
Override indicates you have changed the display from the default settings.

13. Set the Color to **Green** for the Existing phase by selecting the color button. Click **OK**.

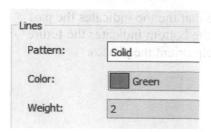

Set the Color to **Blue** for the Demolished phase.
Set the Color to **Magenta** for the New phase.
Change the colors for both Projection/Surface and Cut.

Phase Status	Projection/Surface		Cut	
	Lines	Patterns	Lines	Patterns
Existing	————————		————————	Hidden
Demolished	— — — — — —		— — — — — —	Hidden
New		Override...	————————	
Temporary	....		....	/////

14. Select the **Phase Filters** tab.

Project Phases | **Phase Filters** | Graphic Overrides

15.

	Filter Name	New	Existing	Demolished	Temporary
1	Show All	By Category	Overridden	Overridden	Overridden
2	Show Demo + New	By Category	Not Displayed	Overridden	Overridden
3	Show Previous + Dem	Not Displayed	Overridden	Overridden	Not Displayed
4	Show Previous + New	By Category	Overridden	Not Displayed	Not Displayed
5	Show Previous Phase	Not Displayed	Overridden	Not Displayed	Not Displayed

Note that there are already phase filters pre-defined that will control what is displayed in a view.

16. Click the **New** button on the bottom of the dialog.

New

17.

Project Phases | Phase Filters | Gra

	Filter Name
1	Show All
2	Show Demo + New
3	Show Previous + Dem
4	Show Previous + New
5	Show Previous Phase
6	Show Existing

Change the name for the new phase filter to **Show Existing**.
Show Previous + Demo will display existing plus demo elements, but not new.
Show Previous + New will display existing plus new elements, but not demolished elements.

18. In the New column, select **Overridden**.
In the Existing column, select **Overridden**.
This means that the default display settings will use the new color assigned.
In the Demolished column, select **Not Displayed**.

Project Phases | Phase Filters | Graphic Overrides

	Filter Name	New	Existing	Demolished
1	Show All	By Category	Overridden	Overridden
2	Show Demo + New	By Category	Not Displayed	Overridden
3	Show Previous + Dem	Not Displayed	Overridden	Overridden
4	Show Previous + New	By Category	Overridden	Not Displayed
5	Show Previous Phase	Not Displayed	Overridden	Not Displayed
6	Show Existing	Overridden	Overridden	Not Displayed

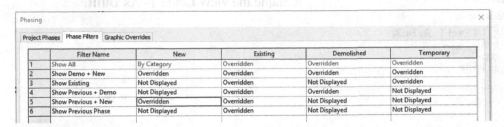

19. Use Overridden to display the colors you assigned to the different phases.
    Verify that in the Show Previous + Demo phase New elements are not displayed.
    Verify that in the Show Previous + New phase Demolished elements are not
    displayed.
    Click **Apply** and **OK** to close the Phases dialog.

20. 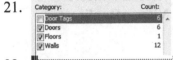 Window around the entire floor plan.
    Select the **Filter** button.

21. 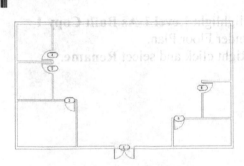 Uncheck **Door Tags**.
    Tags and annotations are not affected by phases.
    Click **OK**.

22. | Phasing | |
    | --- | --- |
    | Phase Created | As-Built |
    | Phase Demolished | None |

    Set the Phase Created to **As-Built**.

23. 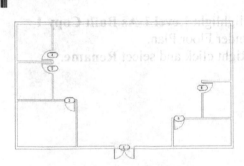 Note that the view changes to display in Green.
    This is because we set the color Green to denote existing elements.

24. 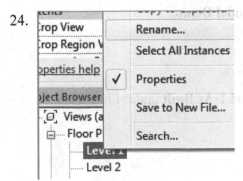 Next, we create three Level 1 floor plan views for each phase.
    Highlight **Level 1** under Floor Plan.
    Right click and select **Rename**.

25. 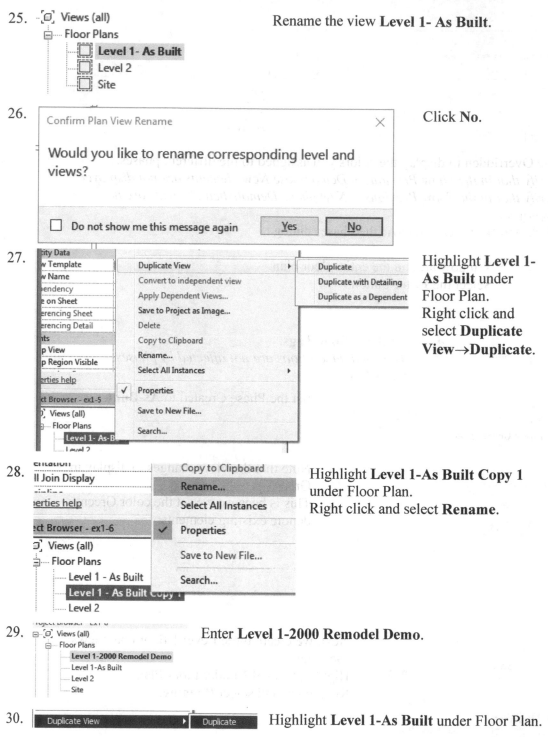 Rename the view **Level 1- As Built**.

26. Click **No**.

27. Highlight **Level 1-As Built** under Floor Plan. Right click and select **Duplicate View→Duplicate**.

28. Highlight **Level 1-As Built Copy 1** under Floor Plan. Right click and select **Rename**.

29. Enter **Level 1-2000 Remodel Demo**.

30. Highlight **Level 1-As Built** under Floor Plan.

31. Right click and select **Duplicate View→Duplicate**.

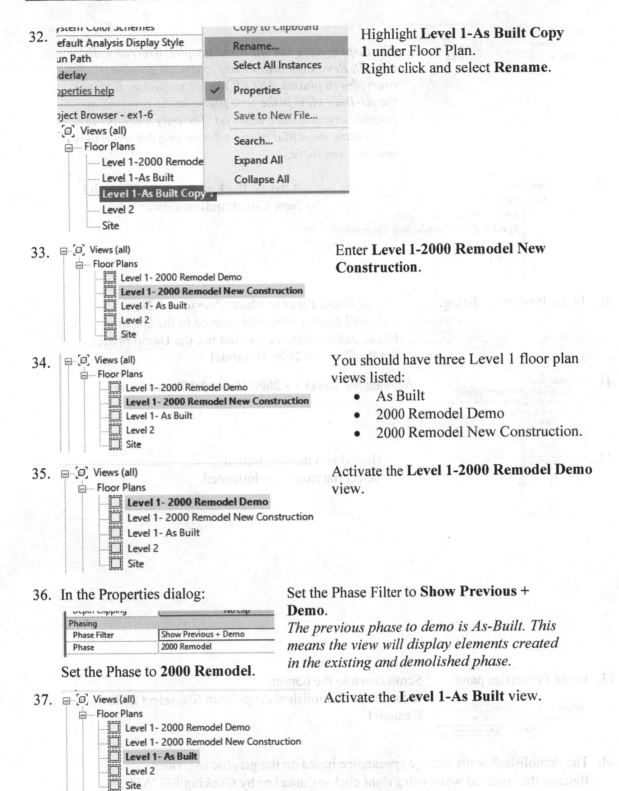

32. Highlight **Level 1-As Built Copy 1** under Floor Plan.
Right click and select **Rename**.

33. Enter **Level 1-2000 Remodel New Construction**.

34. You should have three Level 1 floor plan views listed:
- As Built
- 2000 Remodel Demo
- 2000 Remodel New Construction.

35. Activate the **Level 1-2000 Remodel Demo** view.

36. In the Properties dialog:

Set the Phase Filter to **Show Previous + Demo**.
The previous phase to demo is As-Built. This means the view will display elements created in the existing and demolished phase.

Set the Phase to **2000 Remodel**.

37. Activate the **Level 1-As Built** view.

38. In the Properties dialog:
Set the Phase Filter to **Show All**.
Set the Phase to **As-Built**.

Phasing	
Phase Filter	Show All
Phase	As-Built

The display does not show the graphic overrides. By default, Revit only allows you to assign graphic overrides to phases AFTER the initial phase. Because the As-Built view is the first phase in the process, no graphic overrides are allowed. The only work-around is to create an initial phase with no graphic overrides and go from there.

39.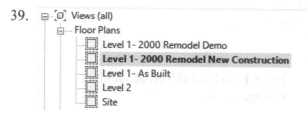

Activate the **Level 1-2000 Remodel New Construction** view.

40. In the Properties dialog:

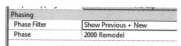

Set the Phase Filter to **Show Previous + New**. This will display elements created in the Existing Phase and the New Phase, but not the Demo phase. Set the Phase to **2000 Remodel**.

41.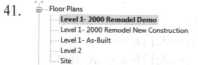

Activate the **Level 1 - 2000 Remodel Demo** view.

42.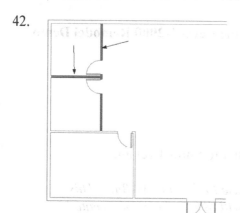

Hold down the Ctrl button.
Select the two walls indicated.

43. In the Properties pane:

Phasing	
Phase Created	As-Built
Phase Demolished	2000 Remodel

Scroll down to the bottom.
In the Phase Demolished drop-down list, select **2000 Remodel**.

44. The demolished walls change appearance based on the graphic overrides.
Release the selected walls using right click→Cancel or by Clicking ESCAPE.

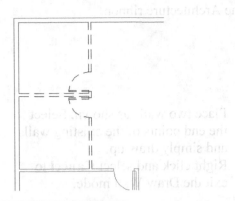

45.

Activate the **Modify** tab on tab on ribbon.
Use the **Demolish** tool on the Geometry panel to demolish the walls indicated.

46.

Note that the doors will automatically be demolished along with the walls. If there were windows placed, these would also be demolished. That is because those elements are considered *wall-hosted*.

Right click and select Cancel to exit the Demolish mode.

47.

This is how the Level 1- 2000 Remodel Demo view should appear.

If it doesn't, check the walls to verify that they are set to Phase Created: As Built, Phase Demolished: 2000 Remodel.

Phasing	
Phase Created	As-Built
Phase Demolished	2000 Remodel

48.
Views (all)
  Floor Plans
    Level 1- 2000 Remodel Demo
    Level 1- 2000 Remodel New Construction
    Level 1- As Built
    Level 2
    Site

Activate the **Level 1 - 2000 Remodel New Construction** view.

49.  Select the **Wall** tool from the Architecture ribbon.

50.

Place two walls as shown. Select the end points of the existing walls and simply draw up.
Right click and select **Cancel** to exit the Draw Wall mode.

51. Select the **Door** tool under the Build panel on the Architecture tab on tab on ribbon.

52.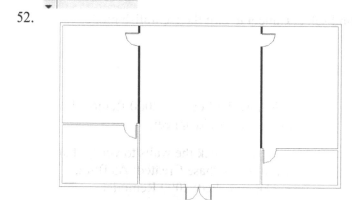

Place two doors as shown. Set the doors 3′ 6″ from the top horizontal wall. Flip the orientation of the doors if needed.

You can Click the space bar to orient the doors before you left click to place.

Note that the new doors and walls are a different color than the existing walls.

53. 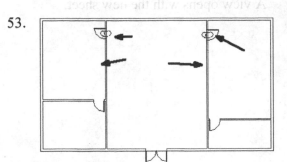 Select the doors and walls you just placed. You can select by holding down the CONTROL key or by windowing around the area.

Note: If Door Tags are selected, you will not be able to access Phases in the Properties dialog.

54. Look in the Properties panel and scroll down to Phasing.

Phasing	
Phase Created	2000 Remodel
Phase Demolished	None

Note that the elements are already set to **2000 Remodel** in the Phase Created field.

55. Switch between the three views to see how they display differently.

56. Remember that Existing should show as Green, Demo as Blue, and New as Magenta.

If the colors don't display correctly in the Level 1 Remodel Demo or Remodel New Construction, check the Phase Filters again and make sure that the categories to be displayed are Overridden to use the assigned colors.

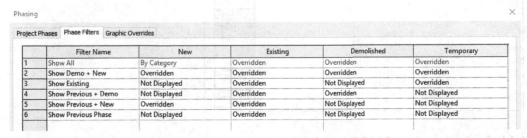

Phasing

Project Phases | Phase Filters | Graphic Overrides

	Filter Name	New	Existing	Demolished	Temporary
1	Show All	By Category	Overridden	Overridden	Overridden
2	Show Demo + New	Overridden	Overridden	Overridden	Overridden
3	Show Existing	Not Displayed	Overridden	Not Displayed	Overridden
4	Show Previous + Demo	Not Displayed	Overridden	Overridden	Not Displayed
5	Show Previous + New	Overridden	Overridden	Not Displayed	Not Displayed
6	Show Previous Phase	Not Displayed	Overridden	Not Displayed	Not Displayed

57. Schedules/Quantities
    Sheets (all)
    New Sheet...
    Fami
    Grou   Browser Organization...
    Revit

Highlight **Sheets** in the Project Browser. Right click and select **New Sheet**.

58. Select titleblocks:

E1 30 x 42 Horizontal : 30x42 Horizontal
None

Click **OK** to accept the default title block.

59.

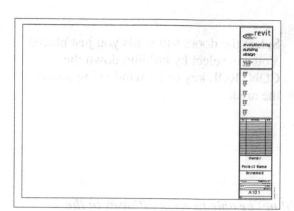

A view opens with the new sheet.

60.

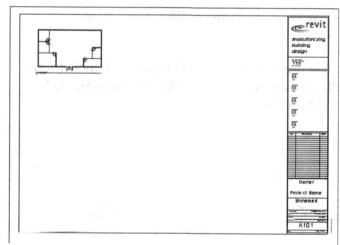

Highlight the Level 1 – As-Built Floor plan.
Hold down the left mouse button and drag the view
onto the sheet. Release the left mouse button to click
to place.

61. A preview will appear on your cursor. Left click to place the view on the sheet.

62. Highlight the **Level 1 - 2000 Remodel Demo** Floor plan.
Hold down the left mouse button and drag the view onto the sheet. Release the left
mouse button to click to place.

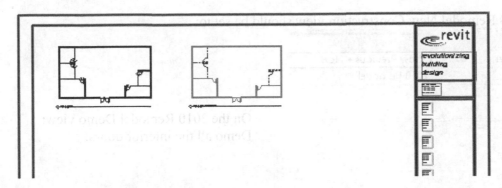

The two views appear on the sheet.

63. Highlight the Level 1 - 2000 New Construction plan.
Hold down the left mouse button and drag the view onto the sheet. Release the left mouse button to click to place.

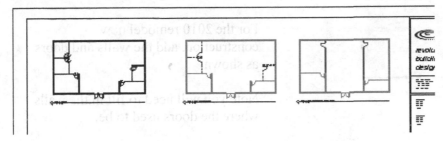

64.   Save as *ex3-23.rvt*.

Challenge Exercise:

Create two more views called Level 1 2010 Remodel Demo and Level 1 2010 Remodel New Construction.

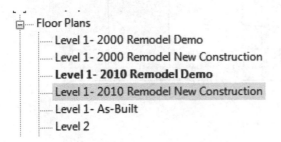

Floor Plans
Level 1- 2000 Remodel Demo
Level 1- 2000 Remodel New Construction
Level 1- 2010 Remodel Demo
Level 1- 2010 Remodel New Construction
Level 1- As-Built
Level 2

Set the Phases and phase filters to the new views.

The 2010 Remodel Demo view should be set to:

Phasing	
Phase Filter	Show Previous + Demo
Phase	2010 Remodel

The 2010 Remodel New Construction view should be set to:

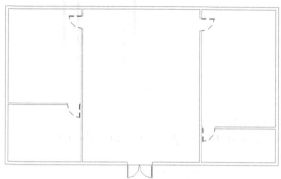

Phasing	
Phase Filter	Show Previous + New
Phase	2010 Remodel

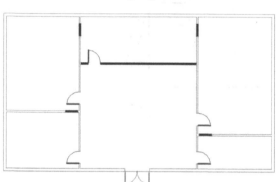

On the 2010 Remodel Demo view: Demo all the interior doors.

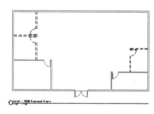

For the 2010 remodel new construction, add the walls and doors as shown.

Note you will need to fill in the walls where the doors used to be.

Add the 2010 views to your sheet.

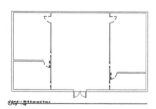

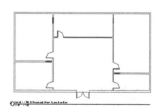

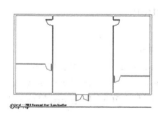

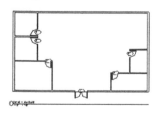

Design Options

Design Options allow you to explore different options for various parts of your project. Design Options work best when you wish to study varying options for small, distinct elements in your building model. The majority of your project should be stable. For example, the building footprint should be decided and the floor plan for the most part should be determined, but you might want to use Design Options to explore different kitchen or bathroom layouts. The building footprint and floor plan are considered part of the Main Model – those elements which are unchanged and not part of the design options.

For each Option set, you must designate a preferred or "primary" option. This is the option which will be shown by default in views. It will also be the option used back in the main model.

Exercise 3-24
Design Options

Drawing Name: **Design_Options**
Estimated Time to Completion: 90 Minutes

Scope
Use of Design Options
Place Views on Sheets
Visibility/Graphics Overrides
Duplicate Views
Rename Views

Solution

1.

Activate the **Manage** ribbon.
Select **Design Options** under the Design Options panel.

2.

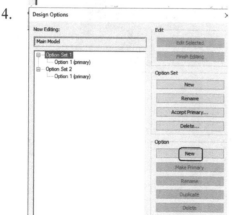

Select **New** under Option Set.
Select **New** a second time.

3.

There should be two Option Sets displayed in the left panel.
Each Option set represents a design choice group. The Option set can have as many options as needed. The more options, the larger your file size will become.

4.

Highlight the **Option Set 1**.
Select the **New** button under Option.
Note that Option Set 1 now has two sub-options.

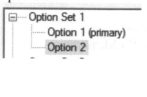

5.

Highlight the **Option Set 2**.
Select the **New** button under Option.
Note that Option Set 2 now has two sub-options.

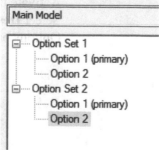

6.

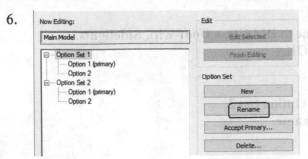

Highlight **Option Set 1**.
Select **Rename**.

7.

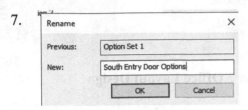

Rename Option Set 1 **South Entry Door Options**.
Click **OK**.

8.

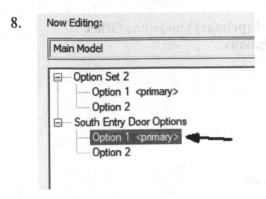

Highlight **Option 1 (primary)** under the South Entry Door Options.
Select **Rename**.

9.

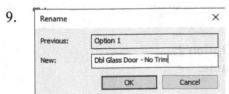

Rename to **Dbl Glass Door - No Trim**.
Click **OK**.

10.

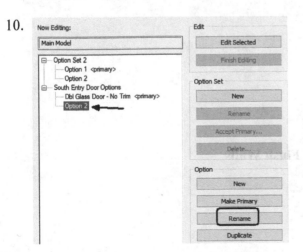

Highlight **Option 2** under the South Entry Door Options.
Select **Rename**.

11.

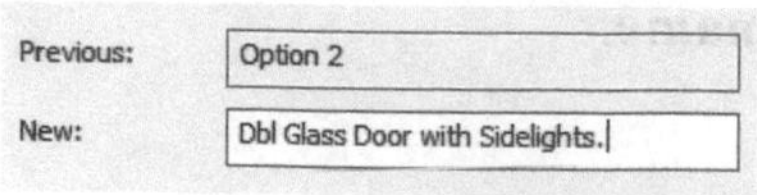

Rename to **Dbl Glass Door with Sidelights**.
Click **OK**.

12.

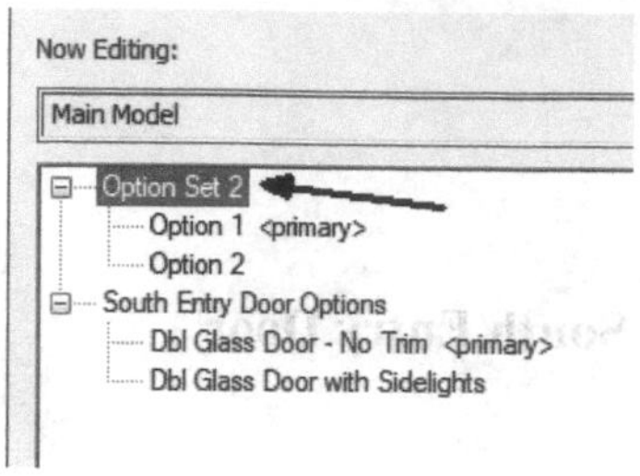

Highlight **Option Set 2**.
Select **Rename**.

13.

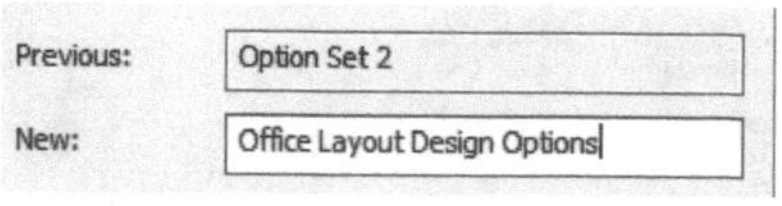

Rename Option Set 2 **Office Layout Design Options**.
Click **OK**.

14.

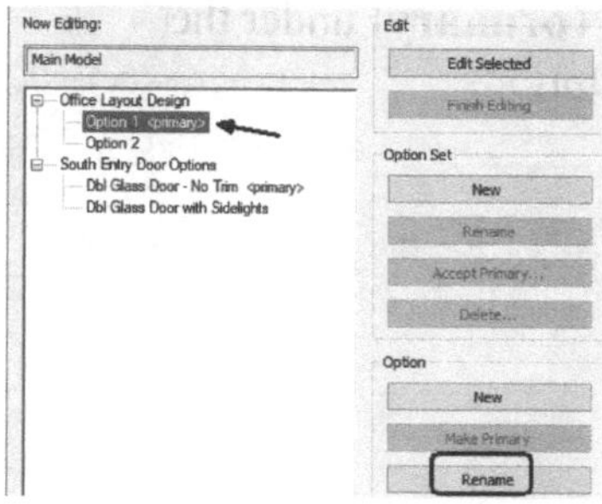

Highlight **Option 1 (primary)** under the **Office Layout Design Options**.
Select **Rename**.

15.

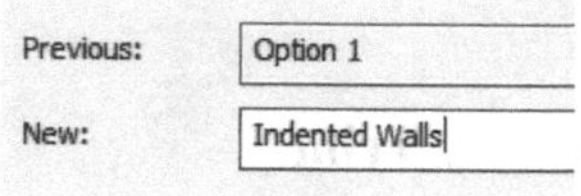

Rename to **Indented Walls**.
Click **OK**.

16.

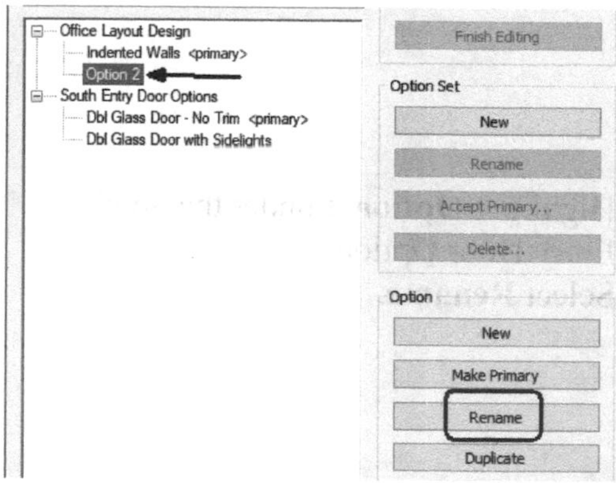

Highlight **Option 2**.
Select **Rename**.

17.

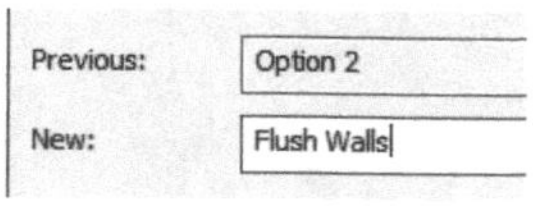

Rename Option 2 **Flush Walls**.
Click **OK**.

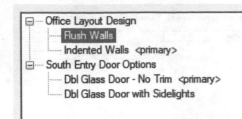

You should have two Option Sets. Each Option Set should have two options.

Notice that Revit automatically sorted the Options and the sets alphabetically.

An Option Set can have as many options as you like, but the more option sets and options, the larger your file size and the more difficult it becomes to manage.

18. Close the Design Options dialog.

19.

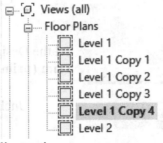

Note in the bottom of the window, you can select which Option set you want active.

20. Using **Duplicate View→Duplicate**, create four copies of the Level 1 view.

```
⊟ [O] Views (all)
  ⊟ Floor Plans
      [ ] Level 1
      [ ] Level 1 Copy 1
      [ ] Level 1 Copy 2
      [ ] Level 1 Copy 3
      [ ] Level 1 Copy 4
      [ ] Level 2
```

Rename the duplicate views:
Level 1 - Office Layout Indented Walls
Level 1 - Office Layout Flush Walls
Level 1 - South Entry Dbl Glass Door – No Trim
Level 1 - South Entry Dbl Glass Door with Sidelights

To rename, highlight the level name and Click F2.

21. Using **Duplicate View→Duplicate**, create two copies of the South Elevation view.
Rename the duplicate views:
South Entry Dbl Glass Door – No Trim
South Entry Dbl Glass Door with Sidelights

22. Activate **Level 1 - South Entry Dbl Glass Door – No Trim**.

23.

Parts Visibility	Show Original
Visibility/Graphics Overrides	Edit...
Graphic Display Options	Edit...
Underlay	None

In the Properties pane:
Select **Edit** Visibilities/Graphics Overrides.

3-137

24. Activate the **Design Options** tab.

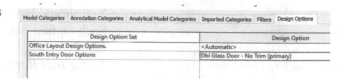

Set **Dbl Glass Door – No Trim** on South Entry Door Options.
Click **Apply** and **OK**.

25. Select the **Design Options** button.

26. Highlight **Dbl Glass Door with Sidelights**.

Click **Edit Selected**.

Click **Close**.

27. Set the Design Option to **Dbl Glass Door - No Trim (primary)**.

28. Uncheck **Active Only**.

29. 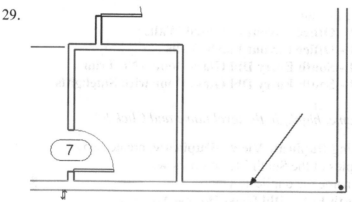 Select the south horizontal wall.

30. Activate the **Manage** ribbon.
Under Design Options, select **Add to Set**.
The selected wall is added to the **Dbl Glass Door - No Trim (primary)** *set.*
We need to add the wall to the set so we can place a door. Remember doors are wall-hosted.

31. 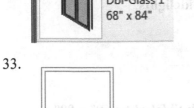 Activate the **Architecture** ribbon.
Select the **Door** tool from the Build panel.

32. 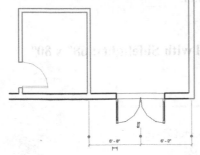 Set the Door type to **Dbl-Glass 1: 68″ x 84″**.

33. Place the door as shown.

34. 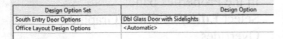 Select the **Design Options** button.

35. Click **Finish Editing**.

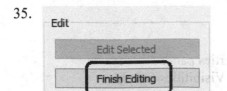

Click **Close**.

36. 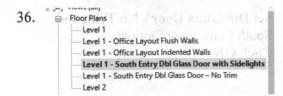 Activate **Level 1 - South Entry Dbl Glass Door with Sidelights**.

37. 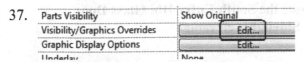 In the Properties pane:
Select **Edit** Visibilities/Graphics Overrides.

38. Activate the **Design Options** tab.

Set **Dbl Glass Door with Sidelights** on South Entry Door Options.
Click **OK**.

39. Select the **Design Options** button.

40.

Highlight **Dbl Glass Door with Sidelights**.

Click **Edit Selected**.

Close the dialog.

41.

Verify the Active Design Option is **Dbl Glass Door with Sidelights**.

42.

Activate the **Architecture** ribbon.
Select the **Door** tool from the Build panel.

43.

Place a **Double-Raised Panel with Sidelights: 68″ x 80″** door as shown.

Double-Raised Panel with Sidelights
68" x 80"

44.

Activate the **South Entry Dbl Glass Door – No Trim** elevation.

45.

In the Properties pane:
Select **Edit** Visibilities/Graphics Overrides.

46. Activate the **Design Options** tab.

Set **Dbl Glass Door - No Trim** on South Entry Door Options.
Click **OK**.

Design Option Set	Design Option
South Entry Door Options	Dbl Glass Door - No Trim (primary)
Office Layout Design Options	<Automatic>

47.

Activate the **South Entry Dbl Glass Door with Sidelights** elevation.

48.

In the Properties pane:
Select **Edit** Visibilities/Graphics Overrides.

49.

Design Option Set	Design Option
South Entry Door Options	Dbl Glass Door with Sidelights
Office Layout Design Options	<Automatic>

Activate the **Design Options** tab.

Set **Dbl Glass Door with Sidelights** on South Entry Door Options.

Click **OK**.

50.

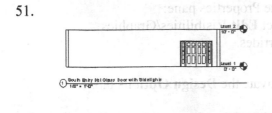

Sheets (all)
— A101 - South Entry Door Option
— A102 - Office Layout Options

Activate the Sheet named **South Entry Door Options**.

51.

Drag and drop the two South Entry Option elevation views on the sheet.

52. Switch to 3D view.

53. Use **Duplicate View→Duplicate** to create two new 3D views.

3D Views
— 3D - South Entry Dbl Glass Door - No Trim
— 3D - South Entry Dbl Glass Door with Sidelights
— {3D}

Rename the views:
3D - South Entry Dbl Glass Door - No Trim
3D - South Entry Dbl Glass Door with Sidelights

54.

3D Views
— 3D - South Entry Dbl Glass Door - No Trim
— 3D - South Entry Dbl Glass Door with Sidelights
— {3D}

Activate **3D - South Entry Dbl Glass Door - No Trim**.

55.

Detail Level	Coarse
Visibility/Graphics Overrides	Edit...
Visual Style	Hidden Line
Graphic Display Options	Edit...

In the Properties pane:
Select **Edit** Visibilities/Graphics Overrides.

56.

Activate the **Design Options** tab.

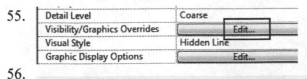

Design Option Set	
South Entry Door Options	Dbl Glass Door - No Trim (primary)
Office Layout Design Options	<Automatic>

Set **Dbl Glass Door - No Trim** on South Entry Door Options.
Click **OK**.

57.

Visibility	Projection/Surface Lines	Halftone
☑ Furniture Tags		☐
☑ Generic Annotations		☐
☑ Generic Model Tags		☐
☑ Grids		☐
☑ Guide Grid		
☑ Keynote Tags		☐
☐ Levels		☐
☑ Lighting Fixture Tags		☐

Model Categories | Annotation Categories | Analytical Model Categories | Imported Categories
☑ Show annotation categories in this view
Filter list: Architecture ⌄

Disable the visibility of **LEVELS** on the Annotation Categories tab.

58.

⊟ 3D Views
  ☐ 3D - South Entry Dbl Glass Door - No Trim
  ☐ **3D - South Entry Dbl Glass Door with Sidelights**
  ☐ {3D}

Activate **3D - South Entry Dbl Glass Door with Sidelights**.

59.

Display Model	Normal
Detail Level	Coarse
Visibility/Graphics Overrides	Edit...
Visual Style	Hidden Line
Graphic Display Options	Edit...

In the Properties pane:
Select **Edit** Visibilities/Graphics Overrides.

60.

Visibility/Graphic Overrides for 3D View: 3D - South Entry Dbl Glass Door with Sidelights

Model Categories | Annotation Categories | Analytical Model Categories | Imported Categories | Filters

Design Option Set	Design Option
Office Layout Design	<Automatic>
South Entry Door Options	Dbl Glass Door with Sidelights

Activate the **Design Options** tab.

Set **Dbl Glass Door with Sidelights** on South Entry Door Options.
Click **OK**.

If you forget which design option you are working in, check the title at the top of the dialog.

61.

Visibility	Projection/Surface Lines	Halftone
☑ Furniture Tags		☐
☑ Generic Annotations		☐
☑ Generic Model Tags		☐
☑ Grids		☐
☑ Guide Grid		
☑ Keynote Tags		☐
☐ Levels		☐
☑ Lighting Fixture Tags		☐

Model Categories | Annotation Categories | Analytical Model Categories | Imported Categories
☑ Show annotation categories in this view
Filter list: Architecture ⌄

Disable the visibility of **LEVELS** on the Annotation Categories tab.

62.

⊟ 🔲 Sheets (all)
  ⊞ **A101 - South Entry Door Options**
  ⋯⋯ A102 - Office Layout Options

Activate the Sheet named **South Entry Door Options**.

63.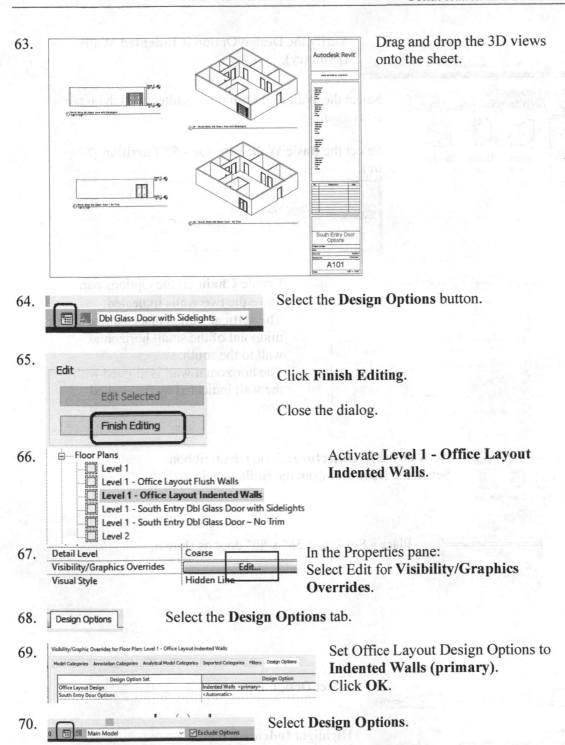

Drag and drop the 3D views onto the sheet.

64. Select the **Design Options** button.

Dbl Glass Door with Sidelights

65.

Edit

Edit Selected

Finish Editing

Click **Finish Editing**.

Close the dialog.

66.

Floor Plans
Level 1
Level 1 - Office Layout Flush Walls
Level 1 - Office Layout Indented Walls
Level 1 - South Entry Dbl Glass Door with Sidelights
Level 1 - South Entry Dbl Glass Door – No Trim
Level 2

Activate **Level 1 - Office Layout Indented Walls**.

67.

Detail Level	Coarse
Visibility/Graphics Overrides	Edit...
Visual Style	Hidden Line

In the Properties pane:
Select Edit for **Visibility/Graphics Overrides**.

68. Design Options

Select the **Design Options** tab.

69. Visibility/Graphic Overrides for Floor Plan: Level 1 - Office Layout Indented Walls

Model Categories Annotation Categories Analytical Model Categories Imported Categories Filters Design Options

Design Option Set	Design Option
Office Layout Design	Indented Walls <primary>
South Entry Door Options	<Automatic>

Set Office Layout Design Options to **Indented Walls (primary)**.
Click **OK**.

70. Main Model ☑ Exclude Options

Select **Design Options**.

71.

Now Editing:
Main Model

Office Layout Design
Flush Walls
Indented Walls <primary>
South Entry Door Options
Dbl Glass Door - No Trim <primary>
Dbl Glass Door with Sidelights

Edit
Edit Selected
Finish Editing

Option Set
New
Rename

Highlight **Indented Walls**.

Click **Edit Selected**.

Close the dialog.

72. Verify the Design Option is **Indented Walls (primary)**.

73. 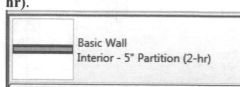 Select the **Wall** tool from the Architecture tab on tab on ribbon.

Select the **Basic Wall: Interior - 5" Partition (2-hr)**.

74. Enable **Chain** on the Options bar.
75. 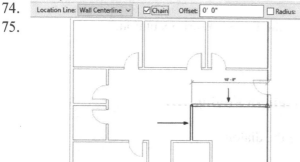 Place the two walls indicated.
The vertical wall is placed at the midpoint of the small horizontal wall to the south.
The horizontal wall is aligned with the wall indicated by the dashed line.

76. Activate the **Architecture** tab on tab on ribbon.
Select the **Door** tool from the Build panel.

77. 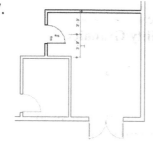 Place a **Sgl Flush 36" x 80"** door as shown.

78.  Select **Design Options**.

79. 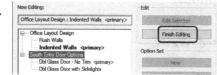 Highlight **Indented Walls**.

Click **Finish Editing**.

Close the dialog.

80. 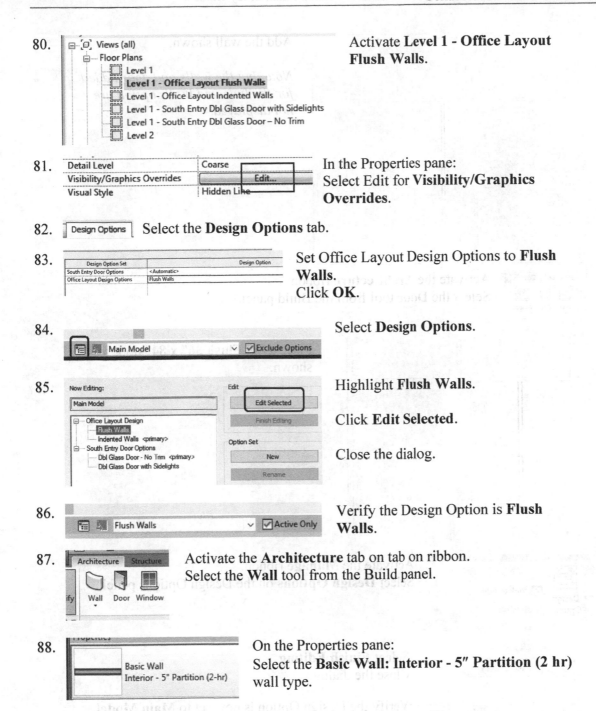 Activate **Level 1 - Office Layout Flush Walls**.

81. In the Properties pane:
Select Edit for **Visibility/Graphics Overrides**.

82. Select the **Design Options** tab.

83. Set Office Layout Design Options to **Flush Walls**.
Click **OK**.

84. Select **Design Options**.

85. Highlight **Flush Walls**.

Click **Edit Selected**.

Close the dialog.

86. Verify the Design Option is **Flush Walls**.

87. Activate the **Architecture** tab on tab on ribbon.
Select the **Wall** tool from the Build panel.

88. On the Properties pane:
Select the **Basic Wall: Interior - 5″ Partition (2 hr)** wall type.

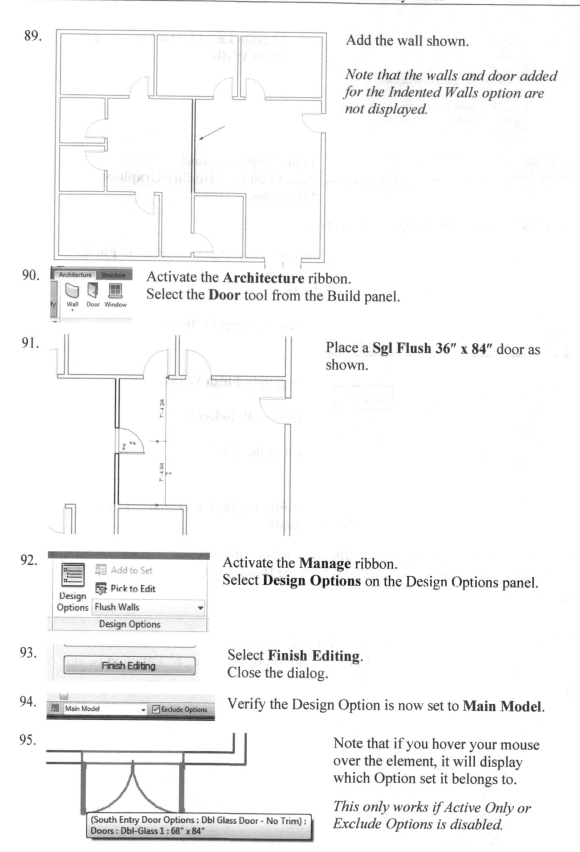

89. Add the wall shown.

Note that the walls and door added for the Indented Walls option are not displayed.

90. Activate the **Architecture** ribbon.
Select the **Door** tool from the Build panel.

91. Place a **Sgl Flush 36″ x 84″** door as shown.

92. Activate the **Manage** ribbon.
Select **Design Options** on the Design Options panel.

93. Select **Finish Editing**.
Close the dialog.

94. Verify the Design Option is now set to **Main Model**.

95. Note that if you hover your mouse over the element, it will display which Option set it belongs to.

This only works if Active Only or Exclude Options is disabled.

(South Entry Door Options : Dbl Glass Door - No Trim) :
Doors : Dbl-Glass 1 : 68" x 84"

96. 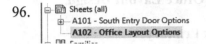 Activate the Sheet named **Office Layout Options**.

97. Drag and drop the two Office Layout options onto the sheet.

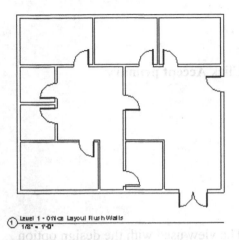 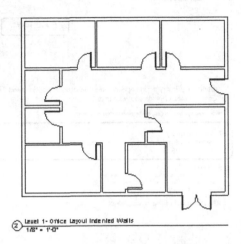

98. Activate the Manage ribbon.
Select **Design Options**.

99. *Let's assume that the client decided they prefer the flush walls option.*

Highlight the **Flush Walls** option.

100. Select **Make Primary**.

Note that (primary) is now next to Flush Walls.

If you see an error message, you can click to ignore it.

101.

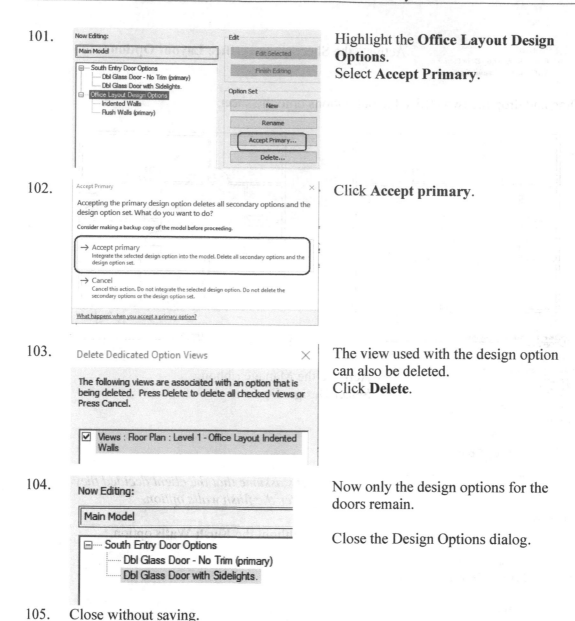

Highlight the **Office Layout Design Options**.
Select **Accept Primary**.

102.

Click **Accept primary**.

103.

The view used with the design option can also be deleted.
Click **Delete**.

104.

Now only the design options for the doors remain.

Close the Design Options dialog.

105. Close without saving.

Exercise 3-25
Design Options Practice Question

Drawing Name: Design_Options_Question.rvt
Estimated Time to Completion: 5 Minutes

Scope
Design Options
Properties

Solution

1.   3D Views
       Approach
       From Yard
       Kitchen
       Living Room
       Living Room - ISO
       Section Perspective
       Solar Analysis
       {3D}

Activate the {3D} view under 3D Views.

2.   Roof
       Metal (primary)
       Timber
       Tile

What is the volume of the roof in Design Option – Timber?

3.   Timber ☑ Active Only

Select the **Timber** design option.
Enable **Active Only**.

4.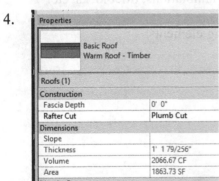

Select the roof.
Go to the Properties panel.

Scroll down.

What is the volume of the roof?

Properties	
Basic Roof Warm Roof - Timber	
Roofs (1)	
Construction	
Fascia Depth	0' 0"
Rafter Cut	Plumb Cut
Dimensions	
Slope	
Thickness	1' 1 79/256"
Volume	2066.67 CF
Area	1863.73 SF
Identity Data	

5.

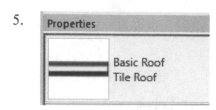

If you switch design options, does the volume of the roof change?

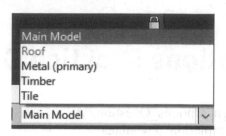

Answer this question:

When should you use phasing as opposed to design options?

Copy/Monitor

Use the Copy/Monitor tool when the following statements are true:

- All of the team members are using Revit software.

- Each team member must be informed of changes to levels, grids, and other elements.

- The teams will link models to work on the same building design. Each team maintains its own edition of the model and uses Revit software to develop the design for their discipline. Each model is linked to the other models to share information about changes to monitored elements in the building design.

- The team should have a project manager who dictates the shared coordinate to be used by all team members and controls CAD standards for annotations, titleblocks, etc.

Copy/Monitor can be used to monitor changes to the following elements:

- Levels
- Grids (but not multi-segment grids)
- Columns
- Walls
- Floors
- Openings
- MEP Fixtures

Exercise 3-26

Monitoring a Linked File

Drawing Name: **i_multiple_disciplines.rvt**
Estimated Time to Completion: 40 Minutes

Scope

Link a Revit Structure file.
Monitor the levels in the linked file.
Reload the modified Structure file.
Perform a coordination review.
Create a Coordination Review report.

Solution

1. If you have an Autodesk online Account,
    the sign in for the user account is used
    for your user name. If you don't want to
    use the name displayed by the Autodesk
    account, you need to change your profile
    to display a different name. This name is
    used in comments for revisions, etc.

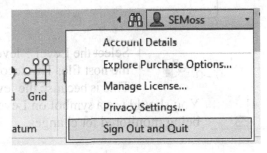

2. Open the *i_multiple_disciplines.rvt* file.

3. 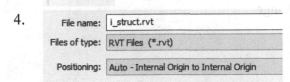 Activate the **Insert** ribbon.
    Select the **Link Revit** tool on the Link panel.

4. File name: i_struct.rvt
    Files of type: RVT Files (*.rvt)
    Positioning: Auto - Internal Origin to Internal Origin

    Locate the *i_struct* file.
    Set the Positioning to **Auto-Internal
    Origin to Internal Origin**.
    Click **Open**.

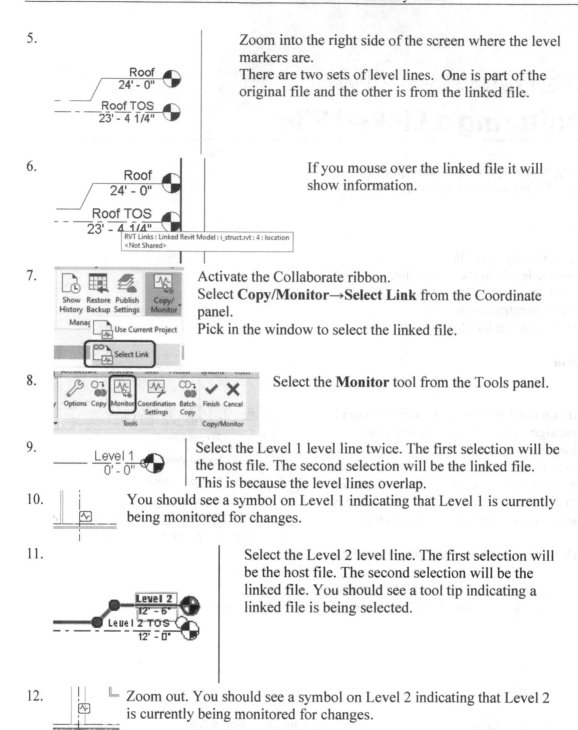

5. Zoom into the right side of the screen where the level markers are.
There are two sets of level lines. One is part of the original file and the other is from the linked file.

6. If you mouse over the linked file it will show information.

RVT Links : Linked Revit Model : i_struct.rvt : 4 : location <Not Shared>

7. Activate the Collaborate ribbon.
Select **Copy/Monitor→Select Link** from the Coordinate panel.
Pick in the window to select the linked file.

8. Select the **Monitor** tool from the Tools panel.

9. Select the Level 1 level line twice. The first selection will be the host file. The second selection will be the linked file. This is because the level lines overlap.

10. You should see a symbol on Level 1 indicating that Level 1 is currently being monitored for changes.

11. Select the Level 2 level line. The first selection will be the host file. The second selection will be the linked file. You should see a tool tip indicating a linked file is being selected.

12. Zoom out. You should see a symbol on Level 2 indicating that Level 2 is currently being monitored for changes.

13.

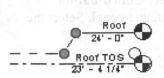

Select the Roof level line.
The first selection will be the host file.
The second selection will be the linked file.
You should see a tool tip indicating a linked file is being selected.

14.

Zoom out. You should see a symbol on the Roof level indicating that it is currently being monitored for changes.

15.

Warning
Elements already monitored

If you try to select elements which have already been set to be monitored, you will see a warning dialog.
Simply close the dialog and move on.

16.

✓ ✗
Finish Cancel
Copy/Monitor

Select **Finish** on the Copy/Monitor panel.

17.

Manage
Links

Activate the **Insert** ribbon.
Select **Manage Links** from the Link panel.

18.

Revit	CAD Formats	DWF Markups	Point Clouds

Linked File	Status	Reference Type	
i_struct.rvt	Loaded	Overlay	

Select the **Revit** tab.
Highlight the *i_struct.rvt* file.

19.

[Reload From...]

Select **Reload From**.

20.

File name:	i_struct_revised
Files of type:	RVT Files (*.rvt)

Locate the *i_struct_revised* file.
Click **Open**.

21.

Warning - can be ignored
Instance of link needs Coordination Review

Show   More Info   Expand >>

A dialog will appear indicating that the revised file requires Coordination Review.
Click **OK**.

22.

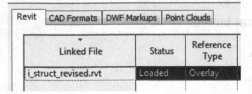

Revit	CAD Formats	DWF Markups	Point Clouds

Linked File	Status	Reference Type
i_struct_revised.rvt	Loaded	Overlay

Note that the revised file has replaced the previous link.
This is similar to when a sub-contractor or other consultant emails you an updated file for use in a project.
Click **OK**.

23. 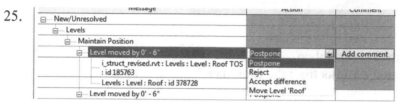 Activate the **Collaborate** ribbon. Select **Coordination Review→Select Link** from the Coordinate panel. Select the linked file in the drawing window.

24. 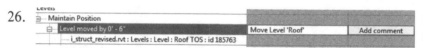 A dialog appears. Expand the notations so you can see what was changed.

Message	Action
New/Unresolved	
Levels	
Maintain Position	
Level moved by 0' - 6"	Postpone
i_struct_revised.rvt : Levels : Level : Roof TOS : id 185763	
Levels : Level : Roof : id 378728	
Level moved by 0' - 6"	Postpone
i_struct_revised.rvt : Levels : Level : Level 2 TOS : id 32567	
Levels : Level : Level 2 : id 9946	

25. 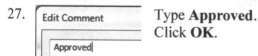 Highlight first change. Select **Move Level Roof** from the drop-down list.

26. 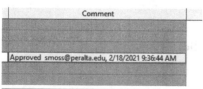 Select the **Add Comment** button.

27. 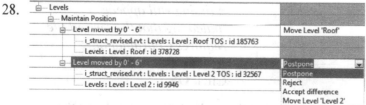 Type **Approved**. Click **OK**.

Note that Revit automatically adds your user name and a time stamp.

Approved smoss@peralta.edu, 2/18/2021 9:36:44 AM

28. 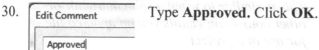 Highlight the second change. Select **Move Level 'Level 2'**.

29. Add comment Select the **Add Comment** button.

30. Edit Comment Type **Approved**. Click **OK**.

Approved

31. Select **Create Report** on the bottom left of the dialog.

32. Browse to your exercise folder. Name the file *review*. Click **Save**.

33. Click **OK** to close the Coordination Review dialog box.

   If you close the Coordination Review dialog before you create the report and then re-open it to create the report, the report will be blank because all the issues will have been resolved.

34. Locate the report you created and double click on it to open.

35. This report can be emailed or used as part of the submittal process.

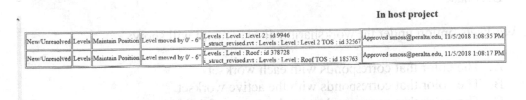

Revit Coordination Report

In host project

New/Unresolved	Levels	Maintain Position	Level moved by 0' - 6"	Levels : Level : Level 2 : id 9946 i_struct_revised.rvt : Levels : Level : Level 2 TOS : id 32567	Approved smoss@peralta.edu, 11/5/2018 1:08:35 PM
New/Unresolved	Levels	Maintain Position	Level moved by 0' - 6"	Levels : Level : Roof : id 378728 i_struct_revised.rvt : Levels : Level : Roof TOS : id 185763	Approved smoss@peralta.edu, 11/5/2018 1:08:17 PM

36. Close the file without saving.

Practice Exam

1. Worksharing allows team members to:
   A. Share families from different projects
   B. Share views from different projects
   C. Work on the same parts of a project simultaneously
   D. Work on different parts of the same project

2. Set Phase Filters for a view in the:
   A. Properties pane
   B. Design Options
   C. Manage ribbon
   D. Project Browser

3. The Coordination Review tool is used when you use:
   A. Worksets
   B. Linked Files
   C. Interference Checking
   D. Phase

4. The Worksets option under the worksharing display menu will show:

   A. The color that corresponds with each workset
   B. The color that corresponds with the active workset
   C. The color that corresponds with the owner of the active workset
   D. The color of all worksets that can be edited (checked out).

5. True/False
   You can ignore review warnings, but you cannot ignore an error.

6. The Coordination Review is useful for:

   A. Checking what monitored elements were changed in the host file
   B. Aligning the origin of one project with a linked project
   C. Performing multiple copy operations from a linked project
   D. Copying a large number of elements from one project to another

7. When you link to a CAD drawing:

    A. You bring in an image of the CAD file which cannot be updated.
    B. You automatically import any external references used by that CAD file.
    C. If the CAD file is modified, you will see any changes when you reload the file.
    D. The CAD file is added to the Revit project as decal.

8. You are using worksets. When you make changes to the local file, these changes:

    A. Cannot be saved to the Central Model.
    B. Can only be saved to the Central Model if the changes are minor.
    C. Are reflected in the Central Model only after synchronizing.
    D. Are automatically reflected in the Central Model.

9. When reviewing errors, why isn't using the Remove Constraints button the best method to resolve the error?

    A. It might not actually remove the constraints
    B. It may add more constraints
    C. It's too easy a solution
    D. It may remove elements that are required in the design

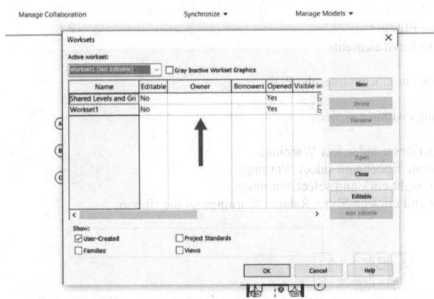

10. What will appear in the Owner field for Workset1 if you change the Editable field to Yes?

    A. The User name you set in the Options will appear in the Owner field
    B. The original creator of the workset will be assigned as the Owner
    C. Workset1 will not be assigned an Owner, but it will still be editable
    D. The User you select from the drop-down list will be assigned the owner

11. I choose Import CAD to bring in an AutoCAD drawing file with the floor plan. My vendor just emailed me that he made some changes to the floor plan. What now?

    A. An asterisk will appear next to the drawing file name indicating there are unsaved changes
    B. The drawing will automatically update
    C. A warning will pop up the next time I reopen the Revit project advising me to reload the drawing
    D. The changes are not reflected in my Revit project, and I do not see any warnings or indications that the file may have changed.

12. When creating a local version of a centralized model, what happens if you select the Detach from Central option when opening the file?

    A. You will have a read-only version of the model and can't make changes but can review.
    B. Changes you make will automatically synchronize with the Central Model.
    C. Changes you make to the local model won't be synchronized and the local file will now be independent of the Central model.
    D. Your username will not be assigned to any worksets.

13. You import a PDF into your Revit project, but you are unable to snap to any elements in the PDF. You should:

    A. Enable Snap on the ribbon
    B. Convert the PDF to Revit elements
    C. Reload the PDF
    D. Just do the best you can tracing over the PDF

14. To review any warnings within a project:

    A. Go to the Manage ribbon and select Warnings.
    B. Go to the Collaborate ribbon and select Warnings.
    C. Select an element, right click and select Warnings.
    D. Select an element and click on Show Related Warnings on the ribbon.

15. Which icon should be used to manage the display settings for worksets?

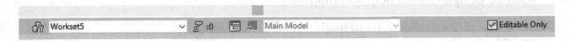

16. Editable Only is enabled on the status bar. This means:

    A. Only elements in the Main Model design option can be edited.
    B. Only elements in the host file can be edited.
    C. Only elements in the active workset can be edited.
    D. Only elements in the worksets with editing enabled can be selected or edited.

17. The Phase of a view is set to New Construction and the Phase Filter is set to Show Previous + New. What will control the graphic display of Existing elements?

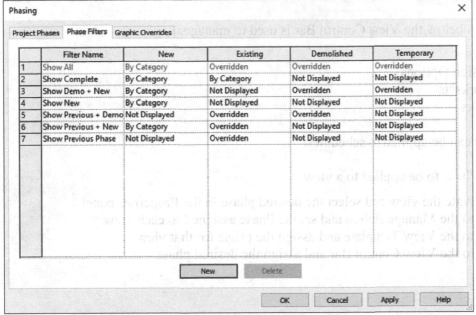

    A. The settings for that category in the Visibility/Graphic Overrides dialog.
    B. Existing elements will not be visible in the view.
    C. The Existing settings on the Graphic Overrides tab.
    D. Object Styles for the New Construction phase.

18. A designer wants to check for clashes in a project. Which tool should be used?
    A. Run Interference Check
    B. Copy/Monitor
    C. Review Clashes
    D. Coordination Review

19. What do design options help accomplish?

    A. Explore various design alternatives in the same model.
    B. Combine elements together to easily place multiple times.
    C. Create layers or subcomponents from individual elements.
    D. Change the display of elements based on the design.

20. T/F Phases allow you to display different versions of a model.

21. T/F Phase filters changes the graphic display of a view depending on the phase.

22. This section below the View Control Bar is used to manage:

    A. Phases
    B. Design Options
    C. Worksets
    D. Links

23. T/F Phases can be applied to schedules.

24. To set the Phase to be applied to a view:

    A. Activate the view and select the desired phase in the Properties panel
    B. Go to the Manage ribbon and set the Phase assigned to each view
    C. Go to the View Template and assign the phase for that view
    D. Go to the View Control Bar and assign the desired phase

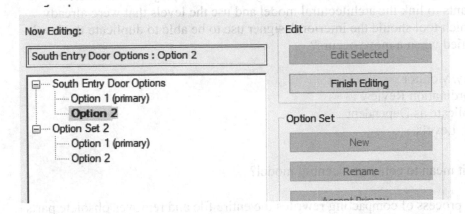

25. You click on Finish Editing. This:
    A. Deletes that design option
    B. Moves elements from one design option to another
    C. Closes the design option you are working in
    D. Incorporates the design option into the main model

26. A designer has a linked CAD file from a Civil Engineer. The designer wants to use the coordinate system setup in the CAD file. Which tool can be used to update the Revit project with the coordinates from the CAD file?

    A. Location
    B. Transfer Project Standards
    C. Acquire Coordinates
    D. Publish Coordinates

27. A designer is working with a Civil Engineer that is using CAD. The designer receives as DWG file and wants to use it in Revit. How should the designer handle the CAD file so that any future updates to the DWG file will be visible in Revit?

    A. Import the DWG file into the Revit project.
    B. Link the RVT file to the CAD file.
    C. Link the DWG file to the Revit project.
    D. Import the RVT file into the CAD file.

28. What is the phase filter setting for a typical demolition plan?

    A. Show Demo + New
    B. Show Demo
    C. Show Previous + Demo
    D. Show Previous + New

29. An interior designer and architect are working on the same project. The interior designer wants to link the architectural model and use the levels that were already created. Which tool should the interior designer use to be able to duplicate the levels and be notified when a move occurs?

    A. Copy/Monitor
    B. Coordination Review
    C. Duplicate as Dependent
    D. Link Levels

30. What does it mean to compact a central model?

    A. The process of compacting rewrites the entire file and removes obsolete parts in order to save space. This reduces the file size.
    B. During the compacting process, unused families and family types are removed from the project. This reduces the file size.
    C. It is the same thing as zipping a file. It allows users to compact all elements of a model into one file so you can send it to other members of the design team.
    D. Compact is another word for archive. It allows users to archive models at milestone submittals.

31. A designer is exporting a model to DWG. The designer needs to export the model based on the internal origin. Where should the designer click to adjust this setting?

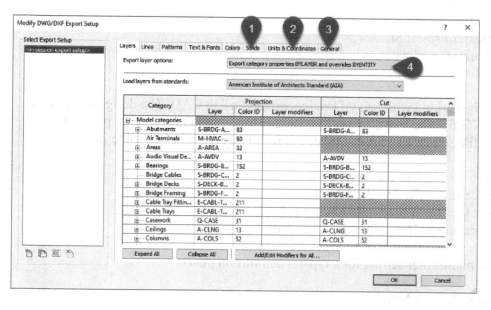

    A. 1
    B. 2
    C. 3
    D. 4

32. A designer is exporting a Revit sheet to a DWG file. She disables the Export views on sheets and links as external references as shown. There are two views on the sheet. What will happen to those views when you open the DWG file in AutoCAD?

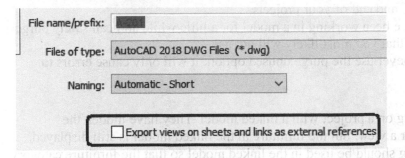

A. They will be excluded from the export.
B. They will appear on the layout tab, but not on the model tab.
C. They will appear on the model tab, but not the layout tab.
D. They will appear on the model tab and on the layout tab.

33. Which of the following is an accurate description of design options?
    A. Design options can only be used to explore options that are related to the building envelope. Duplicate with detailing should be used when interior options are desired.
    B. Design options are created by doing a Save As of the Revit model and creating different files to track each design option. This allows users to work independently on each option.
    C. With design options, a team can develop, evaluate, and redesign building components and rooms within a single project file.
    D. Design options do not work with models that use more than one phase.

34. Once worksharing has been enabled, what is the next step when using this feature and sharing models with others?

    A. Set up permissions
    B. Share a local copy
    C. Create a central model
    D. Purge elements

35.   When is a good time to use the purge tool?

      A. Right after you create the project to remove the unnecessary elements that are located in the project template.
      B. At the middle and end of your projects.
      C. After you have been working in a model for a little while and can safely purge out elements that you most likely will not use.
      D. You should never use the purge unused option; it will only cause errors to occur.

36.   A designer is working on a project with a linked model. They have hidden the furniture category for a view, but the furniture in the linked model is still displayed. Which display setting should be used in the linked model so that the furniture category is hidden in the linked model?

      A. Custom
      B. By linked view
      C. By host view
      D. By category

Answers:
1) D; 2) A; 3) B; 4) A; 5) T; 6) A; 7) C; 8) C; 9) D; 10) A; 11) D; 12) C; 13) A; 14) A; 15) C; 16) D; 17) C: 18) A; 19) A; 20) F; 21) T; 22) B; 23) T; 24) A; 25) C; 26) C; 27) C; 28) C; 29) A; 30) B; 31) B; 32) D; 33) C; 34) C; 35; B; 36) A

Project Standards and Setup

This lesson addresses the following exam questions:

- Configure user interface settings
- Manage project settings
- Customize the Project Browser
- Configure print and export settings

The Project Browser organizes the views, schedules, and sheets of the current project. To display a different view, expand a group, and double-click the view title. Display the model in a single view, or open several views, and click Tile on the View tab to see the views at the same time.

When you select an element in the drawing, you can modify its properties on the Properties palette. For example, use the Type Selector to change the door type from single to double, and that change is reflected in all of the views. Use view controls in the drawing area to manipulate the model and see your changes.

The tools you need to design the building model are grouped by task on the ribbon. When an element is selected, object-specific tools are available on a contextual tab. You can customize the Quick Access toolbar to display the tools that you use most often. For example, right-click a tool, and click Add to Quick Access toolbar.

The View Control Bar provides options for changing the view display, such as the scale or visual style.

The user interface is designed to simplify your workflow, and with a few clicks, you can customize the interface to support the way that you work.

You can save print settings in a project for later reuse.
For example, you may create one print setup for code review, and create another print setup designed for consultants' pages. The print setups are saved with the project, so they can be used across multiple sessions.

Save Reminders

Autodesk Revit does not allow you to save a project automatically; it only provides automatic reminders about saving your document. The reason the software works this way is twofold. Reason 1: Once you save your current project, you cannot perform an undo. So, if you are experimenting with your design work and the software autosaves, you are stuck. Reason 2: Because some building projects are large, an autosave can take several seconds. This can disrupt your thought process and some users find that interruption frustrating.

You can set how often to be reminded or "nagged" about saving in Options.

The Save Reminder setting is independent of which project you are working in.

Exercise 4-1

Set Save Reminders

Drawing Name: **basic project.rvt**
Estimated Time to Completion: 5 Minutes

Scope
Set Save Reminders

Solution

1.

Go to File.

Click on **Options**.

2.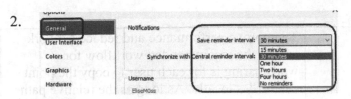

Highlight General.

Select how often you would like to be reminded to save.

Click **OK**.

3. Close without saving.

File Locations

Project Template files	Specifies the template files to list when creating a new project.
Default path for User files	Specify the default path for Revit to use when storing project files.
Default path for family template files	Specify the path for templates and libraries. This is useful if your company has set up templates and stored them on a network or the cloud.

Root path for point clouds	To improve performance and reduce network traffic, the recommended workflow for worksharing is for each user to copy the point cloud files locally. As long as the relative path to the local copies of the point cloud files is the same for each user, the link will remain valid when you synchronize with central. In some cases, it may be easiest to store point cloud files in the root directory.
System analysis workflows	Specify the workflow files to list on the Systems Analysis dialog for use by OpenStudio. Default files are provided for Annual Building Energy Simulation and HVAC Systems Loads and Sizing.
Places 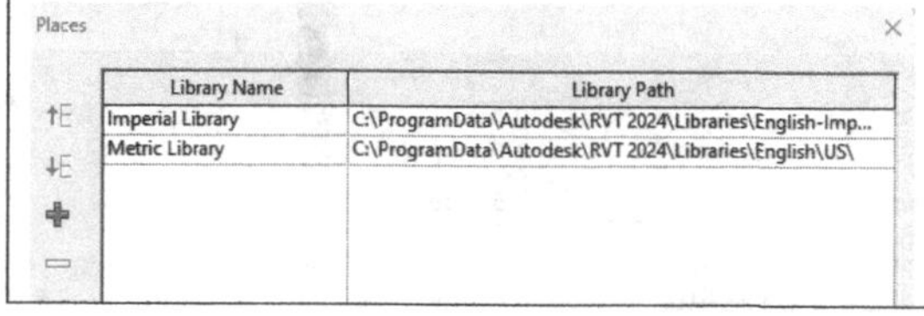	By default, Places defines where the libraries for loadable families can be located. If your company has developed a content library, you can add it here.

Exercise 4-2

Set File Locations

Drawing Name: Close all open files
Estimated Time: 5 minutes

Scope

Options
File Locations

Solution

1. Close all open files or projects.

2.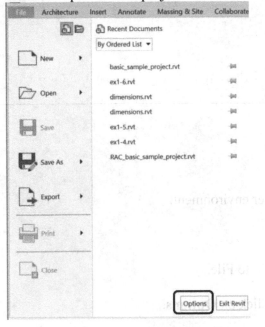

Go to the **File Menu**.

Select the **Options** button at the bottom of the window.

3.   File Locations   Select the **File Locations** tab.

4. Default path for user files:
   C:\Users\Elise\Documents   Browse...

In the **Default path for user files** section, pick the **Browse** button.

5.

> **Default path for user files:**
>
> C:\Revit 2024 Certification Guide\Revit 2024

Navigate to the local or network folder where you will save your files. When the correct folder is highlighted, pick **Open**. Your instructor or CAD manager can provide you with this file information.

Click **OK**.

I recommend to my students to bring a flash drive to class and back up each day's work onto the flash drive. That way you will never lose your valuable work. Some students forget their flash drive. For those students, make a habit of uploading your file to Google Drive, Dropbox, Autodesk Docs, or email your file to yourself.

Changes made to the ribbon are independent of the project you are working in.

Exercise 4-3
Configure Tabs

Drawing Name: **basic_project.rvt**
Estimated Time to Completion: 5 Minutes

Scope
Modify the tab display to customize the user environment.

Solution

1.

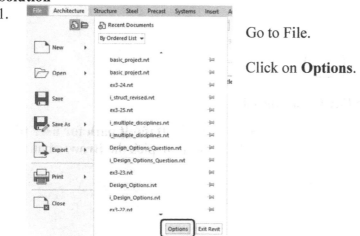

Go to File.

Click on **Options**.

2.

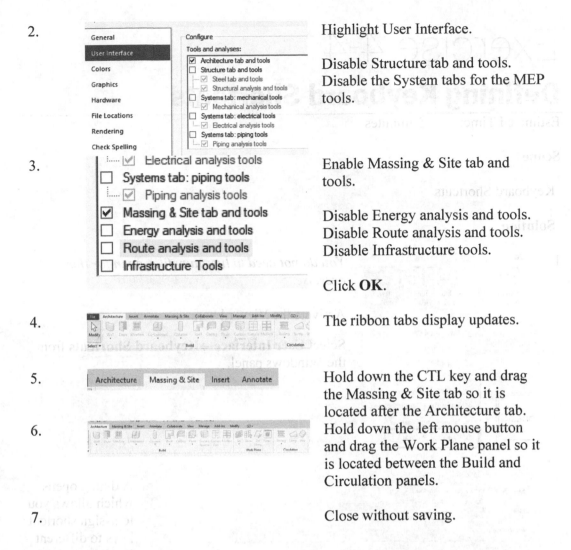

Highlight User Interface.

Disable Structure tab and tools.
Disable the System tabs for the MEP tools.

3.

Enable Massing & Site tab and tools.

Disable Energy analysis and tools.
Disable Route analysis and tools.
Disable Infrastructure tools.

Click **OK**.

4.

The ribbon tabs display updates.

5.

Hold down the CTL key and drag the Massing & Site tab so it is located after the Architecture tab.

6.

Hold down the left mouse button and drag the Work Plane panel so it is located between the Build and Circulation panels.

7.

Close without saving.

Keyboard Shortcuts

In Revit all keyboard shortcuts are made up of at least two keys pressed in sequence. Many commands have default keyboard shortcuts assigned, and many other commands do not have a default shortcut assigned. Commands unassigned a keyboard shortcut can be assigned by you. You also can change the assigned keyboard shortcut to one that makes more sense to you.

Exercise 4-4

Defining Keyboard Shortcuts

Estimated Time:	5 minutes

Scope

Keyboard Shortcuts

Solution

1.

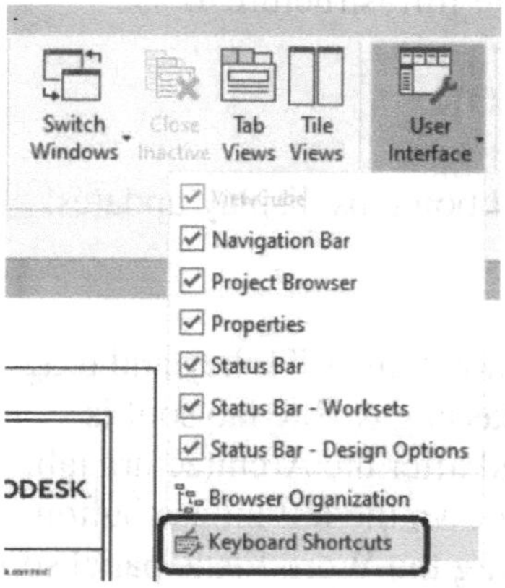

You do not need to have any files open for this exercise.

Activate the View ribbon.

Select **User Interface→Keyboard Shortcuts** from the Windows panel.

2.

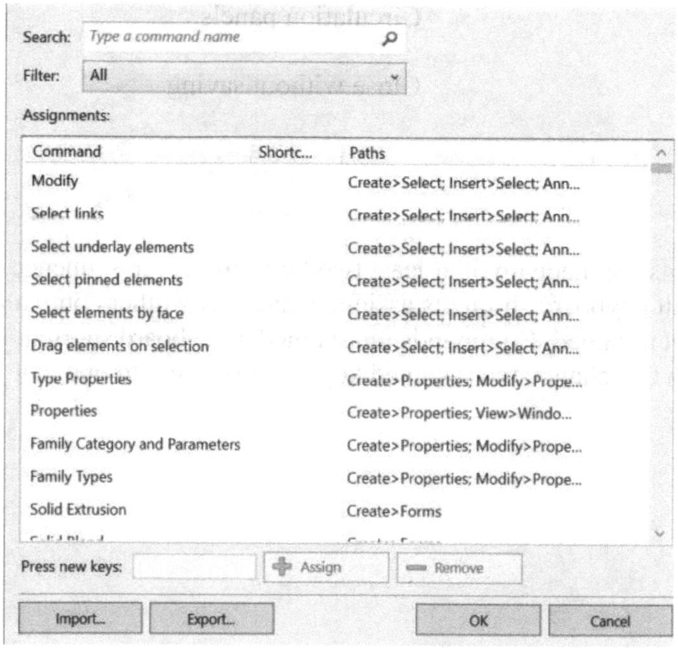

A dialog opens which allows you to assign shortcut keys to different Revit commands.

3. 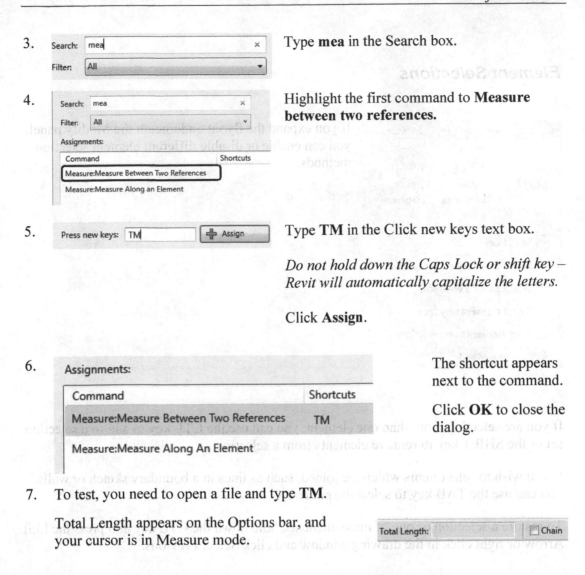 Type **mea** in the Search box.

4. Highlight the first command to **Measure between two references.**

5. Type **TM** in the Click new keys text box.

   Do not hold down the Caps Lock or shift key – Revit will automatically capitalize the letters.

   Click **Assign**.

6. The shortcut appears next to the command.

   Click **OK** to close the dialog.

7. To test, you need to open a file and type **TM**.

   Total Length appears on the Options bar, and your cursor is in Measure mode.

Element Selections

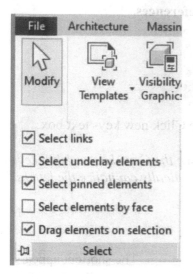

If you expand the flyout underneath the Modify panel, you can enable or disable different element selection methods.

If you are selecting more than one element, you can use the CTL key to add to a selection set or the SHIFT key to remove elements from a selection set.

If you wish to select items which are joined, such as lines in a boundary sketch or walls, you can use the TAB key to select the entire set.

To restore a selection of one of more elements, hold down the CTL key and press the Left Arrow or right click in the drawing window and click Select Previous.

Exercise 4-5
Control Element Selections

Drawing Name: **cubicle layouts.rvt**
Estimated Time to Completion: 20 Minutes

Scope
Practice selecting and de-selecting elements in a project.
Link a Revit File
Bind a Revit File

Solution

1.    Open the Insert ribbon.

    Click **Link Revit**.

2. 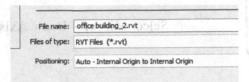   Locate office build_2.rvt.
    Set the Positioning to **Auto – Internal Origin to Internal Origin.**

    Click **Open.**

3. 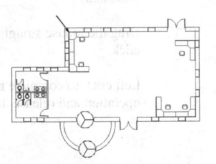   Adjust the position of the linked file.

    Select the linked file and move it so the upper left cubicle fits in the upper left corner as shown.

4.

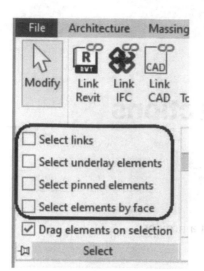

Expand the Modify panel.

Disable all the options except Drag elements on selection.

Try to select the linked file.

5.

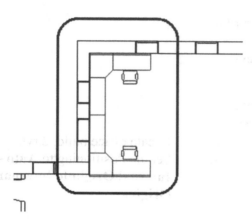

Hold down the CTL key and select the two cubicles and chairs in the upper left corner.

Select **Mirror – Draw Axis**.

6.

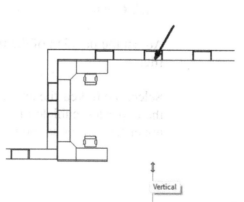

Select the midpoint of the window indicated.

Drag the mouse straight down and click.

Left click to complete the mirror operation and release the selection.

7.

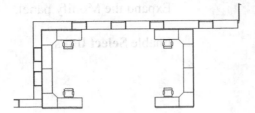
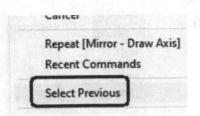

Right click in the drawing window
and click **Select Previous**.

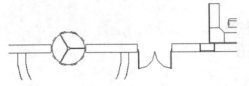

8.

The mirrored elements are
highlighted.

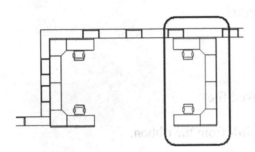

9.

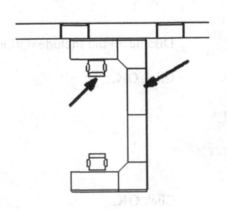

Hold down the SHIFT key and
click on the upper workstation and
chair.

*The SHIFT key removes element
from a selection.*

10.

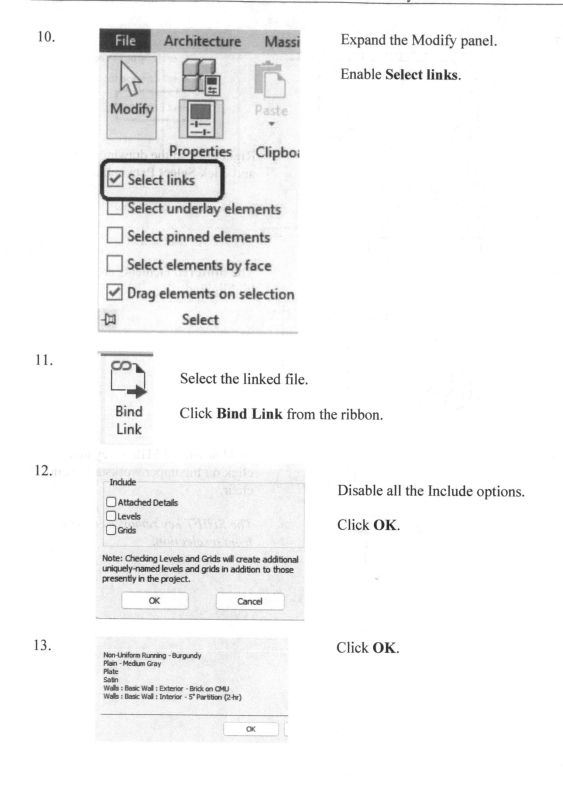

Expand the Modify panel.

Enable **Select links**.

11.

Select the linked file.

Click **Bind Link** from the ribbon.

12.

Disable all the Include options.

Click **OK**.

13.

Click **OK**.

14.  Click **Remove Link**.

15. *The link is inserted as a group.*

Select the link and click **Ungroup** from the ribbon.

16. 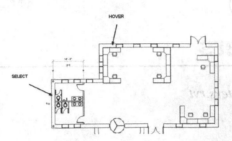 Click to select the indicated wall. Hover the mouse over the wall indicated.

Click the TAB key.

Left click to select the highlighted walls.

17. 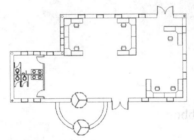 Notice the selected wall, the hovered wall and the connecting wall are all selected.

Left click in the window to release the selection.

18. Close without saving.

Project Parameters

Project parameters are specific to a project, such as Project Name or address. They may appear in schedules or title blocks but not in tags.

Project parameters can be used to organize the browser and to standardize information used across sheets.

Exercise 4-6
Project Parameters

Drawing Name: *project_parameters.rvt*
Estimated Time: 20 minutes

Scope

Project Browser
Project Parameters
Browser Organization

Solution

1.  | Manage | Select the **Manage** ribbon.

2.    Select the **Project Parameters** tool.

Project
Parameters

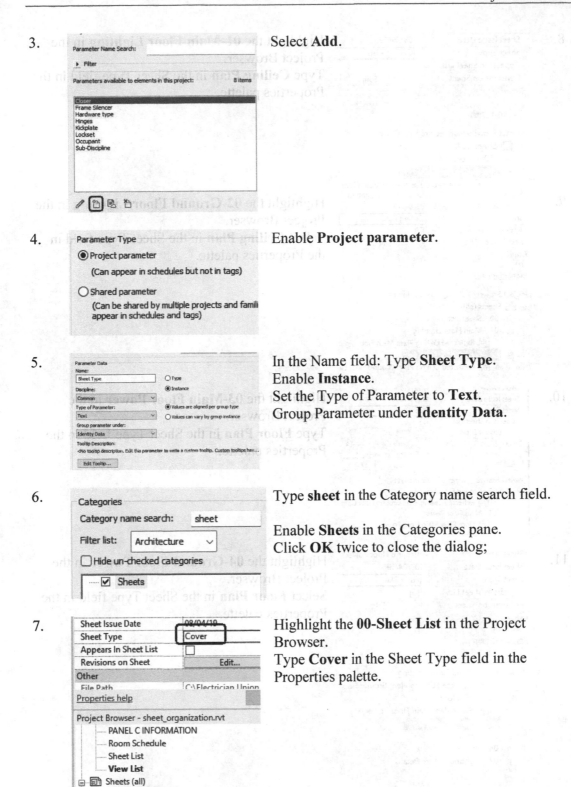

3. Select **Add**.

4. Enable **Project parameter**.

5. In the Name field: Type **Sheet Type**.
Enable **Instance**.
Set the Type of Parameter to **Text**.
Group Parameter under **Identity Data**.

6. Type **sheet** in the Category name search field.

Enable **Sheets** in the Categories pane.
Click **OK** twice to close the dialog;

7. Highlight the **00-Sheet List** in the Project Browser.
Type **Cover** in the Sheet Type field in the Properties palette.

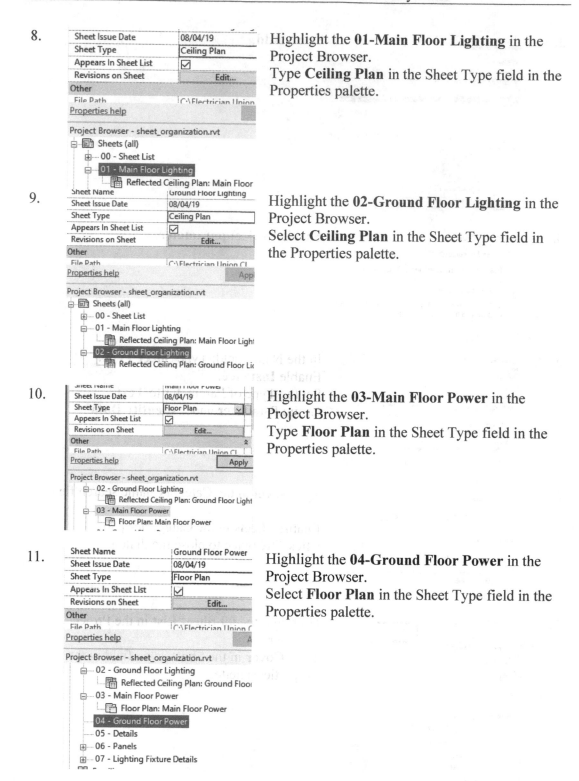

8. Highlight the **01-Main Floor Lighting** in the Project Browser.
Type **Ceiling Plan** in the Sheet Type field in the Properties palette.

9. Highlight the **02-Ground Floor Lighting** in the Project Browser.
Select **Ceiling Plan** in the Sheet Type field in the Properties palette.

10. Highlight the **03-Main Floor Power** in the Project Browser.
Type **Floor Plan** in the Sheet Type field in the Properties palette.

11. Highlight the **04-Ground Floor Power** in the Project Browser.
Select **Floor Plan** in the Sheet Type field in the Properties palette.

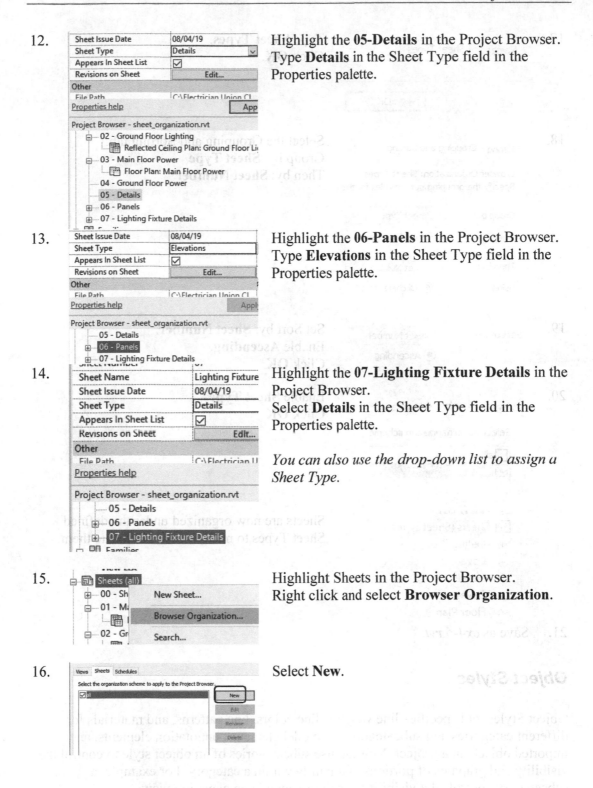

12. Highlight the **05-Details** in the Project Browser. Type **Details** in the Sheet Type field in the Properties palette.

13. Highlight the **06-Panels** in the Project Browser. Type **Elevations** in the Sheet Type field in the Properties palette.

14. Highlight the **07-Lighting Fixture Details** in the Project Browser.
Select **Details** in the Sheet Type field in the Properties palette.

You can also use the drop-down list to assign a Sheet Type.

15. Highlight Sheets in the Project Browser.
Right click and select **Browser Organization**.

16. Select **New**.

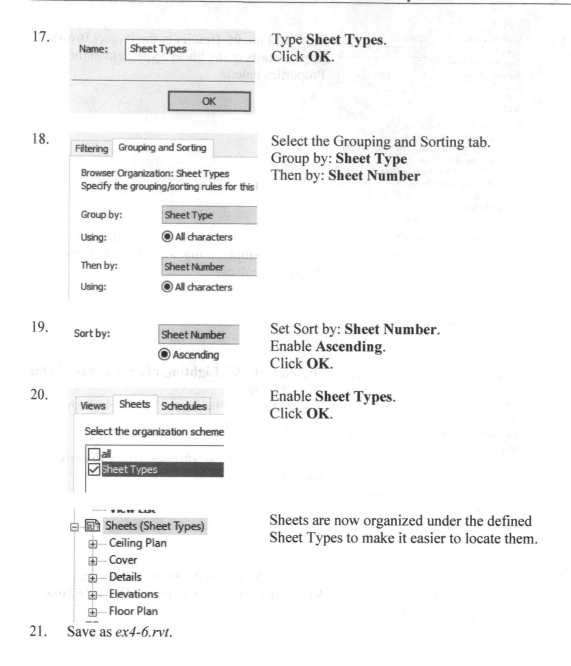

17.   Type **Sheet Types**.
Click **OK**.

18.   Select the Grouping and Sorting tab.
Group by: **Sheet Type**
Then by: **Sheet Number**

19.   Set Sort by: **Sheet Number**.
Enable **Ascending**.
Click **OK**.

20.   Enable **Sheet Types**.
Click **OK**.

Sheets are now organized under the defined Sheet Types to make it easier to locate them.

21.   Save as *ex4-6.rvt*.

Object Styles

Object Styles tool specifies line weights, line colors, line patterns, and materials for different categories and subcategories of model elements, annotation elements, and imported objects in a project. You can use subcategories of an object style to control the visibility and graphics of portions of a family within a category. For example, a subcategory controls the visibility of a door swing in an elevation view.

You can override project object styles on a view-by-view basis.

Exercise 4-7

Configure Object Styles

Drawing Name: *office building.rvt*
Estimated Time: 15 minutes

Scope

Visibility/Graphics
Object Styles

Users have the ability to create different objects styles to comply with company standards.

Solution

1.   -- — Elevations (Building Elevation)
         |---- ☐ East
         |---- ☐ North
         |---- ☐ **South**
         |---- ☐ West

Open the South elevation.

2.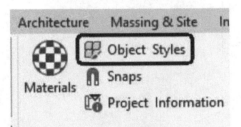

Open the Manage ribbon.

Click **Object Styles**.

3.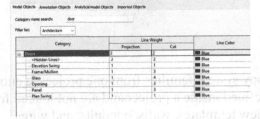

Type **door** in the Category name search field.

Hold down the CTL key to select all the door elements.

Set the color to **Blue**.

Click **Apply**.

4.

Type **window** in the Category name search field.

Hold down the CTL key to select all the door elements.

Set the color to **Magenta**.

Click **OK**.

The display on the elevation updates.

5. Switch to Level 1 floor plan.

Notice that the object styles are applied independently of the view.

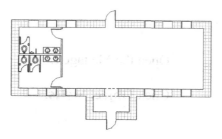

6. Save as *ex4-7.rvt*.

Transfer Project Standards

Transfer Project Standards is used to copy system families from one project to another. System families are elements that are defined inside a project and are not loaded from an external file. System families include view templates, walls, conduits, and wires.

You need to have the project open that you want to import into as well as the project you want to transfer from. If the project you are transferring from has a linked file, you can also import families from the linked file without opening that file.

Items which can be copied from one project to another include:

- Family types (including system families, but not loaded families)
- Line weights, materials, view templates, and object styles
- Mechanical settings, piping, and electrical settings
- Annotation styles, color fill schemes, and fill patterns
- Print settings

You may get a prompt during the transfer process alerting that an item being transferred already exists in the new project. You may opt to overwrite, ignore or cancel for the existing elements. You do not get to pick and choose which elements you overwrite if more than one item has been selected to be transferred.

Exercise 4-8
Transfer Project Standards

Drawing Name: *object styles.rvt, office building_2.rvt*
Estimated Time: 5 minutes

Scope
Transfer Project Standards
Object Styles

A company has developed object styles to mimic the layer colors required by the local county permit board. The CAD designer decides to use the Transfer Project Standards tool to assign the new object styles to their current project.

Solution

1. Open the *object styles.rvt* and *office building_2.rvt* files.
    Open the **Level 1** floor plan for *office building_2.rvt*.

2. Collaborate   View   Manage   Add-Ins   Mo   Open the Manage ribbon.

    roject Parameters   🗄 Transfer Project Standards   Click **Transfer Project**
    hared Parameters   🗄 Purge Unused   **Standards**.
    lobal Parameters   🗄 Project Units
    Settings

3.

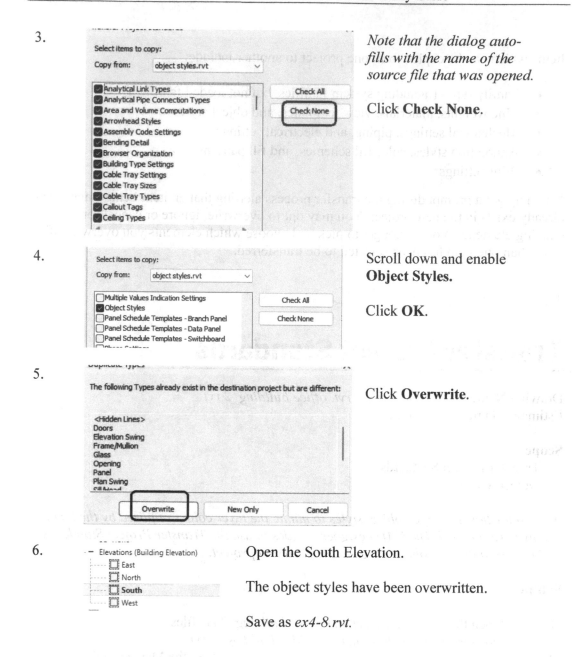

Note that the dialog auto-fills with the name of the source file that was opened.

Click **Check None**.

4.

Scroll down and enable **Object Styles.**

Click **OK**.

5.

Click **Overwrite**.

6.

Open the South Elevation.

The object styles have been overwritten.

Save as *ex4-8.rvt*.

Browser Organization

Many users find it more productive to organize or sort the project browser using view parameters.

Use the Browser Organization tool to group and sort views, sheets, and schedules/quantities in the way that best supports your work. You can specify six levels of grouping. Within groups, items are sorted in ascending or descending order of a selected property.

By default, the Project Browser displays all views (by view type), all sheets (by sheet number and sheet name), and all schedules and quantities (by name).

In addition to grouping and sorting views, you can limit the views that display in the Project Browser by applying a filter. This approach is useful when the project includes many views, sheets, and schedules/quantities, and you want the Project Browser to list a subset. You can specify up to 3 levels of filtering.

The combination of criteria used to filter, group, and sort items in the Project Browser is an organization scheme. You can create separate schemes for views, sheets, and schedules/quantities in the project. Create as many schemes as needed to support different phases of your work.

When creating an organization scheme, you can use any properties of the view, sheet, or schedule/quantity as the criteria for grouping, sorting, and filtering. You can also use project parameters and shared parameters for grouping and filtering.

When working on a project, you can quickly change the organization schemes applied to the Project Browser at any time. Switch between organization schemes whenever needed based on your current work.

Exercise 4-9

Browser Organization

Drawing Name: **browser.rvt**
Estimated Time to Completion: 20 Minutes

Scope
Creating a new browser organization schema.
Using parameters in the project browser.

Solution

1.

Highlight the Views in the Project Browser.

Right click and select **Browser Organization**.

2.

Notice that you can organize Views, Sheets, and Schedules. Each of these can be organized a different way, if you like.

Currently the views are organized by Scale.

Click **Edit**.

3.

No filtering has been applied to this organization schema.

Click the **Grouping and Sorting** tab.

4.

Notice that views are grouped by Type (floor plan, ceiling plan, elevation, etc.), then by View Scale.

They are sorted by level.

5.

Click **OK**.
Place a check next to **Discipline**.

Then click **Edit**.

6.

Click the **Grouping and Sorting** tab.

7.

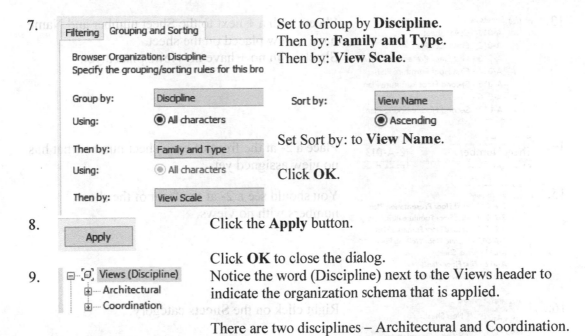

Set to Group by **Discipline**.
Then by: **Family and Type**.
Then by: **View Scale**.

Set Sort by: to **View Name**.

Click **OK**.

8. Apply

Click the **Apply** button.

Click **OK** to close the dialog.

9.

Notice the word (Discipline) next to the Views header to indicate the organization schema that is applied.

There are two disciplines – Architectural and Coordination.

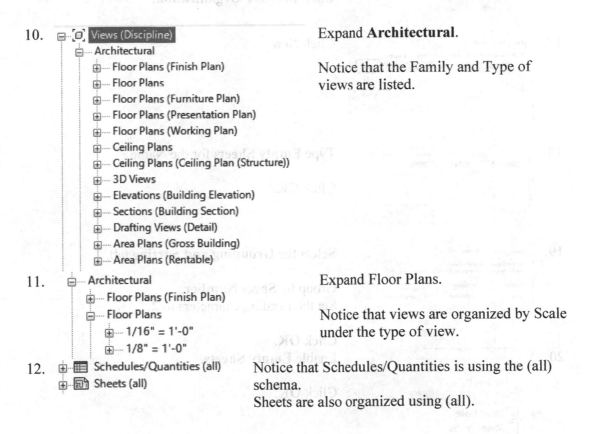

10.

Expand **Architectural**.

Notice that the Family and Type of views are listed.

11.

Expand Floor Plans.

Notice that views are organized by Scale under the type of view.

12.

Notice that Schedules/Quantities is using the (all) schema.
Sheets are also organized using (all).

13. 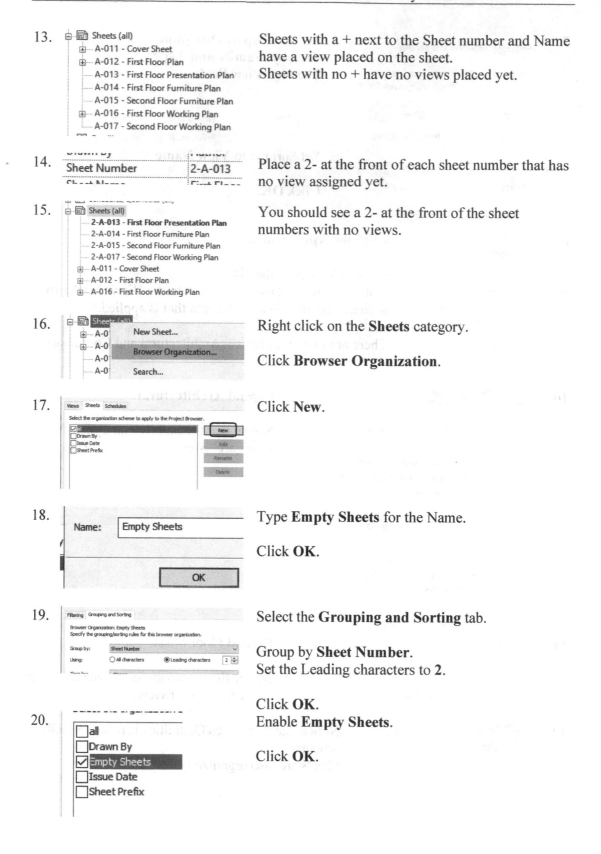 Sheets with a + next to the Sheet number and Name have a view placed on the sheet.
Sheets with no + have no views placed yet.

14. Place a 2- at the front of each sheet number that has no view assigned yet.

15. You should see a 2- at the front of the sheet numbers with no views.

16. Right click on the **Sheets** category.

Click **Browser Organization**.

17. Click **New**.

18. Type **Empty Sheets** for the Name.

Click **OK**.

19. Select the **Grouping and Sorting** tab.

Group by **Sheet Number**.
Set the Leading characters to **2**.

Click **OK**.
Enable **Empty Sheets**.

20. Click **OK**.

21.

```
Sheets (Empty Sheets)
  2-
    2-A-013 - First Floor Presentation Plan
    2-A-014 - First Floor Furniture Plan
    2-A-015 - Second Floor Furniture Plan
    2-A-017 - Second Floor Working Plan
  A-
    A-011 - Cover Sheet
    A-012 - First Floor Plan
    A-016 - First Floor Working Plan
```

All the sheets that are missing views are now separated into their own category.

As you add a view to each sheet, you can edit the Sheet Number and have it move into the category of sheets with views.

Many companies set up templates where the title blocks and sheets are set up for each type of drawing sheet needed for a project. By organizing the sheets this way, you can quickly see which sheets still need to be worked on.

22.

Save as *ex4-9.rvt*.

Search

A search feature is available in Revit to make it easier to select from a long list of options. You can search in the following functional areas:

- Type Selector

- Properties palette (drop-down lists for instance property values)

- Dialogs (drop-down lists for values in dialog tables, category name search for long lists of categories)

- Callout, section, elevation, or detail view creation with the Reference Other View option selected

- Project Browser

In order to perform a search for text or tags, you have to create a schedule and then organize it to locate the desired item.

Exercise 4-10
Perform Effective Searches

Drawing Name: **search.rvt**
Estimated Time to Completion: 5 Minutes

Scope
Search the Project Browser
Use the Type Selector search

Solution

1.

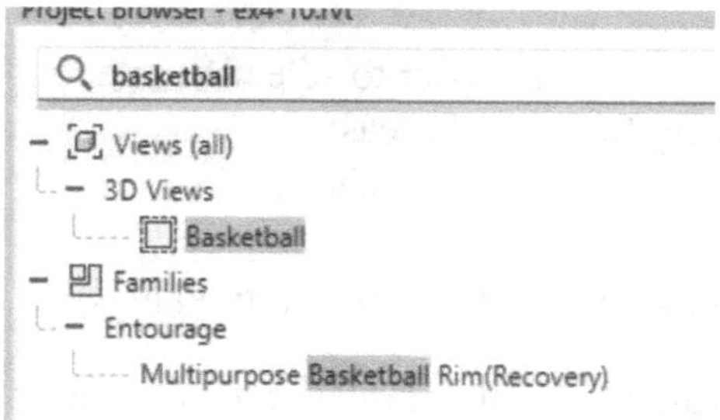

Type **basketball** in the search field of the browser.

Open the **3D Basketball** view.

2.

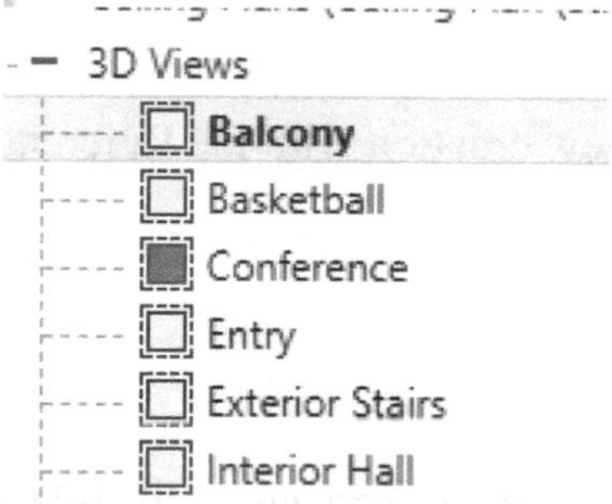

Open the **3D Balcony** view.

3.

Select one of the tables.

In the Type Selector search field, type **table**.

Select the **Table – Communal – Round w/Chairs -60" Diameter.**

Click to release the selection.

4.

The table and chairs are replaced.

Save as *ex4-10.rvt*

Print Setups

To streamline the printing process, you can create and save print setups for different types of print jobs. You can also create and save sets of views/sheets, so it is easy to reprint a set later.

Printed output in Revit is What You See Is What You Get (WYSIWYG), with a few exceptions:
- The background color for the print job is always white.
- By default, reference planes, work planes, crop boundaries, unreferenced view tags, and scope boxes do not print. To include them in the print job, in the Print Setup dialog, clear the corresponding Hide options.
- The print job includes elements that have been hidden from a view using the Temporary Hide/Isolate tool.
- Line weights modified by the Thin Lines tool print at their default line weight.

You can save print settings in a project for later reuse.

For example, you may create one print setup for code review, and create another print setup designed for consultants' pages. The print setups are saved with the project, so they can be used across multiple sessions.

Note: You can transfer print setups to another project using Transfer Project Standards. Click Manage tab ➤ Settings panel ➤ (Transfer Project Standards), and in the Select Items To Copy dialog, select Print Settings.

You can also change saved print settings, revert changes, and rename or delete print settings.

To access the Print Setup dialog, click File tab ➤ Print ➤ (Print Setup).

Exercise 4-11
Configure Print Sets

Drawing Name: **firestation.rvt**
Estimated Time to Completion: 10 Minutes

Scope
Configure a Print Set

Solution

1.

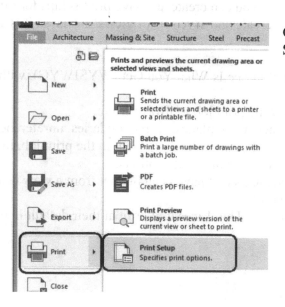

Go to **File→Print→Print Setup**.

2.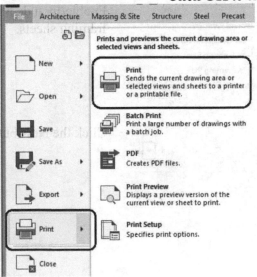

Set the Printer to **Adobe PDF**.
Set the Paper Size to **Arch D**.
Set the Orientation to **Landscape**.
Set the Paper Placement to **Center.**
Set Zoom to **Fit to Page.**
Set Colors to **Color.**

Click **Save As.**

3. Name: Adobe PDF_ArchD_Color

Set the Name to **Adobe PDF_ArchD_Color**.

Click **OK**.
Click **OK** to close the dialog.

4.

Go to **File→Print→Print**.

5.

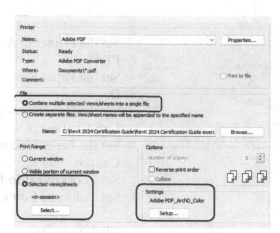

Enable **Combine multiple selected views/sheets into a single file**.
Enable **Selected views/sheets**.
Verify that the Settings is set to the **Adobe PDF_ArchD_Color** setup.

Click **Select** under **Selected views/sheets**.

6.

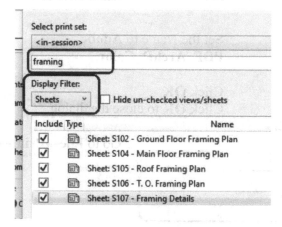

Type **framing** in the search field.

Select **Sheets** in the Display Filter.

Enable/Include all the framing sheets.

7.

Click the **New** button.

8.

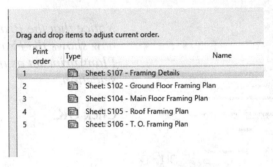

Name the print set **Framing Plans.**

Place a check in front of all the displayed sheets.

Click **Edit print order**.

9.

Drag and drop the Framing Details sheet so it is Sheet 1.

Click **OK**.

10.

Click **Save current set**.

Click **Select** at the bottom of the dialog box.

11.

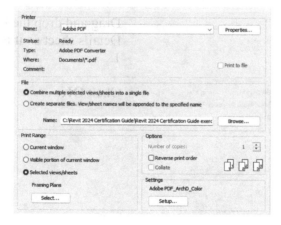

Notice that the Print Range now shows the Framing Plans print set as the active selection.

Click **OK**.

12.

File name: framing plans
Save as type: PDF files (*.PDF)

Browse to the folder where you want to store the file.
Name the file *framing plans*.
Click **Save.**

13.   Save the file as *ex4-11.rvt.*

Saving Library Files

As you become more familiar with Revit, you will work in projects that contain Revit families which you may want to re-use in a future project. You may want to create a folder where you store your favorite or standard Revit loadable families so it is easy to locate them and share them with other team members. If the Revit content was created prior to 2016, it may take a little more work to create it as a loadable and shareable file.

Exercise 4-12
Create Library Files

Drawing Name: **library.rvt, firestation.rvt**
Estimated Time to Completion: 10 Minutes

Scope
Edit Family
Save a Family
Copy and paste a family.

Solution

1.

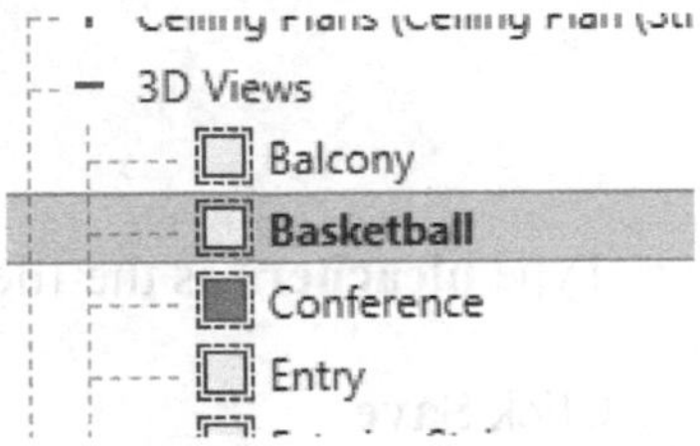

Open *library.rvt*.

Open the **3D Basketball** view.

2.

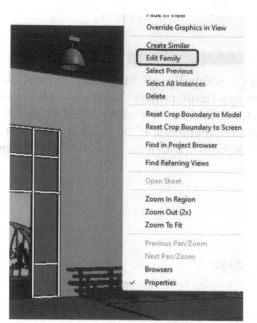

Select the bleachers.

Right click and select **Edit Family**.

3.

Go to **File→Save As→Family**.

4.

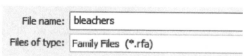

Type **bleachers** as the file name.

Click **Save**.

5.

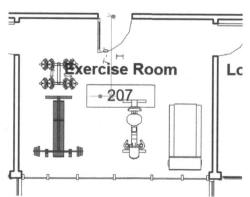

Close the file.
Open *firestation.rvt*.
Open the Main Floor floor plan.
Zoom into the Exercise Room.

Select the Lifting Bench.
If you right click, you will see that the Edit Family option is grayed out. This is because this family was created pre-2016.

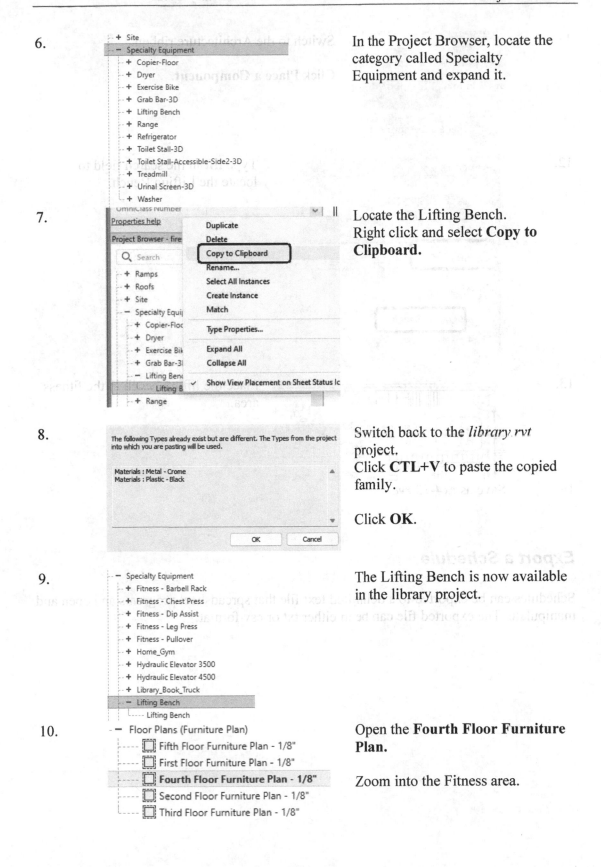

6. In the Project Browser, locate the category called Specialty Equipment and expand it.

7. Locate the Lifting Bench. Right click and select **Copy to Clipboard.**

8. Switch back to the *library.rvt* project.
Click **CTL+V** to paste the copied family.

Click **OK**.

9. The Lifting Bench is now available in the library project.

10. Open the **Fourth Floor Furniture Plan.**

Zoom into the Fitness area.

11. Switch to the Architecture ribbon.

Click **Place a Component**.

12. Type **lift** in the search field to locate the Lifting Bench.

13. Place the lifting bench in the fitness area.

14. Save as *ex4-12.rvt*.

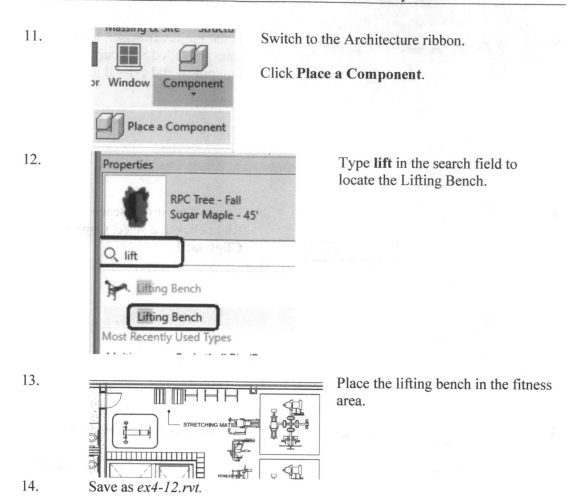

Export a Schedule

Schedules can be exported to a delimited text file that spreadsheet programs and open and manipulate. The exported file can be in either txt or csv format.

Exercise 4-13

Export a Schedule

Drawing Name: schedule_export.rvt [m_ schedule_export.rvt]
Estimated Time: 5 minutes

Scope

Schedules

Solution

1.   Open *schedule_export.rvt [m_ schedule_export.rvt]*.

2. 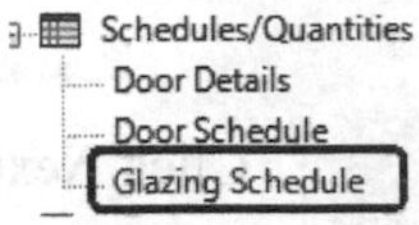   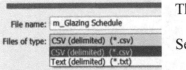   Activate the **Glazing Schedule**.

3.   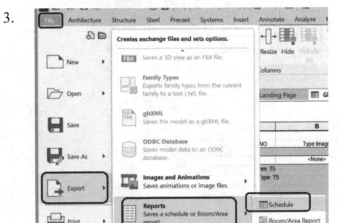

Go to the Applications Menu.

Select **File → Export →
Reports → Schedule**.

*You will need to scroll down to
see the Reports option.*

4.

The schedule can be saved as a csv or txt file.

Select *.txt.

Browse to your exercise folder.
Click **Save**.

5. 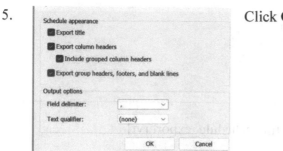 Click **OK**.

6. Launch **Excel**.

7. Open   Select **Open**.

8. Text Files   Set the file types to **Text Files**.

9.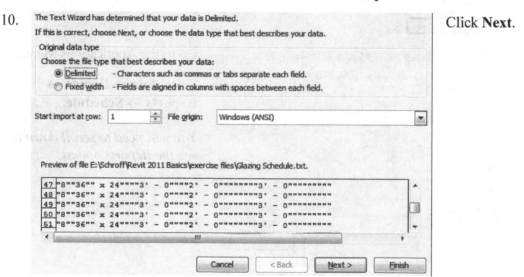

Browse to where you saved the file and select it. Click **Open**.

10. Click **Next**.

11.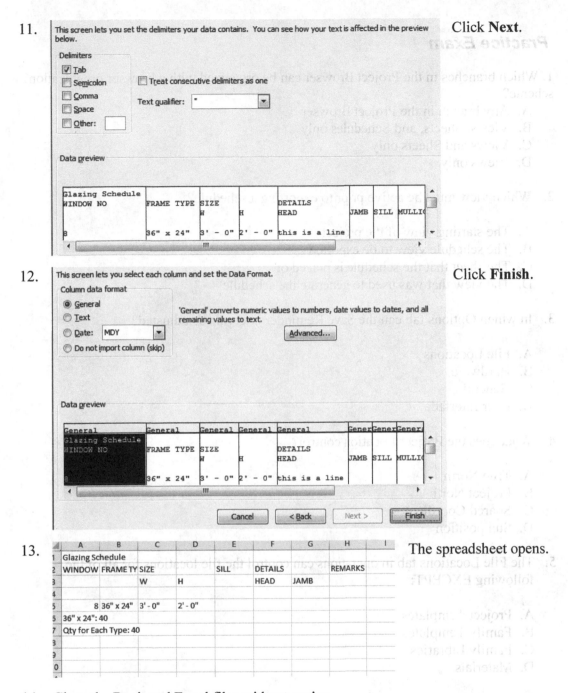

Click **Next**.

12. Click **Finish**.

13. The spreadsheet opens.

14. Close the Revit and Excel files without saving.

Practice Exam

1. Which branches in the Project Browser can be organized with a browser organization scheme?
    A. Any branch in the Project Browser
    B. Views, Sheets, and Schedules only
    C. Views and Sheets only
    D. Views only

2. Which view must be active prior to exporting a schedule?

    A. The starting view of the project
    B. The schedule view to be exported
    C. The sheet that the schedule is placed on
    D. The view that was used to generate the schedule

3. In which Options tab can the Save Reminder intervals be adjusted?

    A. File Locations
    B. Hardware
    C. General
    D. User Interface

4. What does the Project Location control?

    A. True North
    B. Project North
    C. Shared Coordinates
    D. Sun position

5. The File Locations tab in operations can control the file locations for all of the following EXCEPT:

    A. Project Templates
    B. Family Templates
    C. Family Libraries
    D. Materials

6. T/F You cannot adjust the position of panels in the ribbon.

7. Shortcut keys must use:

   A. A combination of two keys
   B. A combination of the CTL key and one other key
   C. A single key
   D. A Function key and one other key

8. A CAD designer has created a selection set of elements and copied them. They escape out of the command and perform a different task. They realize they need to create another copy of the previous selection set. To restore the previous selection set, they:

   A. Hold down the CTL key and click the LEFT arrow.
   B. Right click in the display window and click SELECT PREVIOUS.
   C. Go to the view where the original selection set was created and start selecting the elements again.
   D. Use the GROUP command to define the selection set.

9. To search the Project Browser for a specific sheet name:

   A. Type the keyword for the search in the search field of the Project Browser.
   B. Go to the Manage ribbon and click the Find tool.
   C. Create a sheet list schedule and search the schedule.
   D. Go to the Manage ribbon and click Select By ID.

10. A print set may be used to:

   A. Control the print settings, such as page size and colors, as well as what sheets/views are included
   B. Control the print settings for the printer, but not what sheets/views are included
   C. Batch print from more than one project
   D. Control the size/scale of the views used on the sheets

11. To copy object styles from one project to another, use:

   A. View Templates
   B. Transfer Project Standards
   C. Copy/Paste
   D. Save the project with the desired object styles as a template

12. An architectural designer is creating a browser organization scheme to help organize the views. The available options to group do not meet their needs. What should they do?

    A. Create a project parameter and assign it to the Views category
    B. Click Project Information on the Manage ribbon and create a new parameter
    C. Add a new sort by parameter in the User Interface options in the Options dialog
    D. Create a global parameter

13. Where is the save location of workshared local copies controlled?

    A. Worksets dialog
    B. Options dialog→Collaborate
    C. Options dialog→File Locations
    D. Synchronize with Central dialog

14. Which two items can a project parameter be applied to? (Select two)

    A. System families
    B. Project Locations
    C. Keyboard shortcuts
    D. Loadable families

15. What should be created when the same set of sheets needs to be printed and exported to PDF multiple times?

    A. Organization scheme
    B. Template
    C. Export setup
    D. Print Set

Answers:
    1)   B; 2) B; 3) C; 4) D; 5) D; 6) F; 7) A; 8) A; 9) A; 10) B; 11) A; 12) C; 13) A & D; 14) D

Information Analysis

This lesson addresses the following exam questions:

- Create and manage schedules
- Create Calculated and Combined Parameters
- Create a Note Block
- Create and Edit Area Schemes
- Create sun and shadow studies

Schedules and Linked Models

All fields that are available for elements in the host project are available for elements in linked models. The behavior of some fields changes when you add elements from linked models to a schedule. For example, the Family, Type, Family and Type, Level, and Material parameters become read-only for elements in both the host and linked models. You also cannot filter a schedule by the Family, Type, Family and Type, Level, or Material parameters.

Exercise 5-1
Including Linked Elements in a Schedule

Drawing Name: **schedules.rvt, exam rooms**
Estimated Time to Completion: 15 Minutes

Scope

Include linked elements in a schedule

Solution

1.

Open *schedules.rvt*.

Open the **Working Ground Floor** plan view.

2.

Open the Insert ribbon.

Click **Link Revit**.

3.

Select the *exam rooms.rvt* file.

Set the Positioning to **Auto – Center to Center**.

Click **Open**.

4.

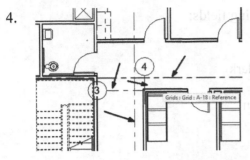

Select the ALIGN tool to adjust the position of the linked file.

Use the grids to locate the linked file.

5.

Grid 3 should be aligned to A-18.
Grid 4 should be aligned to A-19.

6.

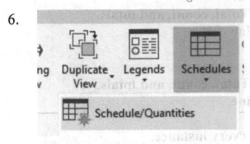

Go to **View→Schedule/Quantities**.

7.

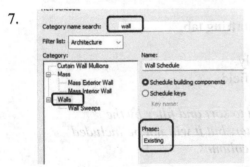

Type **wall** in the category name search field.

Highlight **Walls**.

Enable **Schedule building components**.

Set the Phase to **Existing**.

Click **OK**.

8.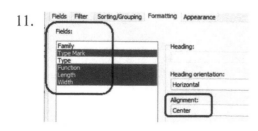

Add the following fields:

- Family
- Type Mark
- Type
- Function
- Length
- Width

Enable **Include elements in links**.

You can use the parameter name search to quickly locate the desired parameters.

9.

Select the **Sorting/Grouping** tab.

Sort by **Function**.
Enable Footer: **Total, count, and totals**.
Enable **Blank line**.

Sort by **Type**.
Enable Footer: **Total, count, and totals**.
Enable **Blank line**.

Enable **Itemize every instance**.

10.

Select the **Formatting** tab.

Highlight Family.
Enable Hidden field.

This allows you to sort and filter by the Family parameter, but it will not be included in the schedule columns.

11.

In the Formatting tab:
Hold down the CTL key.
Select **Type Mark, Function, Length, and Width** so they are highlighted.

Change the Alignment to **Center**.

12.

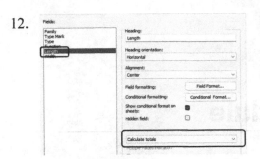

In the Formatting tab:

Highlight the Length field.
Select **Calculate totals**.

13.

Click the Appearance tab.

Set the Title text to: **1/4″ Arial**.
Set the Header text to: **1/16″ Arial**
Set the Body text to: **3/32″ Arial**

Click **OK**.

14.

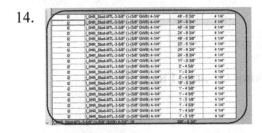

Scroll down the schedule.

The walls with Type Mark l2 are used in the linked model.

Save as *ex5-1.rvt*.

Exercise 5-2
Apply a Phase to a Schedule

Drawing Name: **schedules.rvt**
Estimated Time to Completion: 20 Minutes

Scope

Apply a phase to a schedule
Create a view template for a schedule
Apply a view template to a schedule
Create a new sheet.
Add schedule views to a sheet

Solution

1.
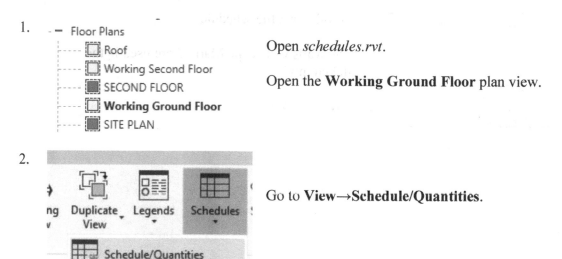

Open *schedules.rvt*.

Open the **Working Ground Floor** plan view.

2.

Go to **View→Schedule/Quantities**.

3.

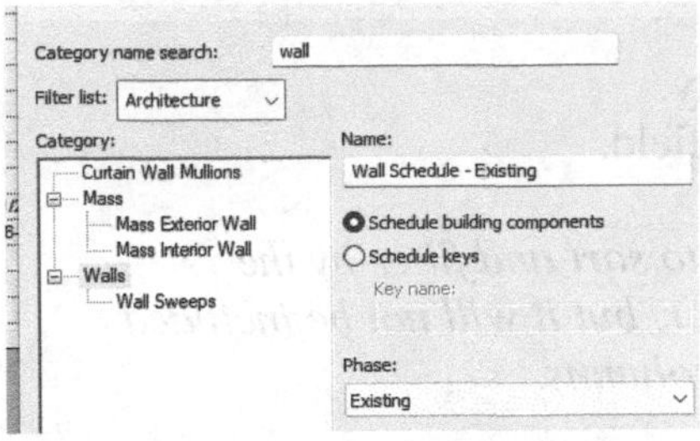

Type **wall** in the category name search field.

Highlight **Walls**.

Change the Name to **Wall Schedule – Existing.**

Enable **Schedule building components**.

Set the Phase to **Existing**.

Click **OK**.

4.

Add the following fields:

- Family
- Type Mark
- Type
- Function
- Length
- Width

You can use the parameter name search to quickly locate the desired parameters.

5.

Select the **Sorting/Grouping** tab.

Sort by **Function**.
Enable Footer: **Total, count, and totals**.
Enable **Blank line**.

Sort by **Type**.
Enable Footer: **Total, count, and totals**.
Enable **Blank line**.

Enable **Itemize every instance**.

6.

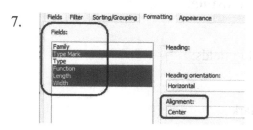

Select the **Formatting** tab.

Highlight **Family**.
Enable **Hidden** field.

This allows you to sort and filter by the Family parameter, but it will not be included in the schedule columns.

7.

In the Formatting tab:
Hold down the CTL key.
Select **Type Mark, Function, Length, and Width** so they are highlighted.

Change the Alignment to **Center**.

8.

In the Formatting tab:

Highlight the Length field.
Select **Calculate totals**.

9.

Click the Appearance tab.

Set the Title text to: **1/4″ Arial**.
Set the Header text to: **1/16″ Arial**
Set the Body text to: **3/32″ Arial**

Click **OK**.

10.

On the View ribbon:

Click **Create Template from Current View**.

11.

Type **Wall Schedule – Phases** in the Name field.

Click **OK**.

12.

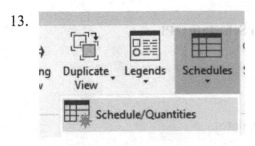

Disable **V/G Overrides Design Options**.
Disable **Phase Filter**.

Click **OK**.

13.

Go to **View→Schedule/Quantities**.

14.

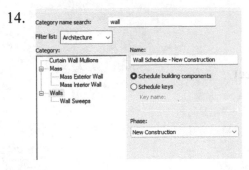

Type **wall** in the category name search field.

Highlight **Walls**.

Change the Name to **Wall Schedule – New Construction.**

Enable **Schedule building components**.

Set the Phase to **New Construction**.

Click **OK**.

15. 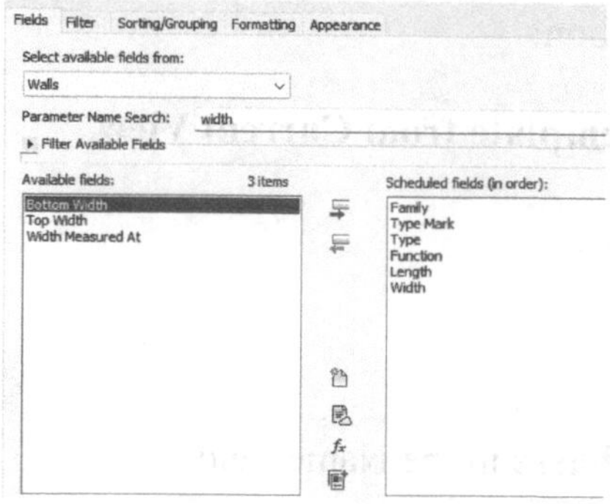 Add the following fields:

- Family
- Type Mark
- Type
- Function
- Length
- Width

You can use the parameter name search to quickly locate the desired parameters.

Click **OK.**

16. 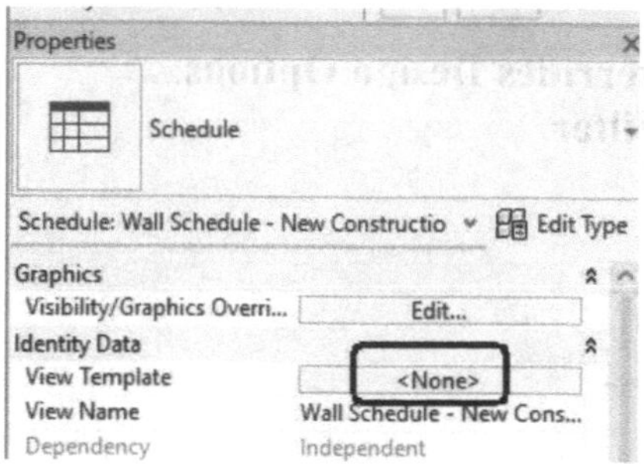 Click the **<None>** button next to View Template in the Properties panel.

17. 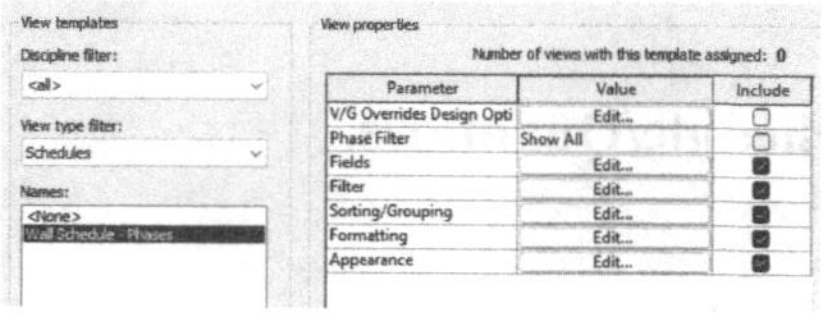 Highlight **Wall Schedule – Phases**.

Click **OK.**

18. 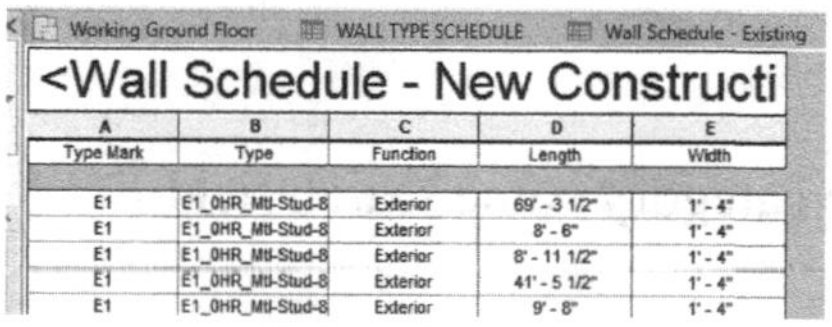 The schedule updates.

19. 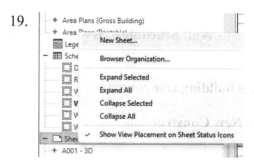 Add a **New Sheet**.

20.

Select titleblocks:

A 8.5 x 11 Vertical
B 11 x 17 Horizontal
C 17 x 22 Horizontal
D 22 x 34 Horizontal
E1 30 x 42 Horizontal : E1 30x42 Horizontal
E1 30 x 42 Horizontal-Cover : E1 30x42 Horizontal
E 34 x 44 Horizontal
None

Select the **E 34 x 44 Horizontal** titleblock.

Click **OK**.

21.

Drag and drop the existing and new construction wall schedules onto the sheet.

You can split a schedule by clicking on it and then clicking on the split icon located on the middle right side of the schedule.

22.

Save as *ex5-2.rvt*.

Exercise 5-3

Apply a Design Option to a Schedule

Drawing Name: **kitchen design options.rvt**
Estimated Time to Completion: 15 Minutes

Scope

Apply a design option to a schedule
Apply a filter to a schedule
Duplicate a schedule view

A kitchen designer has created two design options to present to their client. The client would like to know the cost of each design. The designer creates a schedule for each design option.

Solution

1.

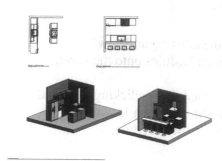

Open *kitchen design options.rvt*.

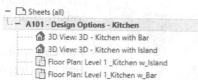

Open the **A101- Design Options Kitchen** sheet.

2.

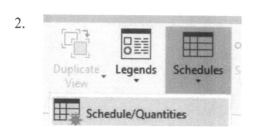

Go to the **View** ribbon.

Click **Schedule/Quantities**.

3.

Highlight **Multi-Category**.

Type **Kitchen Remodel – Island**.

Set the Phase to **New Construction.**

Click **OK**.

4.

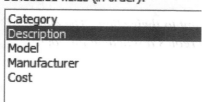

Add the following fields:

- Category
- Description
- Model
- Manufacturer
- Cost

5.

In the Filter tab:
Filter by: Category does not equal **Walls**
And Category does not equal **Floors**

This will filter out walls and floors from the schedule.

6.

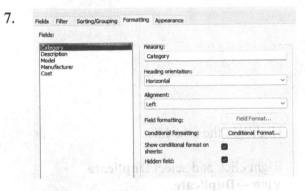

In the Sorting/Grouping tab:
Sort by **Category**.
Enable **Footer**.
Select **Title, count, and totals**.

Enable **Grand totals**.
Set it to **Totals only**.

Enable **Itemize every instance**.

7.

In the Formatting tab:

Highlight **Category**.

Enable **Hidden field**.

8.

In the Formatting tab:

Highlight **Cost.**
Set the Alignment to **Right.**
Select **Calculate totals.**

Click **OK.**

9.

The schedule opens.

Click **Edit** to launch the Visibility/Graphics dialog.

10.

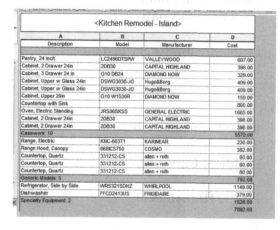

Set the Design Option to **Kitchen w_island.**

Click **OK.**

It appears the materials for the island kitchen remodel will cost around $8,000.

11.

Highlight the schedule.

Right click and select **Duplicate View→Duplicate.**

12.

Rename the schedule **Kitchen Remodel – Bar**.

13.

Click **Edit** next to Visibility/Graphics Overrides.

14.

Design Option Set	
Option Set 1	Kitchen w_Bar <primary>

Set the Design Option to **Kitchen w_Bar.**

Click **OK**.

15.

<Kitchen Remodel - Bar>			
A	**B**	**C**	**D**
Description	Model	Manufacturer	Cost
Cabinet, 4 drawer 24in	DSB3DA24-RM	HUGO&BORG	639.85
Cabinet, 4 drawer 24in	DSB3DA24-RM	HUGO&BORG	639.85
Cabinet, 4 drawer 24in	DSB3DA24-RM	HUGO&BORG	639.85
Cabinet, Single Door 18in	33A W1830R	PROJECT SOURCE	89.98
Cabinet, Double Sink	323 SB30	DIAMOND EXPRESS	459.00
Kitchen Bar- Casework	custom	PROJECT SOURCE	250.00
Cabinet, Single Door 18in	33A W1830R	PROJECT SOURCE	89.98
Casework: 7			2808.51
Bar Stool	LEM STOOL	LAPALMA	1200.00
Bar Stool	LEM STOOL	LAPALMA	1200.00
Bar Stool	LEM STOOL	LAPALMA	1200.00
Bar Stool	LEM STOOL	LAPALMA	1200.00
Furniture: 4			4800.00
Range Hood, Canopy	31T/XP11(75)	WINFLO	269.00
Countertop, Quartz	331212-CS	allen + roth	45.00
Countertop, Quartz	331212-CS	allen + roth	60.00
Kitchen Bar Counter	331212-CS	allen + roth	270.00
Countertop, Quartz	331212-CS	allen + roth	45.00
Countertop, Quartz	331212-CS	allen + roth	45.00
Countertop, Quartz	331212-CS	allen + roth	270.00
Countertop, Quartz	331212-CS	allen + roth	45.00
Generic Models: 8			1049.00
Sink w Faucets	QT-812-BL	KARRAN	256.46
Plumbing Fixtures: 1			256.46
Refrigerator, Side by Side	WRS321SDHZ	WHIRLPOOL	1149.00
Dishwasher	FFCD2413US	FRIGIDAIRE	379.00
Gas Range w Stove	NX60A6511SS	SAMSUNG	829.00
Specialty Equipment: 3			2357.00
			11270.97

It appears the materials for the kitchen with bar remodel will cost around $11,000.

16.

Save as *ex5-3.rvt.*

Exercise 5-4

Using Combined Parameters in a Schedule

Drawing Name: door schedule2.rvt [m_door schedule2.rvt]
Estimated Time: 10 minutes

Scope

Combined Parameters
Schedules

Solution

1.

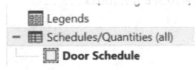

Open the **Door Schedule** view.

2.

Highlight the **SIZE** header.

On the ribbon, click **Ungroup**.

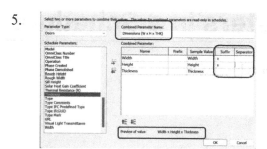

3.
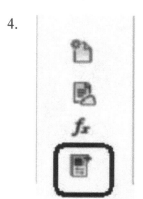

Click **Edit** next to Fields in the Properties pane.

4.

Select **Combine parameters**.

5.

In the Combined Parameter Name field:

Type **Dimensions (W x H x THK)**.

Select the Width, Height, and Thickness parameters from the left pane.

6. In the Suffix column: Type a space, then x, then a space for Width and Height.
Delete the / Separator from all three rows.
Look at the bottom of the dialog and check the preview to see if it looks correct.
Click **OK**.

7.

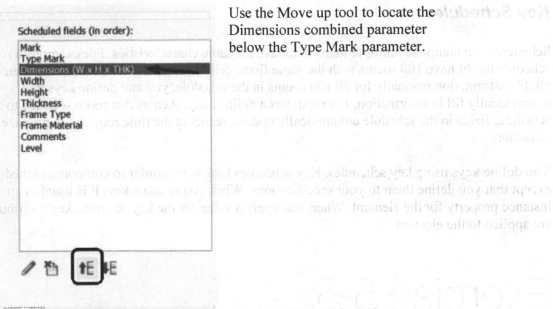

Use the Move up tool to locate the Dimensions combined parameter below the Type Mark parameter.

8.

Select the Formatting tab.

Highlight the **Dimensions combined parameter.**

Set the Alignment to **Center.**

9.

Hold down the CTL key.
Highlight **Width, Height and Thickness.**

Enable **Hidden field.**

Click **OK.**

If you remove the parameters used in the combined parameter, they can't be used in the schedule, so you hide the parameters instead.

10.

The schedule updates.

11.

Save as *ex5-4 [m_ex5-4.rvt].*

Key Schedules

Schedules can comprise multiple items that have the same characteristics. For example, a room schedule might have 100 rooms with the same floor, ceiling, and base finishes. Rather than enter all this information manually for all 100 rooms in the schedule, you can define keys that automatically fill in information. If a room has a defined key, then as that room is added to a schedule, fields in the schedule automatically update, reducing the time required to produce the schedule.

You define keys using key schedules. Key schedules look very similar to component schedules, except that you define them to your specifications. When you create a key, it is listed as an instance property for the element. When you apply a value for the key, then the key's attributes are applied to the element.

Exercise 5-5

Create a Key Schedule

Drawing Name: art gallery.rvt [m_art gallery.rvt]
Estimated Time: 10 minutes

Scope

Schedules
Schedule Keys

Solution

1. Open art gallery.*rvt [m_art gallery.rvt]*.

2. 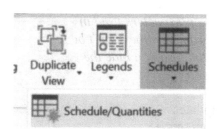 Go to the **View** ribbon.

    Select **Schedules→Schedules/Quantities**.

3.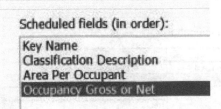

Type **room** in the Category name search field.

Highlight **Rooms** in the Category Panel on the left.

Enable **Schedule Keys**.

In the Name field, type **Space Occupancy Classification**.

In the Key Name field, type **Classification**.

Click **OK**.

4.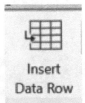

Add the following parameters to the Key Schedule:
- Classification Description
- Area per Occupant
- Occupancy Gross or Net

These are custom project parameters that were added to the exercise file.

Click **OK**.

5. Click **Insert Data Row** on the ribbon.

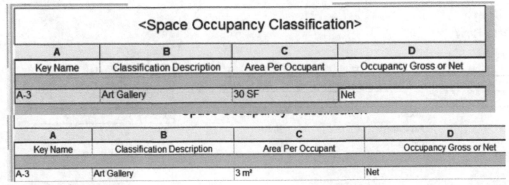

<Space Occupancy Classification>			
A	B	C	D
Key Name	Classification Description	Area Per Occupant	Occupancy Gross or Net
A-3	Art Gallery	30 SF	Net

A	B	C	D
Key Name	Classification Description	Area Per Occupant	Occupancy Gross or Net
A-3	Art Gallery	3 m²	Net

6. Enter the data:

Key Name: A-3
Classification Description: Art Gallery

Area Per Occupant: 30 [3]
Occupancy Gross or Net: Net

7.   Click **Insert Data Row** on the ribbon.

A	B	C	D
Key Name	Classification Description	Area Per Occupant	Occupancy Gross or Net
A-3	Art Gallery	30 SF	Net
B-100	Office	150 SF	Net

<Space Occupancy Classification>			
A	B	C	D
Key Name	Classification Description	Area Per Occupant	Occupancy Gross or Net
A-3	Art Gallery	3 m²	Net
B-100	Office	14 m²	Net

8.   Enter the data:

Key Name: B-100
Classification Description: Office
Area Per Occupant: 150 [14]
Occupancy Gross or Net: Net

9.   Click **Insert Data Row** on the ribbon.

A	B	C	D
Key Name	Classification Description	Area Per Occupant	Occupancy Gross or Net
A-1	Lobby	15 SF	Gross
A-3	Art Gallery	30 SF	Net
B-100	Office	150 SF	Net

<Space Occupancy Classification>			
A	**B**	**C**	**D**
Key Name	Classification Description	Area Per Occupant	Occupancy Gross or Net
A-1	Lobby	1 m²	Gross
A-3	Art Gallery	3 m²	Net
B-100	Office	14 m²	Net

10. Enter the data:

Key Name: A-1
Classification Description: Lobby
Area Per Occupant: 15 [1]
Occupancy Gross or Net: Gross

11. Save as *ex5-5.rvt [m_ex5-5.rvt]*

Calculated Parameters

Formulas can be used when creating a custom family or as a custom parameter in a schedule. You can use formulas in Revit to combine multiple parameters and get a new value based on the parameters used in the formula.

You don't need to have any programming skills in order to write formulas in Revit. You only need to know basic mathematical operation, such as addition, division, and multiplication.

You do need to pay attention to units when creating formulas in schedules. You cannot mix different unit types. You can only create a formula using parameters that are included in the schedule.

Exercise 5-6

Create a Schedule using Formulas

Drawing Name: occupancy_formula.rvt [m_occupancy_formula.rvt]
Estimated Time: 20 minutes

Scope

Schedules
Formulas

Solution

1.   Open *occupancy_formula.rvt [m_occupancy_formula.rvt]*.

2.   Open **Level 1** floor plan.

    – Floor Plans
        ☐ **Level 1**
        ☐ Level 2
        ☐ Site

3.   Select the Room labeled Office 1.

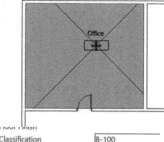

4.   Set the Classification to **B-100**.

    | Classification | B-100 |
    Classification Description   Office
    Occupancy Gross or Net   Net

    Note that the other parameters fill in because they are using the schedule keys.

Dimensions	
Area	54.510 m²
Perimeter	29600.0
Unbounded Height	4000.0
Volume	Not Computed
Computation Height	0.0
Area Per Occupant	14.000 m²

5. Select the Room labeled Office 2.

6. | | |
   |---|---|
   | Classification | B-100 |
   | Classification Description | Office |
   | Occupancy Gross or Net | Net |
   | Occupant | |

   Set the Classification to **B-100**.

7. Hold down the CTL key and select the three rooms labeled Gallery.

   Assign the rooms labeled Gallery to Classification **A-3 Art Gallery**.

8. Select the Lobby.

   Set the Classification to **A-1**.

9. Go to the **View** ribbon.

   Select **Schedules→Schedules/Quantities**.

10.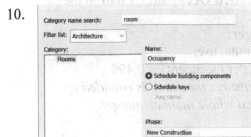

    Type **room** in the Category name search field.

    Highlight **Rooms** in the Category Panel on the left.

    Type **Occupancy** in the Name field.

    Enable **Schedule Building Components**.

    Click **OK**.

11.
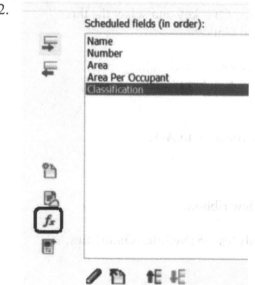

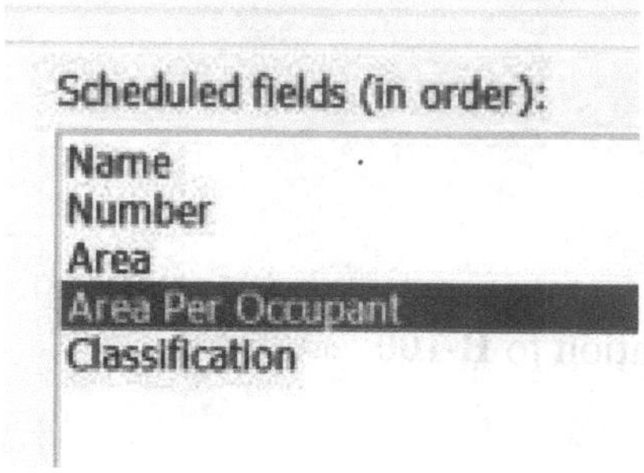

Add the following fields:

- Name
- Number
- Area
- Area Per Occupant
- Classification

12.

Select **Add Calculated Parameter**.

13.

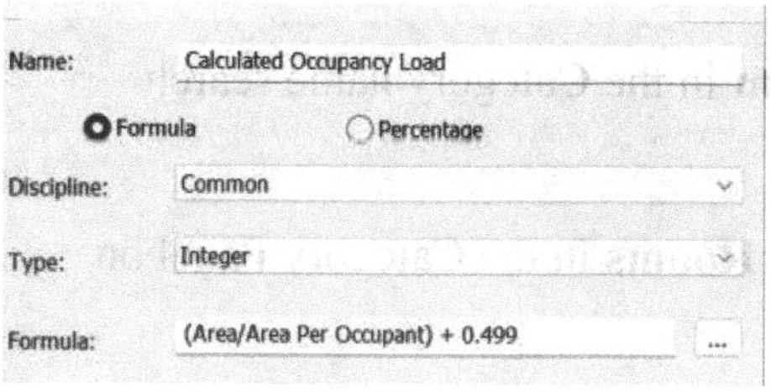

Type **Calculated Occupancy Load** in the Name field.
Select **Integer**.
For the formula, use:
(Area/Area Per Occupant) + 0.499
The 0.499 ensures the value is rounded up to the nearest whole number/integer.

Hint: You can use the ... button to select the Area and Area Per Occupant parameters used in the formula.

Click **OK**.

14.

Select the Filter tab.

Filter by: Area Per Occupant
Has a value.

The restrooms and hallway have no area per occupant assigned as they are considered transient areas. This will omit those rooms from the schedule.

15.

Select the Sorting/Grouping tab.

Sort by **Number**.
Enable **Grand totals**.
Select **Title and totals**.
Enable **Itemize every instance.**

16.

Select the Formatting tab.

Highlight **Area**.
Select **Calculate totals** from the drop down-list.

17.

Heading:

Calculated Occupancy Load

Heading orientation:

Horizontal

Alignment:

Left

Field formatting: Field Format…

Conditional formatting: Conditional Format…

Show conditional format on sheets: ☑

Hidden field: ☐

Calculate totals

Highlight **Calculated Occupancy Load**.
Select **Calculate totals** from the drop down-list.

Click **OK**.

18.

<Occupancy>

Name	Number	Area	Area Per Occupant	Classification	Calculated Occupan
Office	1	662 SF	150 SF	B-100	5
Office	2	444 SF	150 SF	B-100	3
Gallery	6	584 SF	30 SF	A-3	20
Gallery	7	679 SF	30 SF	A-3	23
Lobby	8	535 SF	15 SF	A-1	36
Gallery	9	667 SF	30 SF	A-3	23
Grand total		3572 SF			110

<Occupancy>

Name	Number	Area	Area Per Occupant	Classification	Calculated Occupancy Load
Office	1	55 m²	14 m²	B-100	4
Office	2	40 m²	14 m²	B-100	3
Gallery	6	76 m²	3 m²	A-3	28
Lobby	7	58 m²	1 m²	A-1	42
Gallery	8	82 m²	3 m²	A-3	30
Gallery	9	71 m²	3 m²	A-3	26
Grand total		382 m²			133

The schedule opens.

19. Save as *ex5-6.rvt [m_ex5-6.rvt]*

Note Blocks

A note block (also called an annotation schedule) is a list of construction notes for a drawing. You can create a note block and place it on a sheet to document the building model.

Exercise 5-7
Create a Note Block

Drawing Name: **schedules.rvt**
Estimated Time to Completion: 15 Minutes

Scope

Place generic annotations with the symbol tool.
Create a note block schedule.
Add a note block to a sheet.

Solution

1.
   - Floor Plans
     - Roof
     - Working Second Floor
     - SECOND FLOOR
     - Working Ground Floor
     - SITE PLAN
     - GROUND FLOOR - Overall
     - GROUND FLOOR - Furniture
     - GROUND FLOOR - Department

   Open the **SITE PLAN** view.

2. Symbol

   Open the Annotate ribbon.

   Click the **SYMBOL** tool.

3. general_note
   Note A

   Use the Type Selector to set the symbol to **Note A.**

4.

Place the symbol above the property line.

5.

Switch the Type Selector to Note B.

Place the Note B symbol near the trees in the site plan view.

ESC out of the symbol command.

6.

general_note	
Note A	
Generic Annotations (1)	Edit Type

Select Note A that was placed.

On the Properties palette, click **Edit Type**.

7.

Type Parameters

Parameter	Value
Graphics	
Leader Arrowhead	Arrow Filled 20 Degree
Text	
Note Description	Property line.
ID	A

Note that the symbol has a note description associated with the symbol.

8.

Family:	general_note
Type:	Note B

Type Parameters

Parameter	Value
Graphics	
Leader Arrowhead	Arrow Filled 20 Degree
Text	
Note Description	Landscape area.
ID	B

Select Note B from the Type drop-down.

Read the Note Description.

Click **OK.**

9.

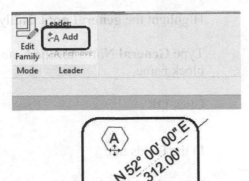

Select **Note A**.

On the contextual ribbon, select **Add**.

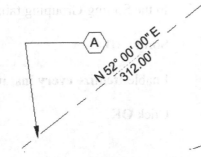

10.

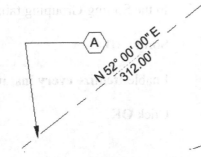

Adjust the leader so it is pointing to the property line.

11.

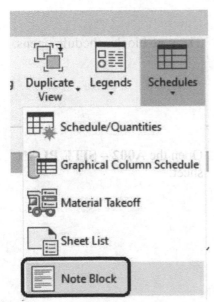

Open the View ribbon.

Go to **Schedules→Note Block**.

12.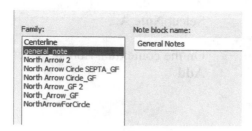

Highlight the **general note** family.

Type **General Notes** for the Note block name.

Click **OK**.

13.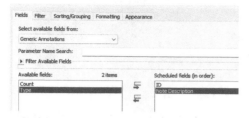

In the Fields tab:

Select the following fields:

- ID
- Note Description

14.

In the Sorting/Grouping tab:

Sort by **ID**.

Enable **Itemize every instance**.

Click **OK**.

The note block schedule opens.

15. Open the **A002 – SITE PLAN** sheet.

16.

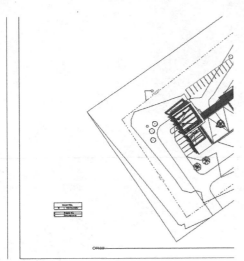

Drag and drop the note block onto the sheet.

Save as *ex5-7.rvt*.

Area Schemes

An area is a subdivision of space within a building model, usually larger than individual rooms. Areas are not necessarily bounded by model elements. Many architects like to start out their conceptual building models by visualizing common and private areas.

Distinguishing the level of privacy that certain spaces demand is an important part of designing a successful and comfortable space. Separating spaces based on public, semi-public and private designations informs the design by grouping spaces with similar demands. Defining levels of desired privacy can be different for each project.

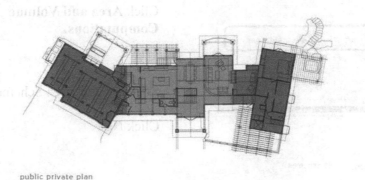

public private plan

For a house there is almost always a simple division between public spaces and private spaces that can be established. Conventionally the bedrooms and bathroom and sometimes office are more or less private and kitchen, living room and dining room are more or less public. For each project, there will be a different attitude toward the relative public/private aspects of these spaces.

Many architects also like to develop transitional spaces/areas as one moves from a public to private area and vice versa.

Exercise 5-8
Area Scheme

Drawing Name: **area_plans.rvt**
Estimated Time to Completion: 30 Minutes

Scope

Create an area scheme
Create an area plan and define area boundaries
Configure areas
Create a color fill legend
Create a project parameter

Solution

1.　

Open the Architecure ribbon.

Expand the Room & Area panel.

Click **Area and Volume Computations**.

2.　

Switch to the Area Schemes tab.

Click **New**.

3.

Change the Name to **Conceptual**.

Click **OK.**

4.

On the Architecture ribbon:

Click **Area Plan**.

5.

Select the **Conceptual** area plan type from the drop-down list.

Highlight **Ground Floor.**

Disable **Do not duplicate existing views**.

This will create a duplicate view of the ground floor using the area plan.

Click **OK**.

6.

Automatically create area boundary lines associated with all external walls?

Click **Yes**.

7.

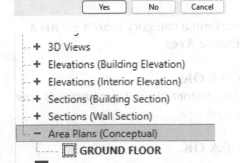

In the Project Browser, there is a new category called Area Plans (Conceptual).

Verify that this is the active view.

8.

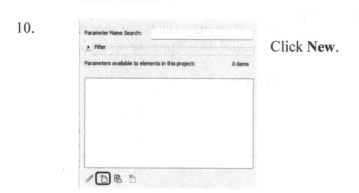

Hold down the CTL key.
Select a property line, a parking space, a tree, a car, and a reference plane.

On the ribbon:

Click **Hide Category**.

9.

Go to the Manage ribbon.

Select **Project Parameters**.

10.

Click **New**.

11.

Enable **Project Parameter**.
Type **Conceptual Area Type**.
Enable **Instance**.
Set the Data Type to **Text.**
Group the parameter under **Identity Data**.

Perform a category search for **area**.
Enable **Area**.

Click **OK**.

12.

The custom project parameter is now listed.

Click **OK.**

13. 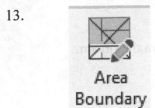 Select **Area Boundary** from the Architecture ribbon.

14.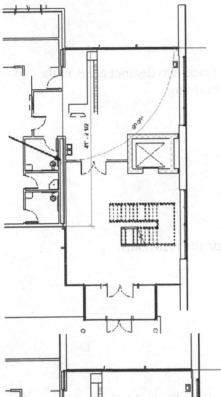

Draw a vertical line to create a new area on the right wing of the building.

15.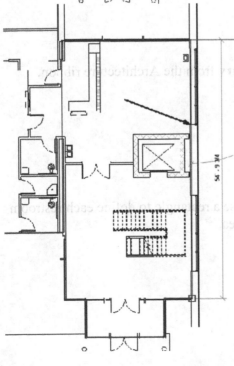

Draw a vertical line to create a closed polygon on the right wing of the building.

16.

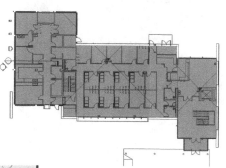

Select the **Area** tool from the ribbon.

17.

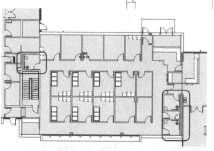

Place two distinct areas in the building.

18.

Locate the restrooms.

19.

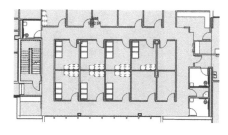

Select **Area Boundary** from the Architecture ribbon.

20.

Use a rectangle to define each restroom area.

21.

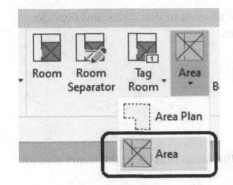

Select the **Area** tool from the ribbon.

22.

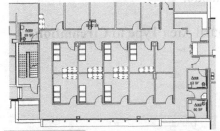

Place an area in each restroom.

23.

Area
Boundary

Select **Area Boundary** from the Architecture ribbon.

24.

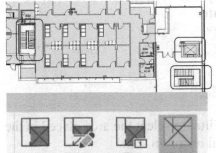

Use a rectangle to define the stairs and elevator areas,

25.

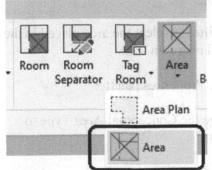

Select the **Area** tool from the ribbon.

26.

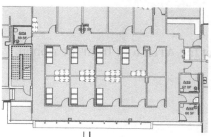

Place an area in the stairs and elevator spaces.

27.

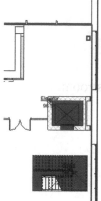

Use Filter to select the areas placed in the stairs and the elevator.

On the Properties palette:

Change the Conceptual Area Type to **Circulation**.

28.

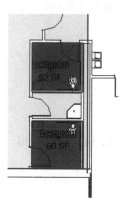

Use Filter to select the areas placed in the two restrooms.

On the Properties palette:

Change the Conceptual Area Type to **Private**.

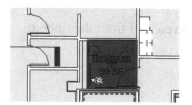

29.

Use Filter to select the area placed in the central restroom.

On the Properties palette:

Change the Conceptual Area Type to **Private**.

30.

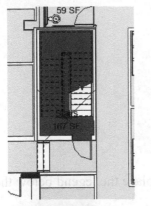

Use Filter to select the area placed in the central stairs.

On the Properties palette:

Change the Conceptual Area Type to **Circulation**.

31.

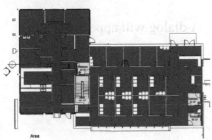

Select the area placed in the main building.

On the Properties palette:

Change the Conceptual Area Type to **Common**.

Identity Data	
Number	1
Name	Area
Image	
Comments	
Conceptual Area Type	Common
IFC Parameters	

32.

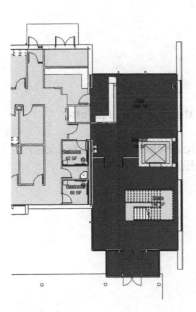

Select the area placed in the right wing of the building.

On the Properties palette:

Change the Conceptual Area Type to **Common**.

33.

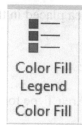

Color Fill
Legend

Color Fill

Switch to the Annotate ribbon.

Select the **Color Fill Legend** tool.

34.

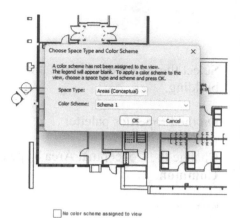

Click to place the legend below the view.

A dialog will appear.

Click **OK**.

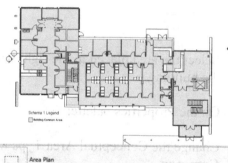

The areas are colored, but they are not differentiated.

35.

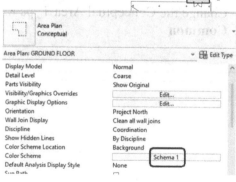

In the Properties palette:

Click **Schema 1** next to Color Scheme.

36.

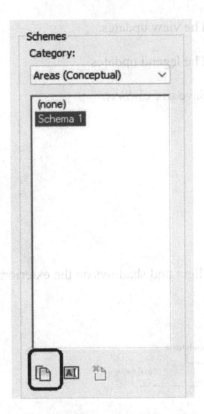

Highlight **Schema 1.**

Click **Duplicate**.

37.

Name: Conceptual

OK

Type **Conceptual** in the Name field.

Click **OK**.

38.

	Value	Visible	Color	Fi
1	Circulation	☑	Green	<Soli
2	Common	☑	RAL 1026	<Soli
3	Private	☑	Magenta	<Soli

Change the Title to **Area**.
Set the Color to **Conceptual Area Type**.

Assign **Green** to Circulation.
Assign **Yellow** to Common.
Assign **Magenta** to Private.

Click **OK.**

39.

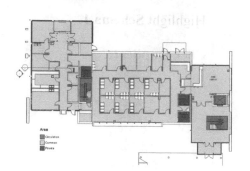

The view updates.

The legend updates.

Save as *ex5-8.rvt*.

Sun and Shadow Studies

Solar studies help you visualize the impact of natural light and shadows on the exteriors and interiors of projects.

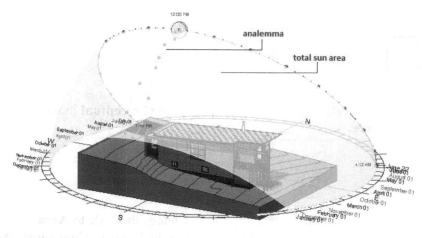

Still mode with sun selected

The sun path is a visual representation of the sun's range of movement across the sky at the geographic location specified for the project.

The sun path displays in the context of the project and allows you to position the sun at any point within its range of movement, between sunrise and sunset, throughout the year. Using the sun path's on-screen controls, you can create solar studies by placing the sun at any point along its daily path, and at any point along its analemma.

To experience the full potential of solar studies, display the sun path, and then use the sun path's context menu to access the Sun Settings dialog whenever you need it. Using the sun path and Sun Settings dialog together gives you the combined advantage of the sun path's highly visual, interactive controls and the dialog's presets and shared settings.

Exercise 5-9
Create a Solar Study

Drawing Name: **solar study.rvt**
Estimated Time to Completion: 10 Minutes

Scope

Adjust sun settings
Specify a project location

Solution

1. 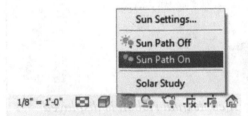 Open the 3D **Solar Study** view.

2.  Enable **Sun Path On** using the Sun button in the View Control bar.

   Zoom out to see the solar control.

3. 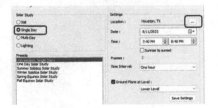 Click **Sun Settings** using the Sun button in the View Control bar.

4. Enable **Single Day**.

   Click the **...** button next to Location.

5.

Set Define Location by **Default City List**.

Select **San Francisco, CA** from the list.

Click **OK**.

6.

Under Presets:

7.

Toggle **Shadows On** in the View Control bar.

8.

Click **Solar Study** on the View Control bar.

9.

A contextual ribbon is displayed.

Click **Play in Loop**.
To stop, click **Play in Loop** again.

10. Save as *ex5-9.rvt*.

Exercise 5-10
Export a Solar Study

Drawing Name: **export_solar_study.rvt**
Estimated Time to Completion: 5 Minutes; It may take several hours to process the exported file, so do this exercise at the end of the day or before a lunch break.

Scope

Create an animation file of a solar study.
Create a multi-day solar study.

Solution

1.   Open the 3D **Solar Study** view.

2.   Select **Sun Settings** from the ribbon.

3.   Enable **Multi-Day**.

4.   Set the Start date to **8/1/2023.**
Set the End date to **11/30/2023**.

Set the Time Interval to **One hour**.

Click **OK**.

The preview updates.

5.

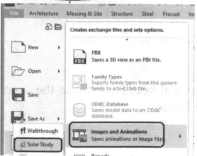

Go to **File→Export→Images and Animations→Solar Study**.

6.

Set the **Visual Style to Shaded with Edges**.

Click **OK**.

7.

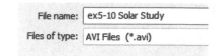

Browse to your homework file.

Save the file as an *.avi file.

8.

Set the Compressor to **Microsoft Video 1**.

Click **OK**.

You will see a process bar at the bottom of the screen to let you know how long it will take.

1%   Processing: Viewer Animator : Playex5-10.rvt - 3D View: Solar Study [18 of 1708] [Elapsed Time : 0:00:01:19]

Practice Exam

1. An architectural designer is creating a wall schedule and wants to include walls from the host model and a linked model. How would this be accomplished?

   A. In the Schedule Properties dialog, on the Sorting/Grouping tab, select Itemize Every Instance.
   B. In the Schedule Properties dialog, on the Fields tab, select Include Elements in Links.
   C. In the Schedule Properties dialog, on the Filter tab, filter by the Host model and linked model.
   D. In the Schedule Properties dialog, on the Formatting tab, select Use Project Settings

2. What is included in a note block schedule?

   A. Generic Annotations
   B. Key notes
   C. Text notes
   D. CAD blocks

3. Which tool is used to create a key schedule?

   A. Key Schedule
   B. Note Block
   C. Keynote Legend
   D. Schedule/Quantities

4. In order to define an area plan type, first you need to define a(n):

   A. Floor plan
   B. Color scheme
   C. Area scheme
   D. Area boundary

5. What controls the position of the sun?

   A. The Project Location
   B. True North
   C. The time of day
   D. Project Coordinates

6. An architectural designer is working on a room schedule. The designer has grouped the schedule by level and added a footer with Title, count, and totals selected. However, the total area is not showing. What needs to be done to display the total area?

    A. Change the Footer option to Totals only
    B. Set the Area parameter to Calculate totals
    C. Add a calculated parameter for the Area
    D. Enable Grand Totals in the Schedule Properties dialog

7. A note block can be:

    A. A schedule for all the sheets in the project
    B. A schedule for all the materials in the walls
    C. A schedule for all the wall types in the project
    D. A schedule for all the symbols placed in the project

8. An architectural designer creates an area plan. Some of the automatically generated area boundary lines are incorrect. What should the designer do to correct the area boundaries?

    A. Create a new area scheme and then refresh the area plan with the new area scheme.
    B. Use the Area Boundary tool to create the correct area boundaries.
    C. Activate the Room Separator tool and create the correct area boundaries.
    D. Select the area plan and edit the sketch.

9. An architectural designer is working on a door schedule. The width and height parameters are both included in the same column. How was this accomplished?

‹DOOR SCHEDULE›

A	B	C
Level	Mark	Width x Height
Lower Level	1501B	3' - 0" x 8' - 0"
Lower Level	1501C	3' - 0" x 7' - 0"
Lower Level	1502	3' - 0" x 7' - 0"
Lower Level	1503	3' - 0" x 7' - 0"
Lower Level	1504	3' - 0" x 7' - 0"
Lower Level	1505	4' - 0" x 7' - 0"

    A. The columns were grouped together after the schedule was created
    B. A calculated parameter was created
    C. The header was changed in the Formatting tab of the schedule
    D. The Combine parameters tool was used

10.   What must be turned on to perform a solar study?

        A. Lighting
        B. Elevation
        C. Shadows
        D. Sun

11.   A kitchen remodel consultant has created several different design options for their
        client's new kitchen. The client wants to know how much each option will cost. The
        consultant does the following:

        A. Creates phases for each design option and then assigns the phase to different
           schedules.
        B. Creates a schedule for each design option and then assigns the preferred design
           option to the schedule view.
        C. Creates a new project file for each kitchen design. Then, creates a schedule in that
           project. Then, creates a plot for each project file to present the client.
        D. Creates a schedule for the primary/preferred design option. Creates a duplicate of
           the schedule for each design option and then manually edits each duplicate to
           reflect the assigned design option.

12.   When exporting a solar study, Revit will create this file type:

        A. *.mov
        B. *.jpg
        C. *.mp4
        D. *avi

13.

\<DOOR SCHEDULE\>

A	B	C
Level	Mark	Width x Height
Lower Level	1501B	3' - 0" x 8' - 0"
Lower Level	1501C	3' - 0" x 7' - 0"
Lower Level	1502	3' - 0" x 7' - 0"
Lower Level	1503	3' - 0" x 7' - 0"
Lower Level	1504	3' - 0" x 7' - 0"
Lower Level	1505	4' - 0" x 7' - 0"

A designer creates a door schedule as shown using a combined parameter for Column C. They then delete the Width and Height fields, which were hidden. What happens to Column C?

A. Nothing
B. The data in Column C is replaced with ???
C. Column C is deleted
D. A warning dialog appears advising that necessary parameters have been deleted

Answers:
   1)   B; 2) A; 3) A; 4) C; 5) A; 6) D; 7) D; 8) B; 9) D; 10) D; 11) B; 12) D; 13) A

About the Author

Elise Moss has worked for the past thirty years as a mechanical designer in Silicon Valley, primarily creating sheet metal designs. She has written articles for Autodesk's Toplines magazine, AUGI's PaperSpace, DigitalCAD.com and Tenlinks.com. She is President of Moss Designs, creating custom applications and designs for corporate clients. She has taught CAD classes at Laney College, DeAnza College, Silicon Valley College, and for Autodesk resellers. Autodesk has named her as a Faculty of Distinction for the curriculum she has developed for Autodesk products, and she is a Certified Autodesk Instructor. She holds a baccalaureate degree in mechanical engineering from San Jose State.

She is married with two sons. Her older son, Benjamin, is an electrical engineer. Her middle son, Daniel, works with AutoCAD Architecture in the construction industry. Her husband, Ari, has a distinguished career in software development.

Elise is a third-generation engineer. Her father, Robert Moss, was a metallurgical engineer in the aerospace industry. Her grandfather, Solomon Kupperman, was a civil engineer for the City of Chicago.

She can be contacted via email at elise_moss@mossdesigns.com.

More information about the author and her work can be found on her website at www.mossdesigns.com.

<u>Other books by Elise Moss</u>

AutoCAD 2024 Fundamentals
Revit 2024 Basics

9781630575977